motors and transformers

based on the 1993 NEC®

AF807609

AMERICAN TECHNICAL PUBLISHERS, INC.
HOMEWOOD, ILLINOIS 60430

James G. Stallcup

1 2 3 4 5 6 7 8 9 – 92 – 9 8 7 6 5 4 3 2

Printed in the United States of America

Library of Congress Cataloging-in-Publication Data

Stallcup, James G.
Motors and transformers based on the 1993 NEC / James G. Stallcup
p. cm.
Includes index.
ISBN 0-8269-1738-0
1. Electric motors. 2. Electric transformers. 3. National Fire Protection Association. National Electrical Code (1993) I. National Fire Protection Association. National electrical code, 1993. II. Title.
TK2514.S733 1992 92-41117
621.46—dc20 CIP

CONTENTS

INTRODUCTION

MOTORS AND TRANSFORMERS BASED ON THE 1993 NEC® explains and illustrates methods used to design and size electrical components of complete circuits from the service to the driven load. Frequent references to the 1993 National Electrical Code® and careful study of its provisions are required. Copies of the National Electrical Code® 1993 (NFPA No. 70) may be obtained from:

National Fire Protection Association
Batterymarch Park
Quincy, MA 02269

Motor subjects included are: overcurrent protection devices that start and accelerate the driven load; conductors that connect the power source to the motor windings; reduced starters that lower inrush current where locked-rotor current presents problems; protection and selection of motor control conductors used to start, stop, jog, or reverse motors; sizing and selecting of motor disconnecting means using a fused or nonfused disconnect and automatic or non-automatic CB's; sizing raceways, junction boxes, gutters, or wireways that supply an individual motor or group of motors; and troubleshooting procedures.

Transformer subjects included are: sizing transformers to supply loads in buildings; sizing overcurrent protection for primary and secondary sides of transformers; sizing conductors for primary and secondary sides of transformers; sizing taps to be taken from the secondary sides of transformers to supply loads; location and clearance of transformers in buildings; sizing buck-and-boost transformers for branch circuits; sizing capacitors to correct power factor; connections and hookups; and troubleshooting procedures.

MOTORS AND TRANSFORMERS BASED ON THE 1993 NEC® contains 16 chapters. Of these, chapters 1 through 8 discuss motors, and 10 through 14 discuss transformers. Chapter 9 contains four motor tests and chapter 15 contains four transformer tests. Chapter 16 contains eight final tests, covering both motors and transformers. Each text chapter is followed by Review Questions. Tests and Review Questions include a variety of True-False, Multiple Choice, Completion, and Identification questions, and Problems. Answers to all questions and problems are given in the Instructor's Guide. Follow the procedures shown to answer questions and problems.

True-False. *Circle T if the statement if true. Circle F is the statement is false.*

(T) F **13.** Reactor starting is usually installed to start and run motors rated over 600 volts.

Multiple Choice. *Select the response that correctly completes the statement. Write the appropriate letter in the space provided.*

A **1.** A three-wire, single-phase system has two conductors that are ________ volts-to-ground.

A. 120
B. 194
C. 208
D. 240

Completion. *Determine the response that correctly completes the statement. Write the appropriate response in the space provided.*

starting **6.** Shaded-pole motors use a shaded coil to provide ________ torque.

Identification. *Determine the response that correctly matches the given element(s). Write the appropriate response in the space provided.*

Identify the components of the universal motor.

7. single-phase supply
8. armature
9. field pole and windings
10. brushes and commutators

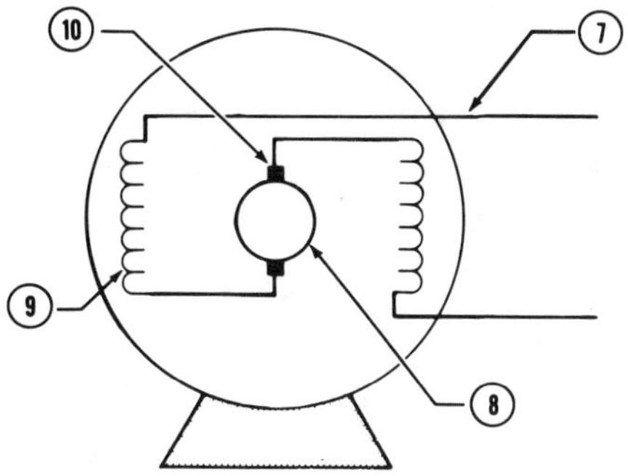

Problems. *Perform the operation indicated. Show all work.*

4. Connect the leads of the six pole motor for delta operation.

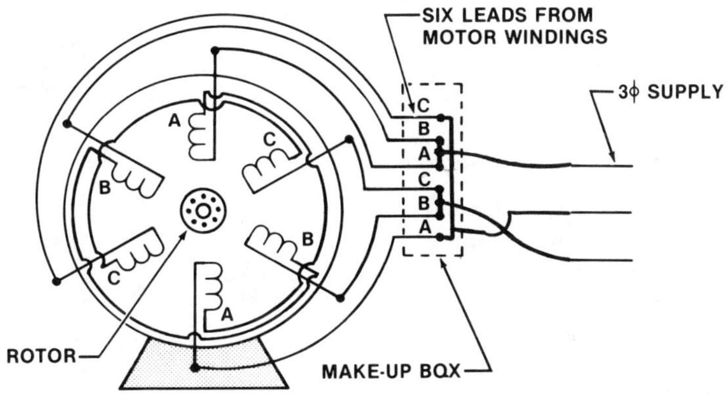

CALCULATIONS

When total wattage or volt-amperage is to be divided by phase-to-phase (three-phase) voltage times 1.732, the following values are substituted:

for 208 volts × 1.732, use 360
for 230 volts × 1.732, use 398
for 240 volts × 1.732, use 416
for 440 volts × 1.732, use 762
for 460 volts × 1.732, use 797
for 480 volts × 1.732, use 831

WATTS—VOLT-AMPERES

In general, within the National Electrical Code® the term *watts* (w) has been superseded by the term volt-amperes (VA) for the computation of loads. However, references to nameplate ratings still reflect the term *watts* on certain loads. This method is used in MOTORS AND TRANSFORMERS BASED ON THE 1993 NEC®.

MOTOR OPERATION

Chapter 1

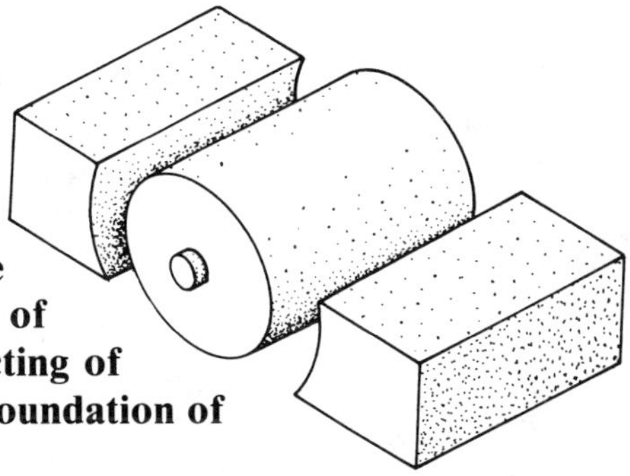

Electricity and magnetism play the major role in the operation of a motor. Electricity furnishes power to the field windings (poles), which in turn induces magnetic lines of force from north to south poles. In the motor, the rotor is connected to the load and drives the load. The rotor cuts the lines of force in its rotation through the fields. The attracting of unlike poles and the repelling of like poles is the foundation of motor operation.

PRINCIPLES OF MAGNETISM

Magnetism is invisible lines of force that exist between the north and south poles of all *permanent magnets* or *electromagnets*. These lines of force, called a *force field* or *magnetic field*, connect the north and south poles of permanent magnets.

Permanent Magnet

The earth is a permanent magnet with lines of force rotating around it from the north pole to the south pole. See Figure 1-1. A piece of soft iron placed within the force field of a permanent magnet becomes magnetized. The force field of the permanent magnet magnetizes the piece of soft iron. The soft iron does not have to touch the permanent magnet to become magnetized. The soft iron then takes on the same characteristics as the permanent magnet. This type of action in motor operation is *induction*.

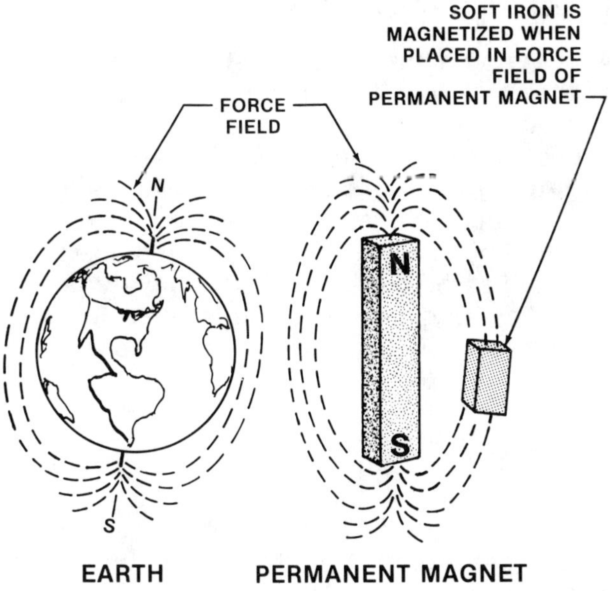

Figure 1-1. The earth and permanent magnets have a force field (lines of force). A piece of soft iron placed in the force field of a permanent magnet becomes magnetized.

Poles Attract or Repel

The poles of two permanent magnets either attract or repel. Unlike poles attract each other; like poles repel each other. If a permanent magnet is suspended from a point by a string, the suspended magnet will rotate by the attracting or repelling end of a second magnet. This is the major principle of motor operation. The rotor rotates through the magnetic field created by the field poles. See Figure 1-2.

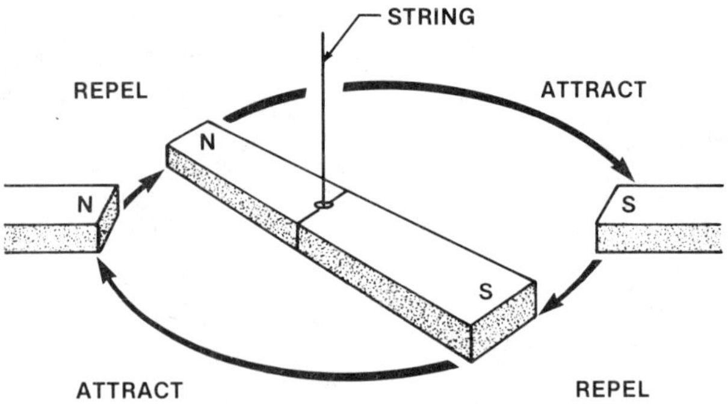

Figure 1-2. Like poles repel; unlike poles attract.

ELECTROMAGNETS

An electromagnet, in its most basic form, is a single conductor wound around a piece of soft iron. A straight conductor carrying current produces a weak magnetic field. A soft iron core with many turns of wire around it forms a coil and produces a strong magnetic field. The number of turns in the coil determines the strength of the coil. A coil with many turns produces a strong magnetic field. A coil with fewer turns produces a weaker magnetic field.

Reversing Poles

Reversing the current flow through an electromagnet changes the polarity of its poles. Current flow in one direction produces a south pole on one end; current flow in the opposite direction produces a north pole. Alternating current changes the poles of an electromagnet from north to south due to the flow of current changing direction. See Figure 1-3.

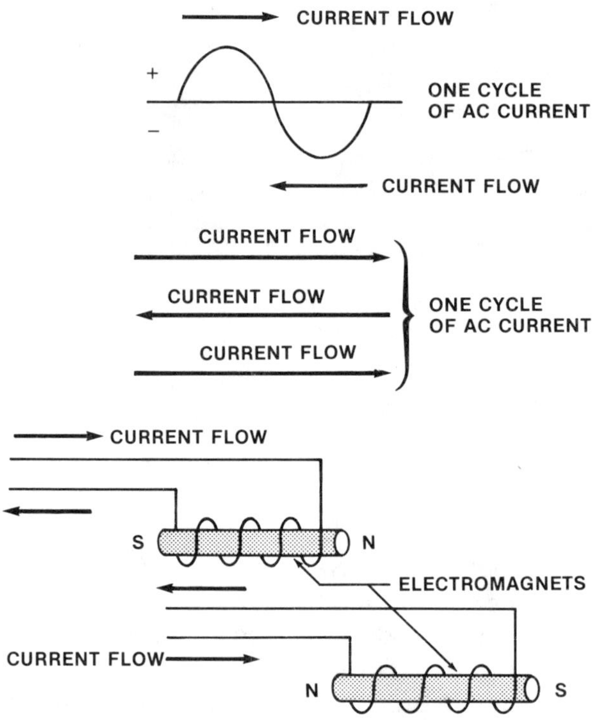

Figure 1-3. Alternating current changes the poles of an electromagnet from a north pole to a south pole by reverse current flow.

Typical Induction Motors

A typical induction motor consists of a fixed stator and a rotating rotor. The stator is cut into thin sections and laminated and assembled in a sandwich-like manner to reduce *eddy current* losses. Eddy currents are the circulating currents induced in the conducting material of the rotor when it cuts through the magnetic flux lines of the magnetic field.

When eddy currents flow in a solid piece of metal, a heating effect is produced. By sandwiching or cutting the metal into sections, the eddy currents flow only in the cut sections. This decreases the heating effect.

The fixed stator is equipped with two or more field poles with wire wound around them and connected together to create two or more electromagnets. When 60 hertz (cycles) of alternating current is supplied to the electromagnets, the poles of the electromagnets reverse their polarity 120 times per second. This occurs every time the current changes direction. *NOTE*: Stator poles and electromagnets are basically the same. See Figure 1-4.

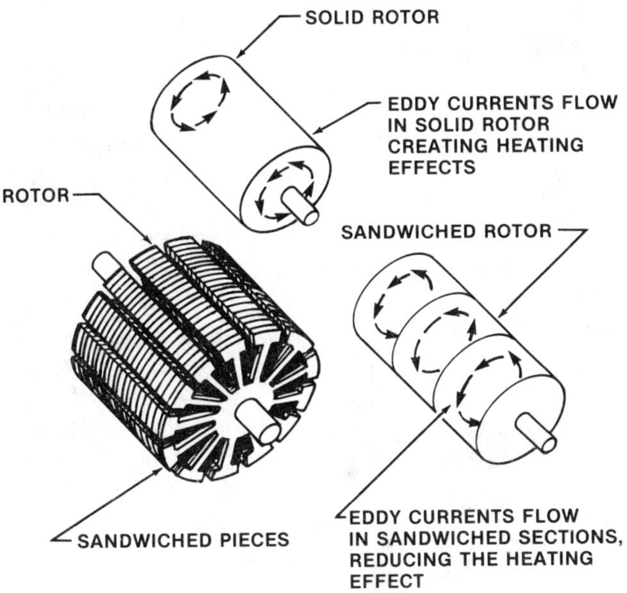

Figure 1-4. A rotor is sandwiched together and laminated to reduce eddy currents.

The rotor is made of slotted sections cut and sandwiched together to reduce eddy current losses. Copper or aluminum bars are embedded in the rotor and welded together by a ring. The ring provides a path for the flow of current through the bars of the rotor. The voltage induced into these bars is low so insulation between the bars and rotor is not required.

The rotor may be placed between two or more stator poles. A 60-hertz AC power source is applied to the stator poles. The magnetic field created by the stator poles builds up and collapses for each alternation. The alternations change the poles of the stator from south to north or north to south. The rotor is pushed and pulled through the rotating magnetic field of the stator poles. See Figure 1-5.

Expanding and collapsing fields of the stator poles induce current into the bars of the rotating rotor. This is accomplished by the rotor cutting the magnetic lines of force created by the stator poles or field windings. *NOTE*: The magnetic field in the rotor is opposite the magnetic field of the stator field poles.

Rotating Magnetic Fields

Induction motors operate on either single-phase, two-phase, or three-phase systems. In a single-phase AC motor the rotating magnetic field is created by splitting the phases and shifting the AC power applied to the stator field poles. The rotor in a single-phase motor must have a means of starting. The field alternates at such a fast rate (60 times a second) that the rotor cannot follow the alternating field. However, the rotor must turn fast enough to catch the rotating field. The rotor can be turned by hand or by a starting winding.

Only one current (phase A) is in a single-phase, 120V motor. This current changes the stator poles from north to south as the current alternates.

In polyphase AC motors, the phase displacement of the different voltages is used. The voltage is either in-phase or out-of-phase. In-phase currents rise and fall simultaneously. Currents 180° out-of-phase have one current rise past zero as the other falls past zero. Currents 90° out-of-phase have one current reaching a peak while the other is at zero.

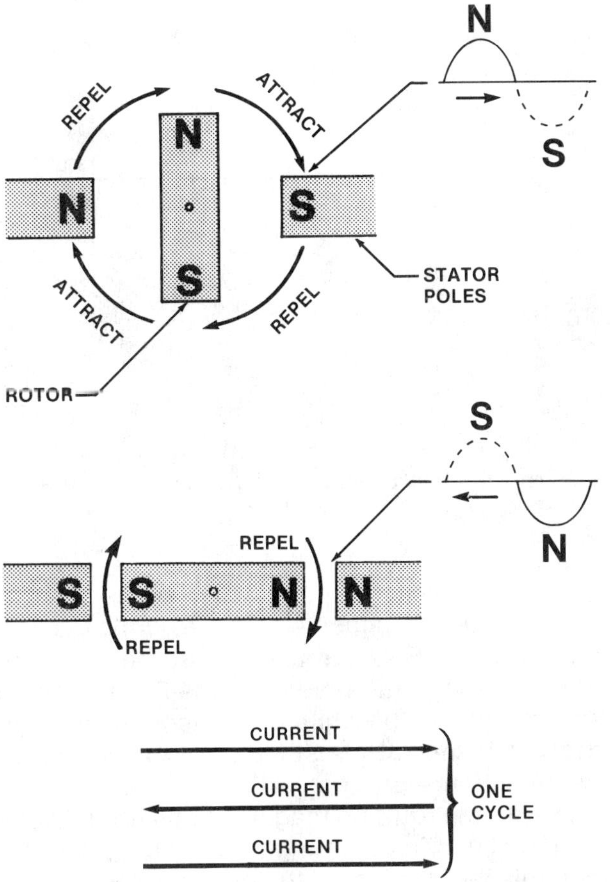

Figure 1-5. Basic operation of an induction motor.

The rotating magnetic field of the stator using two voltages 90° out-of-phase is shown in Figure 1-6. The current flow in phase A at the 0° position is at maximum. The current flow in phase B is at zero. The windings of phase A in the stator will be at maximum value and so will its magnetic field. The magnetic field of phase B windings will be at zero. The currents in both phases are equal values at the 45° position.

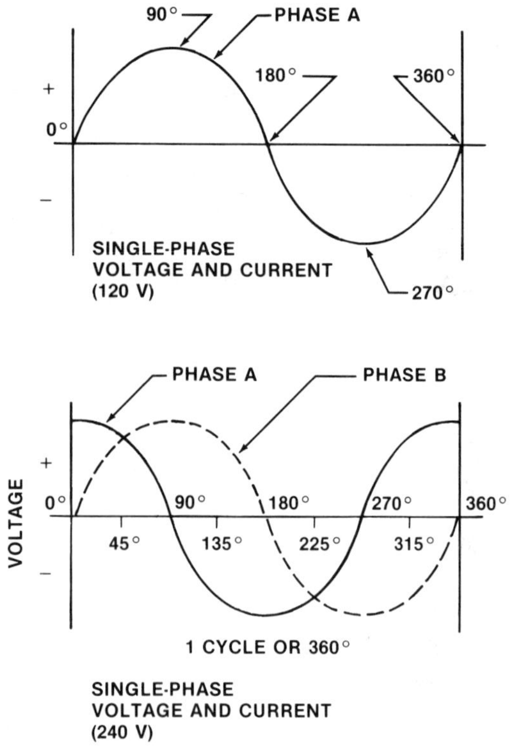

Figure 1-6. Relationship of single-phase currents.

The current in phase A is at zero while the current in phase B is at maximum at the 90° position. The magnetic field of phase B windings is at maximum value while the phase A winding is at zero. At the 135° position, the current flow in each winding is equal. The rotor continues to turn with the rotating magnetic field until it completes 360°. The rotating magnetic field is accomplished by placing the two poles (windings) at right angles to each other in the stator and supplying voltages 90° out-of-phase.

SINGLE-PHASE VOLTAGES AND CURRENTS

A 120-volt, single-phase voltage and current is like one person riding a bicycle. There is only one person with a stroke that will peak and produce power.

A 240-volt, single-phase voltage and current is like two people riding a bicycle. There are two riders providing a power-producing stroke. As one rider leaves the power-producing stroke (peak), the other rider enters the peak and produces power to drive the bicycle. Therefore, a 240-volt, single-phase motor is more efficient than a 120-volt, single-phase motor.

THREE-PHASE VOLTAGES AND CURRENTS

Three-phase voltages and currents operate like three riders on a bicycle. As the first rider leaves the peak stroke, the second rider enters the peak. As the second rider leaves the peak stroke, the third rider enters the peak. Three riders can produce more power than one or two riders. A three-phase motor will produce more power because three different phases are peaking and providing a smooth and continuous power to drive the rotor and load. See Figure 1-7.

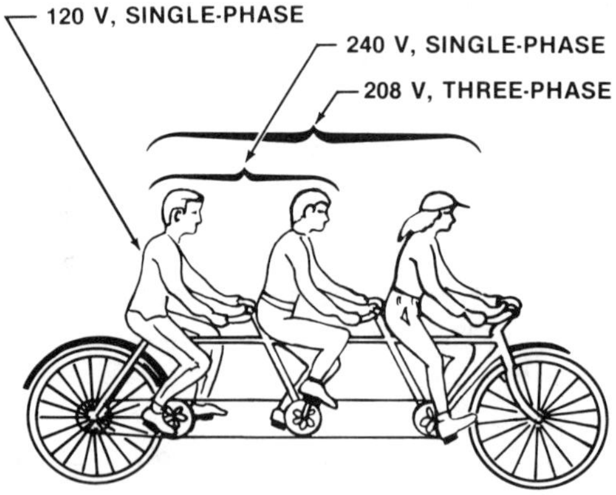

Figure 1-7. Three-phase voltages and currents operate like three riders on a bicycle.

The three phases of voltages and currents supply one of the three separate pairs of poles. In the peak stroke, the first phase delivers the greatest power. As the first phase leaves the peak stroke, the second phase enters the peak, delivering its greatest stroke of power. As the second phase leaves the peak, the third phase enters the peak and the process repeats itself. See Figure 1-8.

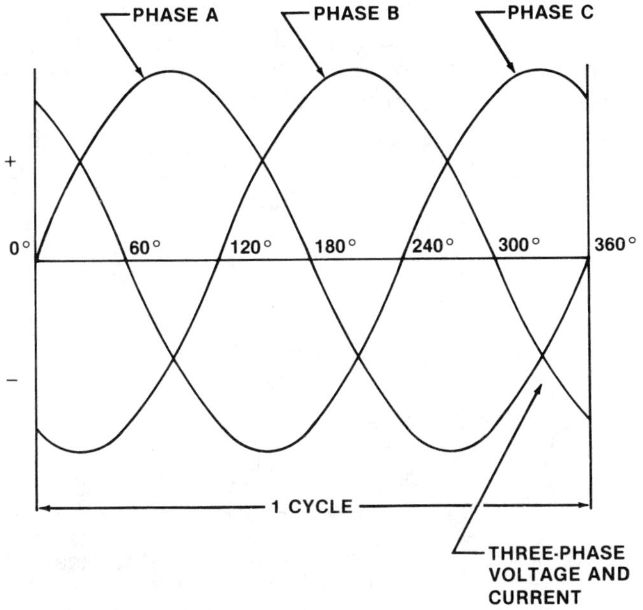

Figure 1-8. The three phases of voltages and currents supply one of the three separate pairs of poles.

Phase 1 has its greatest power stroke and magnetic field between poles 1 and 4. Phase 2 has its greatest power stroke between poles 2 and 5, and Phase 3 has its greatest power stroke between poles 3 and 6. The three phases of voltages and current are displaced 120° on the stator of the motor. See Figure 1-9.

Dual bars installed in the rotor provide the motor with good starting torque and low slip. The outer bars close to the surface have a high-resistance winding. The inner bars have a low-resistance winding. The outer bars provide good starting. The inner bars allow more current to flow at the running speed of the motor. See Figure 1-10.

These bars are designed to allow the percentage of slip desired for each motor. Rotor slip is caused from magnetic lines of force cutting across these bars embedded in the rotor. Current is induced in the rotor when these bars cut

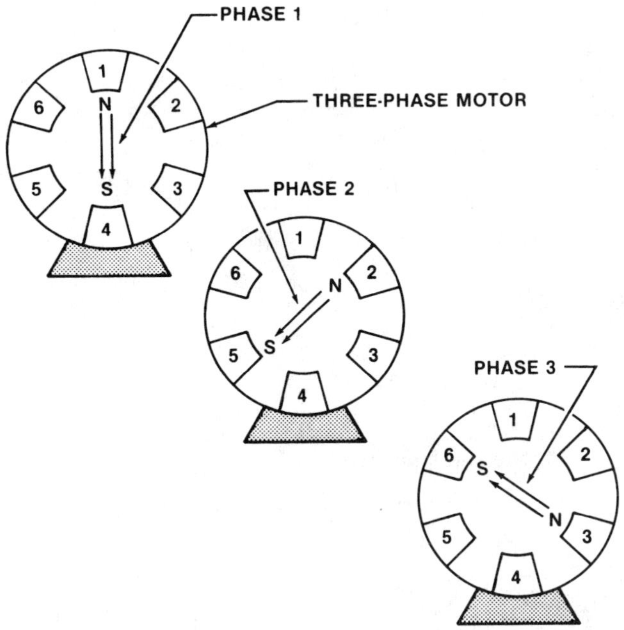

Figure 1-9. The greatest power stroke and magnetic field for Phase 1 occur between poles 1 and 4; for Phase 2, between poles 2 and 5; for Phase 3, between poles 3 and 6.

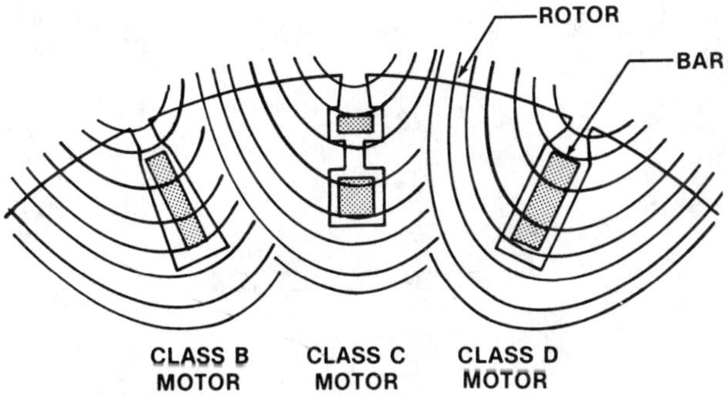

Figure 1-10. Bars in the rotor can produce different classes of motors.

the lines of force created by the alternations of current in the stator field. Note that the rotor will never rotate at the same speed as the alternations of current and rotating field. This difference in rotating speed is the *slip of the motor*. See Figure 1-11.

The more slip a motor has resulting from a driven load, the slower the rotor will turn due to the greater number of lines of force cut during rotation. If a motor is designed to have 5% slip, the rotor will rotate at 5% of the synchronous speed. See Figure 1-12.

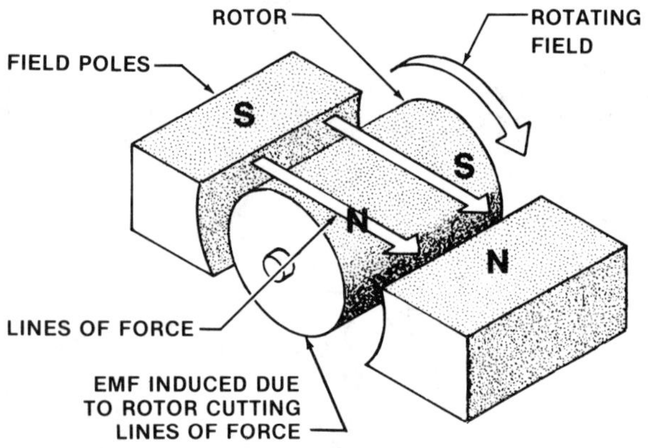

Figure 1-11. The rotor will rotate at a rate less than that of the rotating field.

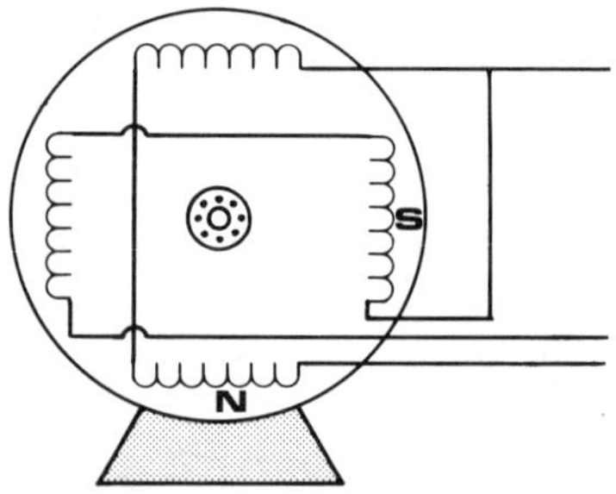

PROBLEM: What is the synchronous speed and actual speed of a four-pole motor operating at 5% slip?

Step 1: $syn\ sp = 120 \times \dfrac{frequency}{no.\ of\ poles}$

Step 2: $syn\ sp = \dfrac{120 \times 60\ \text{cps}}{4} = 1800\ \text{rpm}$

ALTERNATE METHOD

$$syn\ sp = \frac{cps \times cps}{pairs\ of\ poles}$$

$$syn\ sp = \frac{60\ \text{cps} \times 60\ \text{cps}}{2} = 1800\ \text{rpm}$$

Step 3: *syn sp* = 1800 rpm

Step 4: *slip* = 1800 rpm × 5% slip = 90 rpm

Step 5: *actual sp* = 1800 rpm − 90 rpm = 1710 rpm

Answer: **syn sp = 1800 rpm**
actual sp = 1710 rpm

where
syn sp = synchronous speed
actual sp = actual speed
cps = cycles per second
rpm = revolutions per minute

Figure 1-12. Calculating the synchronous and actual speeds of a motor.

As the rotor turns at a faster speed through the field, fewer lines of force are cut. This causes the rotor to slow down due to the loss of current in the rotor. When the rotor is at rest and no lines of magnetic force are cut, the rotor has 100% slip. The motor can pull four to six times its running current with 100% slip. However, when power is applied to the stator poles, the rotor begins to turn through the stator field.

The rotor tries to catch the alternation of current creating the magnetic lines of force between the stator field poles. As the rotor turns, the percentage of slip begins to decrease until the designed amount of slip is reached, which is usually 2% to 5%. See Figure 1-13.

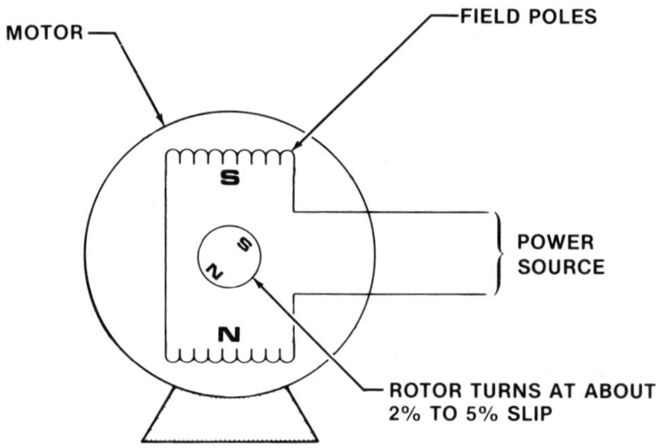

Figure 1-13. The rotor in an induction motor turns at about 2% to 5% slip.

CLASS B MOTORS

Class B motors are designed with a speed regulation based on 2% to 5% slip. Class B motors drive loads such as fans, blowers, and centrifugal pumps.

The Class B motor has a starting torque of about 150% times the full-load torque rating of the motor. The full-voltage starting current is approximately 600% to 725% of the running current (amperes) of the motor.

Class B motors have normal starting current and normal starting torque. They are used on loads that are started and reversed infrequently and are the most used motors in the electrical industry.

CLASS C MOTORS

Class C motors are designed with a speed regulation based on 2% to 5% slip. Class C motors drive hard-to-start loads such as conveyors, compressors, crushers, and reciprocating pumps.

The Class C motor has a starting torque of about 225% of the full-load torque rating of the motor. The full-voltage starting current is approximately 600% to 650% of the running current of the motor. Class C motors have normal starting current and high starting torque.

CLASS D MOTORS

Class D motors are designed with a speed regulation based on 5% to 13% slip and are known as *high-slip motors*. Class D motors drive very hard-to-start loads that are started and reversed frequently, such as cyclical loads, punch presses, cranes, elevators, and hoists.

Class D motors have a starting torque of about 275% of the full-load torque rating of the motor. The full-voltage starting current is approximately 525% to 625% of the running current of the motor. Class D motors have low starting current and high starting torque.

PHASE CONVERTERS

Phase converters utilize 1ϕ power to run electrical equipment requiring 3ϕ power. Two types of phase converters are static and rotary. A *static* phase converter has no moving parts. A *rotary* phase converter has an internal rotor that shall turn before the load is applied. See 455-2.

Branch Circuits—455-6; 455-7

Conductors to a phase converter are sized at 1.25 times the phase converter's nameplate 1ϕ input full load amperage per 455-6. The OCPD is selected based on 1.25 times the phase converter's nameplate 1ϕ input full load amperage per 455-7. The OCPD shall not exceed the 1.25, but shall be equal to or lower than 1.25. For example, a rating of 50 A requires a 50 A OCPD, while a 65 A rating requires a 60 A OCPD. See Figure 1-14. Section 460-7, Ex. 1 allows the next size OCPD above 125% × FLA, 1ϕ input to be used.

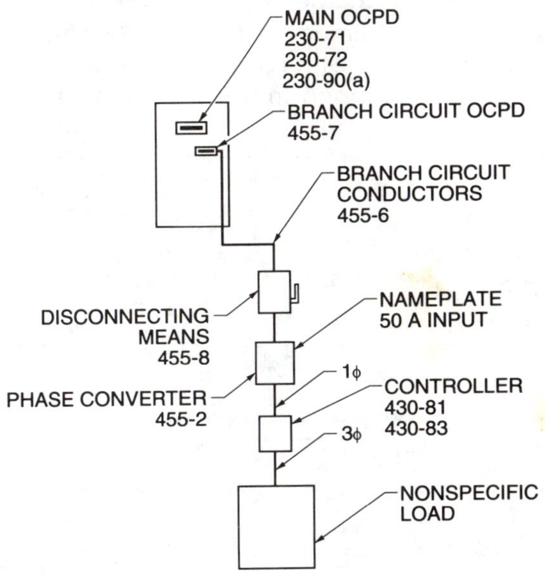

PROBLEM: What size THWN Cu conductors and OCPD are required to supply power to the phase converter?

Conductors

Step 1: *455-4.*
Nameplate = 50 A

Step 2: *455-6.*
50 A × 1.25 = 62.5 A

Step 3: *Table 310-16.*
62.5 A = #6 THWN Cu

Answer: **Conductors = #6 THWN Cu**

OCPD

Step 1: *455-7*, Ex. 1
40 A x 1.25 = 62.5 A

Step 2: *455-7*; *240-6.*
62.5 A = 70 A OCPD

Answer: **OCPD = 70 A**

Figure 1-14. Conductors and OCPDs are selected at 1.25 times the FLC rating of the nonspecific load per 455-6 and 455-7.

Feeder Circuit—455-6, Ex.; 455-7, Ex. 1, 2

The conductors to supply power to two or more phase converters that convert 1ϕ power to 3ϕ power are sized at 2.50 times the 3ϕ amperage of all motors and other loads served. The OCPD is sized at 2.50 times the full load 3ϕ amperage of all motors and other loads. If the percentage does not correspond to a standard size OCPD per 240-6, the next size OCPD above this percentage can be selected per 455-7, Ex. 2. See Figure 1-15.

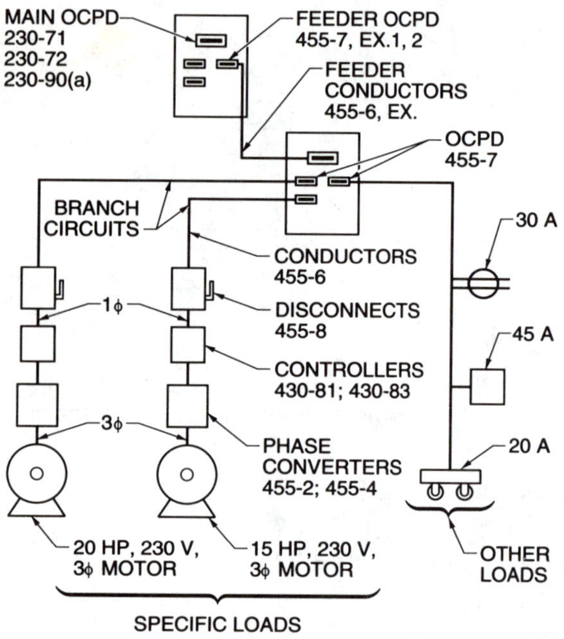

PROBLEM: What size THWN Cu conductors and OCPD are required to supply power to the group of motors and other loads?

Conductors

Step 1: *455-6, Ex.; Table 430-150.*
15 HP = 42 A
20 HP = 54 A

Step 2: *455-6, Ex.*
Motors = 42 A + 54 A × 2.50 = 240 A
Other loads = 30 A + 45 A + 20 A × 2.50 = 237.5 A
Total = 240 A + 237.5 A = 477.5 A

Step 3: *Table 310-16.*
477.5 A = 800 kcmil

Answer: **Conductors = #800 kcmil THWN Cu**

OCPD

Step 1: *455-7, Ex. 1, 2; 240-6.*
477.5 A = 500 A

Answer: **OCPD = 500 A**

Figure 1-15. Conductors and OCPDs are selected at 2.50 times the 3ϕ FLC current rating of all motors and other loads.

REVIEW—CHAPTER 1

Name ____________________ Date ____________

True-False

T F 1. Magnetism is invisible lines of force that exist between the north and south poles of electromagnets.

T F 2. Permanent magnets are not connected by a force field.

T F 3. A piece of soft iron becomes magnetized when placed within the force field of a permanent magnet.

T F 4. The unlike poles of permanent magnets attract each other.

T F 5. The like poles of permanent magnets repel each other.

T F 6. The strength of an electromagnet is determined by the number of turns on its coil.

T F 7. A piece of soft iron passed through the windings of a coil produces a weaker magnetic field.

T F 8. The direction of current flow through an electromagnet determines the polarity of the pole.

T F 9. Direct current reverses polarity without the use of additional components.

T F 10. Alternating current changes the polarity of the poles with each alternation.

T F 11. A typical induction motor includes a stator and rotor.

T F 12. Eddy currents flowing in a solid rotor create a heating effect.

T F 13. The stator of an induction motor is equipped with two or more field poles.

T F 14. Sandwiched and laminated rotors aid in reducing the heating effect created by eddy currents.

T F 15. A field pole that is polarized with a south pole will repel the north pole of the rotor.

Completion

________ 1. The expanding and collapsing fields of the stator poles of an induction motor ________ current into the bars of the rotating rotor.

________ 2. The rotor of an induction motor ________ the magnetic lines of force created by the stator poles to produce torque.

________ 3. The rotor of an induction motor must have an additional ________ to start.

________ 4. In a polyphase AC induction motor, the starting torque is developed by the phase ________ of the different voltages.

________ 5. The voltage of a single-phase, 120-volt motor can be compared to ________ person(s) riding a bicycle.

________ 6. The outer bars of an induction rotor motor provide the motor with a ________ resistance winding.

________ 7. An EMF is ________ into the rotor as it cuts the magnetic lines of force created by the field poles.

________ 8. The rotor will never turn at the same speed as the rotating magnetic field due to the ______ of the induction motor.

________ 9. Actual speed of an induction motor can be calculated by the percentage of ________ designed into the rotor.

________ 10. The rotor will always attempt to catch the ________ of current creating the magnetic lines of force between the field poles.

________ 11. The slip of an induction motor will ________ as the rotor accelerates to its running speed.

________ 12. Class ______ motors are the motors most commonly used to start and drive loads.

________ 13. Class C motors have a(n) ________ starting torque than Class B motors.

________ 14. When the rotor is turning rapidly through the field, ________ lines of force are cut.

________ 15. Rotor slip is produced by the magnetic lines of force cutting across the copper or aluminum ______ embedded in the rotor.

Multiple Choice

________ 1. If 60 hertz AC current is supplied to the field pole windings, the poles will reverse their polarity ______ times per second.
A. 90
B. 100
C. 120
D. 180

________ 2. Currents that are 180° out-of-phase have one current rise past zero as the other current falls past ______.
A. 0
B. 15
C. 20
D. 30

________ 3. A cycle of 120-volt, single-phase voltage and current has ______°.
A. 90
B. 180
C. 270
D. 360

________ 4. The rotating magnetic field of the stator of an induction motor has two voltages ______° out-of-phase.
A. 30
B. 60
C. 90
D. 120

______ **5.** When an induction motor is started, the starting current can be four to ______ times the running current.
A. six
B. eight
C. ten
D. twelve

______ **6.** Class B motors are designed with a slip of ______%.
A. 6
B. 8
C. 2 to 5
D. 2 to 12

______ **7.** Class C motors have ______ torque.
A. medium
B. low
C. poor
D. high

______ **8.** The rotor of a Class D motor is designed to have ______ slip.
A. low
B. poor
C. high
D. no

______ **9.** Class B motors have a starting torque of approximately ______% times full-load torque.
A. 150
B. 175
C. 200
D. 250

______ **10.** Class D motors have a starting torque of approximately ______% times full-load torque.
A. 125
B. 150
C. 175
D. 275

Problems

1. Draw a permanent magnet with a piece of soft iron in the force field. Label the drawing.

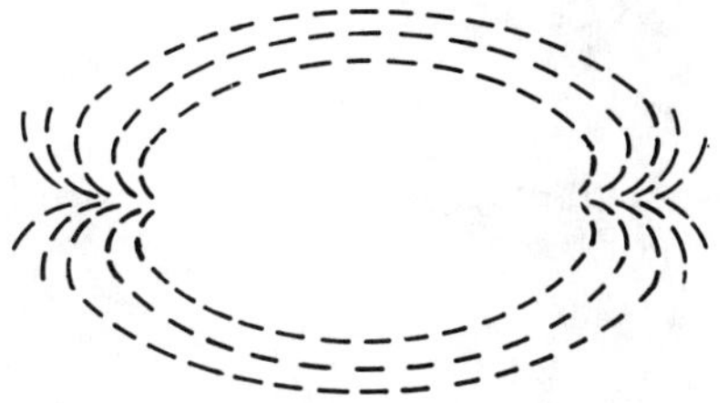

2. Show the basic operation of a motor using permanent magnets to demonstrate rotation.

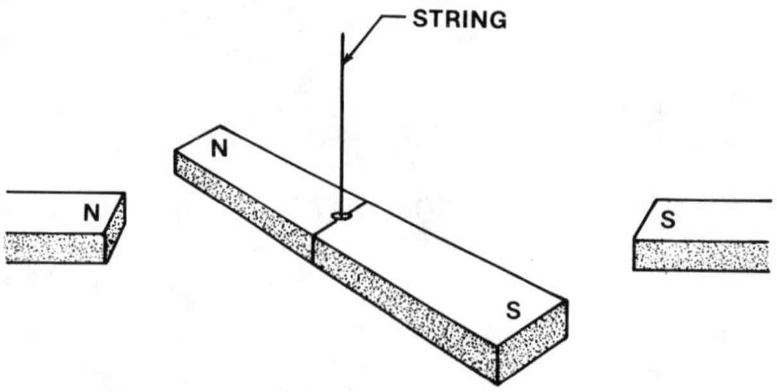

3. Draw arrows to show the current flow of the electromagnet.

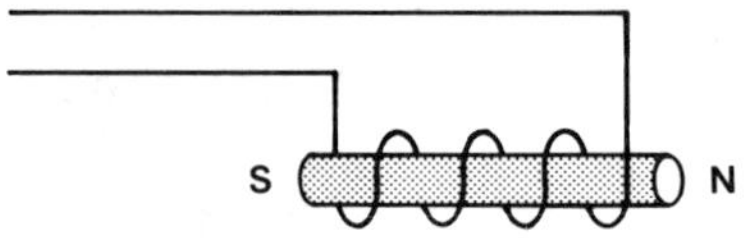

4. What is the synchronous speed of a two-pole induction motor?

5. What is the actual speed of a 3600 rpm induction motor with 10% slip?

6. What is the synchronous speed of a four-pole induction motor?

7. What is the actual speed of an 1800 rpm induction motor with 4% slip?

8. What is the synchronous speed of a six-pole induction motor?

9. What is the actual speed of a 1200 rpm induction motor with 13% slip?

10. What are the actual speeds of an induction motor with synchronous speeds of 3600, 1800, and 1200 rpm operating with 5% slip?

MOTOR TYPES

Chapter 2

Journeyman electricians and maintenance electricians install, connect, and maintain all types of motors. The majority of motors in the United States utilize alternating current (AC). These motors are either single-phase or three-phase and are designed to operate on single or dual voltages based on the connections of the windings of the stator.

Squirrel-cage induction motors are the most widely used motors because they have fewer parts, are less expensive, and require less maintenance than rotor, synchronous, and direct current (DC) motors.

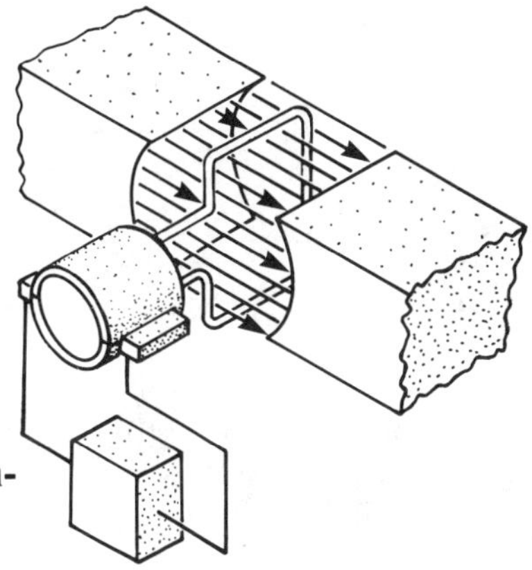

SINGLE-PHASE MOTORS

Single-phase motors operate only on single-phase electrical systems. The motor windings are wound for 120-volt single-phase, 240-volt single-phase, 480-volt single-phase (rarely used), or 120/240-volt single-phase. The most commonly used single-phase motors are

(1) split-phase
(2) capacitor-start
(3) capacitor start-and-run
(4) permanent split-capacitor
(5) shaded-pole
(6) universal

Split-phase Motors

Split-phase motors have two windings on the stator. These are the main (running) winding and the auxiliary (starting) winding. Starting windings are placed about 30° from the running windings to start the motor and produce the torque needed to drive the load.

The starting winding is connected in parallel with the running winding, and a centrifugal switch (starting switch) is placed in series with the starting winding. The starting switch has contacts, which are closed when the rotor is at rest. When the motor is started and begins to accelerate up to its running speed, the contacts open, taking the starting winding out of the circuit at about 75% to 80% of the motor's running speed. See Figure 2-1. When the motor comes up to its running speed, the starting winding drops out of the circuit and the split-phase motor operates as a single-phase induction motor.

Split-phase motors have high inrush current when starting because of the high resistance of the starting winding. Starting windings have small wires with many turns. Running (field) windings have larger size wire with fewer turns. The starting winding, with its smaller wire and greater number of turns, has a greater resistance than the running winding. The running winding has a lower resistance with its larger wire and fewer turns per winding. See Figure 2-2.

If the motor fails to start, either the contacts of the centrifugal switch or the starter winding is burned out. The starting winding is wound on top of the running winding on the stator of the motor. The starting winding has a higher resistance than the running winding because it is wound with smaller wire.

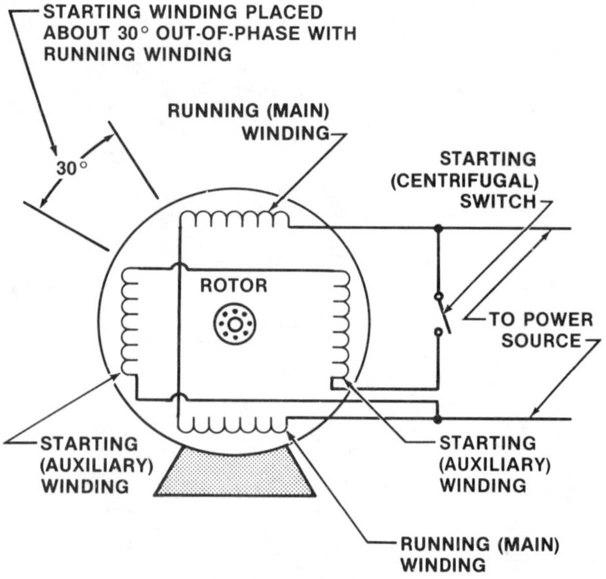

Figure 2-1. Starting windings help start split-phase motors and are removed from the circuit by a centrifugal switch when the motor reaches 75% to 80% of its running speed.

WINDING	RESISTANCE
Running	Low
Starting	High

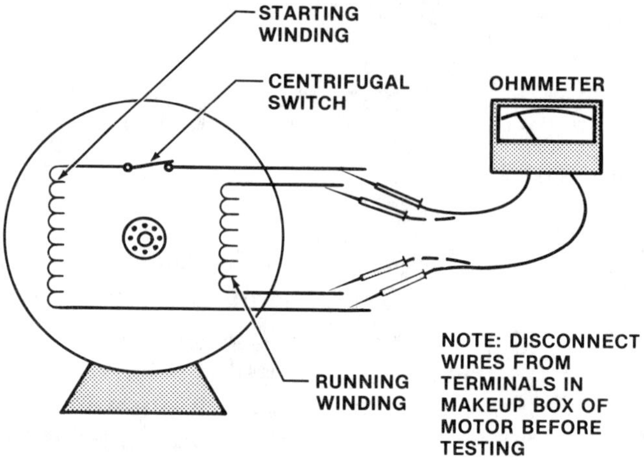

Figure 2-2. An ohmmeter is used to check running windings or starting windings. Running windings have lower resistance than starting windings.

Testing Windings. The following are the three main methods used to check the starting winding if the motor fails to start.

1. Turn the shaft of the motor by hand. If it starts, the trouble is in the starter winding circuit.

2. Disassemble the motor and check the starting winding for an open circuit.

3. Use a test light or an ohmmeter to determine that the winding is complete. Refer to Figure 2-2.

Reversing Rotation. Split-phase motors can be reversed by changing the flow of current through the running or starting windings. When the flow of current is in the same direction in the starting winding as in the running winding, the rotor will rotate in a counterclockwise direction. When the current flow is in the opposite direction in relation to the starting and running windings, the rotor will rotate in a clockwise direction. See Figure 2-3.

Lead Identification. The color coding of starting and running windings varies with the type of motor used. A typical color coding for newer motors is red for T_1, black for T_2, yellow for T_3, and blue for T_4. Older motors are usually tagged. M_1 and M_2 are used for the running winding, and S_3 and S_4 are used to tag the starting winding. R_1 and R_2 may also be used to tag the running winding.

Thermal Protection. Some split-phase motors are equipped with an additional switch to protect the motor from overheating. Overheating can be caused by lack of ventilation when the motor's inlets and outlets are covered with lint or dirt. Overheating can also be caused by high temperatures due to high currents in the windings from a stuck bearing in the motor or on the driven load.

Overload protection devices (bi-metal disks or strips composed of dissimilar metals) are connected in series with the running winding. The overload protector moderates the amount of current flow and temperature rise of the windings. If the current flow and temperature rise exceed the set value, the protector will open the circuit. When the running winding temperature decreases, the protector will connect the power supply and start the motor. *NOTE:* The overload protector will shut the motor down and restart it until the problem is corrected. See Figure 2-4.

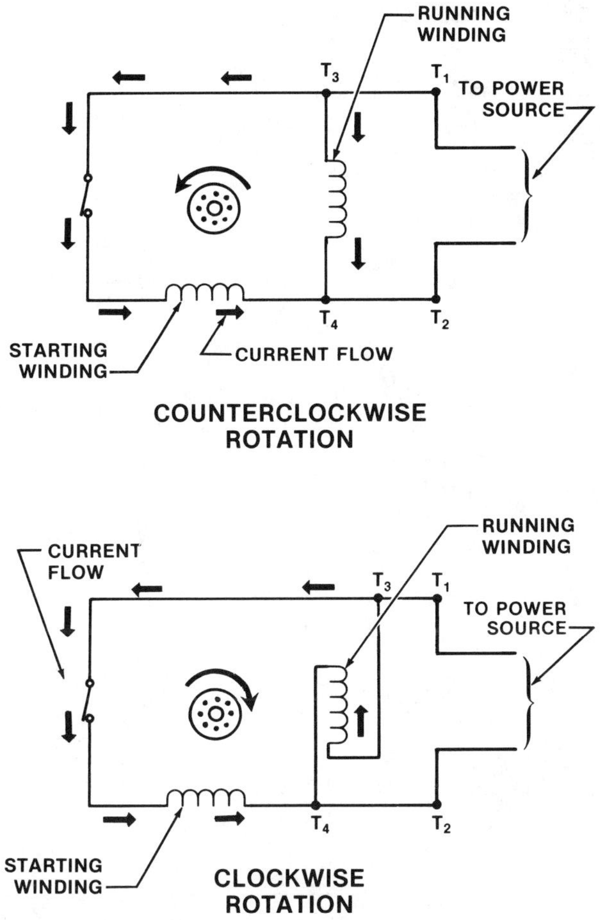

Figure 2-3. Split-phase motors are reversed by changing the direction of current through the starting and running windings.

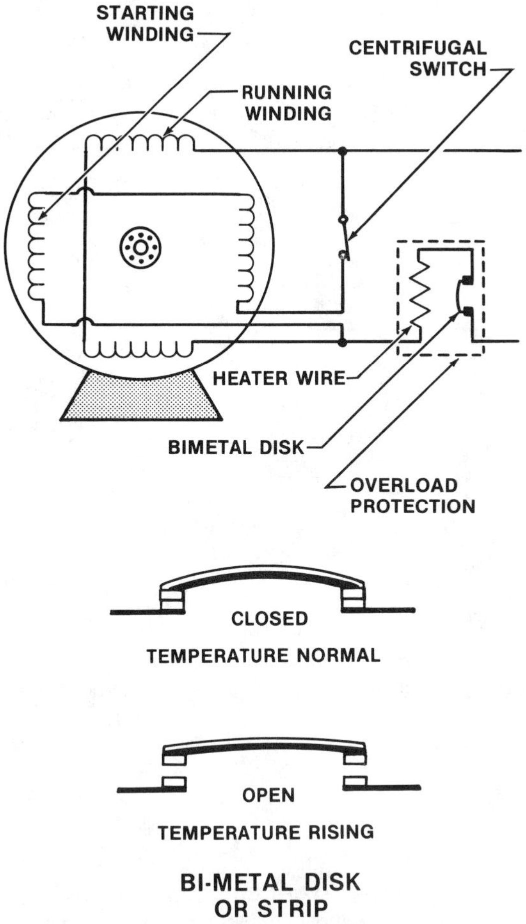

Figure 2-4. Bi-metal disks or strips provide overload protection to the running windings.

Capacitor-start Motors

Capacitor-start motors are equipped with starting and running windings. Also, the motor has a condenser (capacitor) to create a greater starting torque. The capacitor is connected in series with the starting winding and the centrifugal switch.

Current in the starting winding that is released by the capacitor leads the running phase voltage, obtaining a greater angle of displacement between the starting and running windings. This provides increased starting torque for the motor. For example, a split-phase motor with a capacitor has twice as much starting torque as a regular split-phase motor.

High resistance windings in split-phase motors cause the current to be more in phase with line voltage. Therefore, the starting torque of a split-phase motor is not as high as the starting torque of a capacitor-start motor. See Figure 2-5.

Selecting Capacitors from Manufacturer's Chart. Where the existing power factor for motors is known and the power factor is low, the size kVAR capacitors needed to correct the existing power factor can be found in the manufacturer's capacitor calculating chart in Figure 2-6.

The following is an example of how to use this chart. A 60-horsepower, three-phase motor pulls 77 amps at 460 volts. The power factor of the circuit supplying the motor is 70% and is to be increased to 90%. To increase the power factor to 90%, find the operating volt-amps of the motor due to the 70% existing power factor.

SPLIT-PHASE MOTORS	
Without Capacitor	With Capacitor
Line Current 67% Greater	Line Current 67% Less
Starting Torque 50% Less	Starting Torque Twice as Much

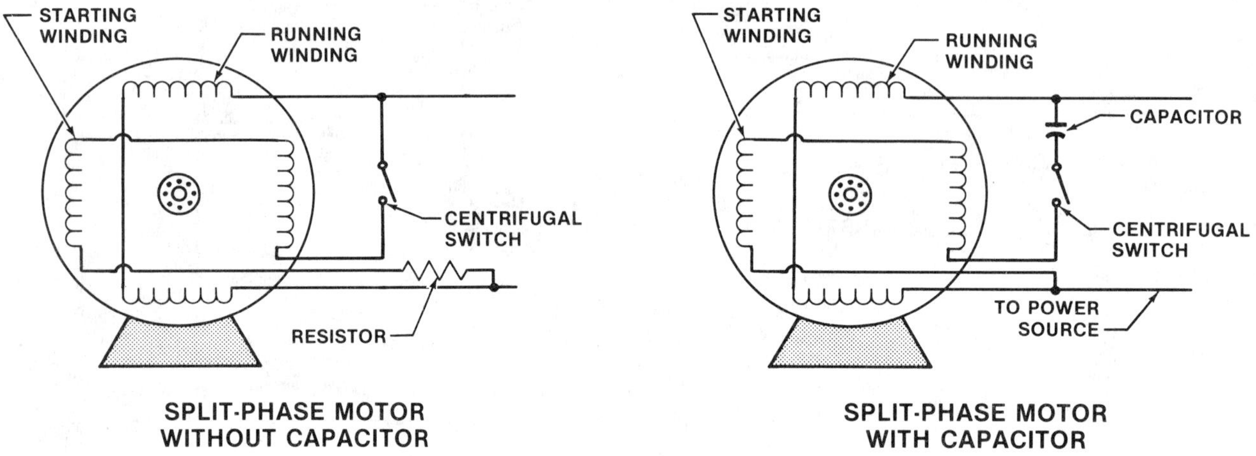

Figure 2-5. Capacitors provide increased torque for easy starting.

$$460 \text{ V} \times 1.732 \times 77 \text{ A} = 61{,}369 \text{ VA}$$

$$61{,}369 \text{ VA} \times 70\% = 42{,}958.3 \text{ VA}$$

Next, refer to the "existing power factor" column of the capacitor calculating chart and select 70%. In the "desired power factor" row, select 90%, which is the power factor to which the circuit is to be corrected. Locate the multiplier by finding the number at which the "existing power factor" (70%) and the "desired power factor" (90%) meet, which is .536.

The multiplier is used as follows:

$$\frac{42{,}958.3 \text{ VA}}{1000} \times .536 = 23.03 \text{ kVAR}$$

See Figure 2-6 for a step-by-step procedure for calculating the kVARs needed to correct power factor.

The size capacitor rated in 23.03 kVARs is selected from any manufacturer's chart. Smaller capacitor rated in microfarads (μf) is found by applying the following formula.

$$\mu f = \frac{159{,}300}{hertz} \times \frac{amps}{volts}$$

For example, the microfarads of a 15-horsepower, 230-volt, three-phase motor pulling 46 amps can be found by the following procedure.

Step 1: $$\mu f = \frac{159{,}300}{hertz} \times \frac{amps}{volts}$$

$$\mu f = \frac{159{,}300}{60} \times \frac{46 \text{ A}}{230 \text{ V} \times 1.732}$$

$$\mu f = 306.9$$

Answer: **306.9 μf**

Testing Capacitors. The starting and running windings can be identified by the same procedure used in Figure 2-2. However, capacitors must be checked by one of three methods: ohmmeter testing, in-line fuse testing, or screwdriver testing. See Figure 2-7.

Reversing Rotation. Capacitor-start motors can be reversed by changing the flow of current through the starting capacitor. Current flowing in the same direction through the capacitor, starting windings, and running windings produces a counterclockwise rotation of the motor. Current flowing through the capacitor and starting windings in one direction and through the running windings in the opposite direction produces a clockwise rotation of the motor. See Figure 2-8.

CAPACITOR CALCULATING CHART

%	80	81	82	83	84	85	86	87	88	89	90	91	92	93	94	95	96
50	0.982	1.003	1.034	1.060	1.086	1.112	1.139	1.165	1.192	1.220	1.248	1.276	1.306	1.337	1.369	1.403	1.442
51	.937	.962	.989	1.015	1.041	1.067	1.094	1.120	1.147	1.175	1.203	1.231	1.261	1.292	1.324	1.358	1.395
52	.893	.919	.945	.971	.997	1.023	1.050	1.076	1.103	1.131	1.159	1.187	1.217	1.248	1.280	1.314	1.351
53	.850	.876	.902	.928	.954	.980	1.007	1.033	1.060	1.088	1.116	1.144	1.174	1.205	1.237	1.271	1.308
54	.809	.835	.861	.837	.913	.939	.966	.992	1.019	1.047	1.075	1.103	1.133	1.164	1.196	1.230	1.267
55	.769	.795	.821	.847	.873	.899	.926	.952	.979	1.007	1.035	1.063	1.090	1.124	1.156	1.190	1.228
56	.730	.756	.782	.808	.834	.860	.887	.913	.940	.968	.996	1.024	1.051	1.085	1.117	1.151	1.189
57	.692	.718	.744	.770	.796	.822	.849	.875	.902	.930	.958	.986	1.013	1.047	1.079	1.113	1.151
58	.655	.681	.707	.733	.759	.785	.812	.838	.865	.893	.921	.949	.976	1.010	1.042	1.076	1.114
59	.618	.644	.670	.696	.722	.748	.775	.801	.828	.856	.884	.912	.939	.973	1.005	1.039	1.077
60	.584	.610	.636	.662	.688	.714	.741	.767	.794	.822	.850	.878	.905	.939	.971	1.005	1.043
61	.549	.575	.601	.627	.653	.679	.706	.732	.759	.787	.815	.843	.870	.904	.936	.970	1.008
62	.515	.541	.567	.593	.619	.645	.672	.698	.725	.753	.781	.809	.836	.870	.902	.936	.974
63	.483	.509	.535	.561	.587	.613	.640	.666	.693	.721	.749	.777	.804	.838	.870	.904	.942
64	.450	.476	.502	.528	.544	.580	.607	.633	.660	.688	.716	.744	.771	.805	.837	.871	.909
65	.419	.445	.471	.497	.523	.549	.576	.602	.629	.657	.685	.713	.740	.774	.806	.840	.878
66	.398	.414	.440	.466	.492	.518	.545	.571	.598	.626	.654	.682	.709	.743	.775	.809	.847
67	.358	.384	.410	.436	.462	.488	.515	.541	.568	.596	.624	.652	.679	.713	.745	.779	.817
68	.329	.355	.381	.407	.433	.459	.486	.512	.539	.567	.595	.623	.650	.684	.716	.750	.788
69	.299	.325	.351	.377	.403	.429	.456	.482	.509	.537	.565	.593	.620	.654	.686	.720	.758
70	.270	.296	.322	.348	.374	.400	.427	.453	.480	.508	.536	.564	.591	.625	.657	.691	.729
71	.242	.268	.294	.320	.346	.372	.399	.425	.452	.480	.508	.536	.563	.597	.629	.663	.701
72	.213	.239	.265	.291	.317	.343	.370	.396	.423	.451	.479	.507	.534	.568	.600	.634	.672
73	.186	.212	.238	.264	.290	.316	.343	.369	.396	.424	.452	.480	.507	.541	.573	.607	.645
74	.159	.185	.211	.237	.263	.289	.316	.342	.369	.397	.425	.453	.480	.514	.546	.580	.618
75	.132	.158	.184	.210	.236	.262	.289	.315	.342	.370	.398	.426	.453	.487	.519	.553	.591
76	.105	.131	.157	.183	.209	.235	.262	.288	.315	.343	.371	.399	.426	.460	.492	.526	.564
77	.079	.105	.131	.157	.183	.209	.236	.262	.289	.317	.345	.373	.400	.434	.466	.500	.538
78	.053	.079	.105	.131	.157	.183	.210	.236	.263	.291	.319	.347	.374	.408	.440	.474	.512
79	.026	.052	.078	.104	.130	.156	.183	.209	.236	.264	.292	.320	.347	.381	.413	.447	.485
80	.000	.026	.052	.078	.104	.130	.157	.183	.210	.238	.266	.294	.321	.355	.387	.421	.459

PROBLEM: The existing circuit to a motor has a power factor of 65%. How many kVARs will it take to correct the power factor to 96%? The motor is 125-horsepower, 460-volt, three-phase.

Step 1: Existing VA

460 V × 1.732 × 156 A = 124,332 VA

Step 2: 124,332 VA × 65% = 80,815.8 VA

Step 3: Find multiplier in chart.

$$\frac{80{,}815.8 \text{ VA}}{1000} \times .878 = 70.96 \text{ kVAR}$$

Answer: **70.96 kVAR**

Figure 2-6. Selecting the kVARs from the capacitor calculating chart based on the existing power factor of the motor to the correct power factor.

Capacitor Start-and-run Motors

Capacitor start-and-run motors are equipped with a starting and a running capacitor. Each capacitor has a different value. The running capacitor is connected in series with the starting winding and in parallel with the starting capacitor. The capacitance of both capacitors are added together in the circuit when the motor is started.

A high starting torque is provided when the motor is started and accelerated up to its running speed. When the motor reaches its running speed, the centrifugal switch opens the circuit and drops out the starting capacitor. The running capacitor is left in the running circuit to provide a higher running torque while the motor is in operation.

Starting capacitors may be *electrolytic* (composed of paper and foil to insulate the capacitor) or *oil-type* (enclosed in oil). Electrolytic starting capacitors are commonly used with motors having low horsepower ratings.

Running capacitors also may be electrolytic

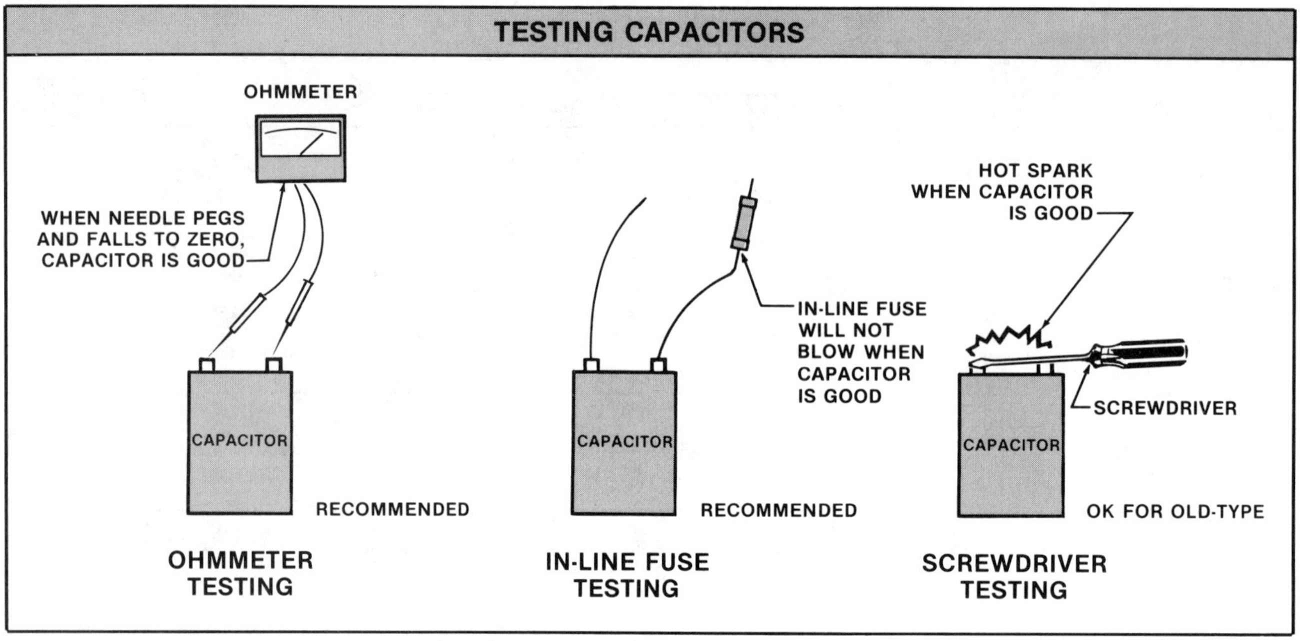

Figure 2-7. Capacitors are tested with an ohmmeter, in-line fuse, or screwdriver.

or oil-type. However, running capacitors are commonly oil-type because they provide a cooler operating range than electrolytic capacitors. See Figure 2-9. Starting and running windings can be identified by the procedure shown in Figure 2-2.

Reversing Rotation. The rotation of the motor can be reversed by following the procedure shown in Figure 2-8. Terminals utilized in reversing motor rotation are found upon removal of the nameplate cover. Reversing switches may also be used.

Permanent Split-capacitor Motors

Permanent split-capacitor motors use a capacitor connected in series with the running winding. The capacitor creates a phase lag in the

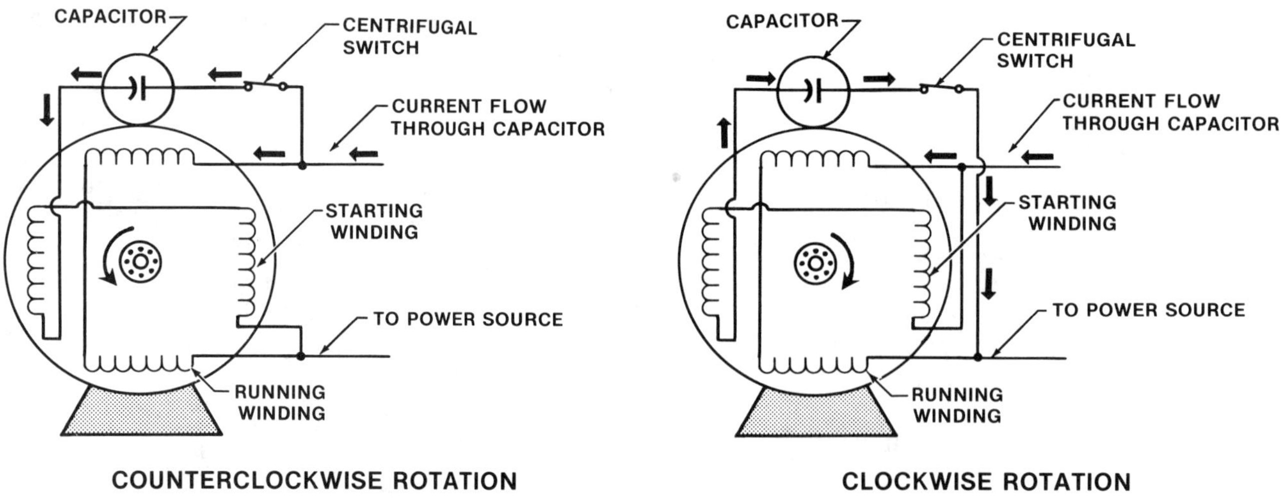

Figure 2-8. Motor rotation is determined by the direction of current flow through the capacitor and windings.

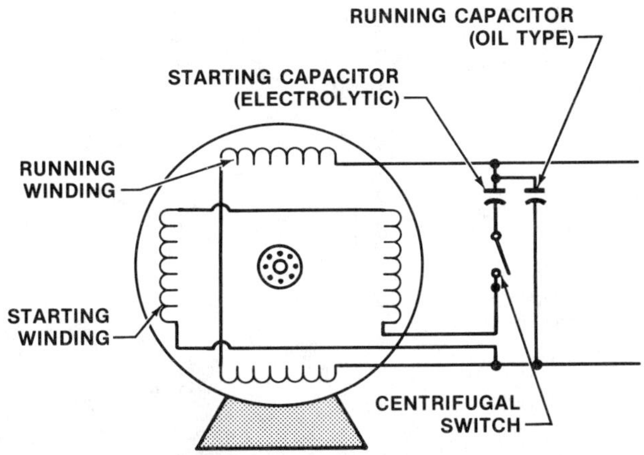

Figure 2-9. Capacitor start-and-run motors are equipped with two capacitors. One capacitor starts the motor and the other runs the motor.

starting winding needed to start the motor and drive the load.

The windings of a permanent split-capacitor motor can be identified by checking the resistance of the starting and running windings. Refer to Figure 2-2. Permanent split-capacitor motors are commonly called *single-value motors.*

The starting winding, with its capacitor, and the running winding remain in the circuit while the motor is in operation. The main difference between a permanent split-capacitor motor and a capacitor-start motor is that no centrifugal switch is needed for the permanent split-capacitor motor. A permanent split-capacitor motor cannot start and drive a load with a heavy starting torque. See Figure 2-10.

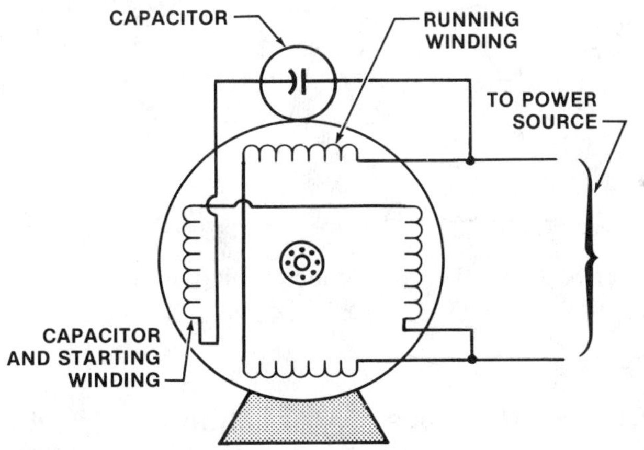

Figure 2-10. Permanent split-capacitor motors do not require a centrifugal switch. The capacitor is never moved from the circuit.

Reversing Rotation. Permanent split-capacitor motors can be reversed by using a reversing switch. The reversing switch changes the connection to the capacitor. In one position of the switch, the capacitor is in series with the starting winding. In the other position of the switch, the capacitor is in series with the field winding. By switching the capacitor, the rotation of the motor is changed from one direction to another.

Terminal leads may also be reversed to change the direction of motor rotation. However, most permanent split-capacitor motors are used in equipment requiring a reversing switch (for example, fans and blower motors). See Figure 2-11.

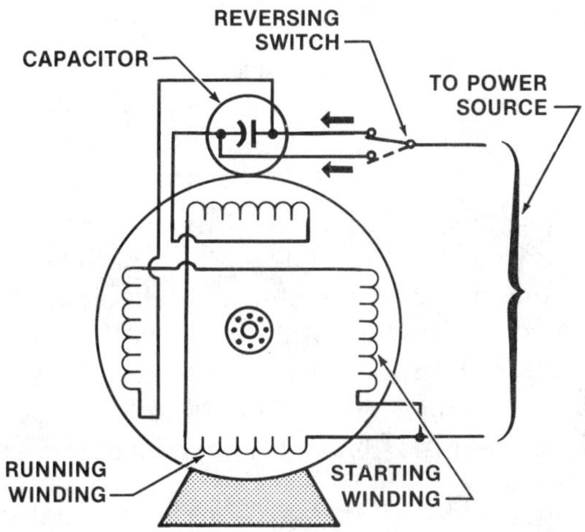

Figure 2-11. Permanent split-capacitor motors are commonly reversed using a reversing switch.

Shaded-pole Motors

The trailing edge of each pole of a shaded-pole motor is wound with a shaded coil. The shaded coil produces the slip to provide the torque required to start and run the load. Shaded-pole motors have very low starting torque and cannot be used with equipment that requires high starting torque. Each field pole is cut with a slot containing the shaded coil. The coil forms a closed circuit (loop) with the running windings wound around each field pole. See Figure 2-12.

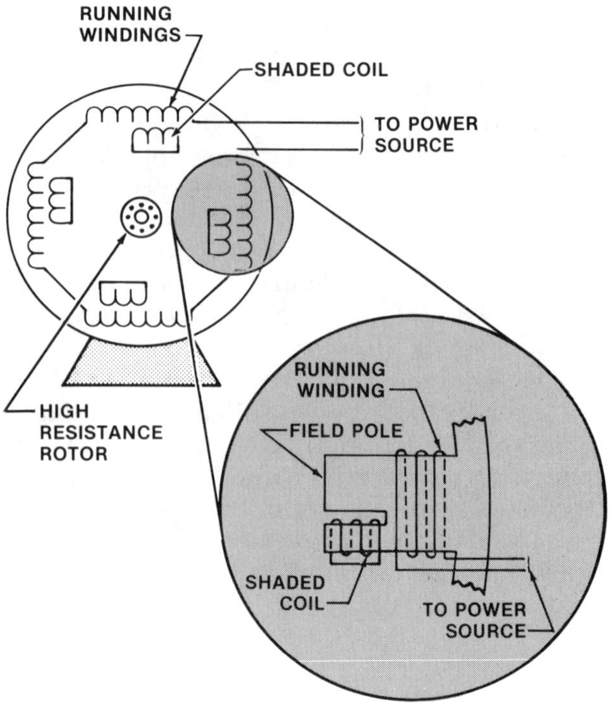

Figure 2-12. The shaded coil of a shaded-pole motor forms a closed circuit with the running windings.

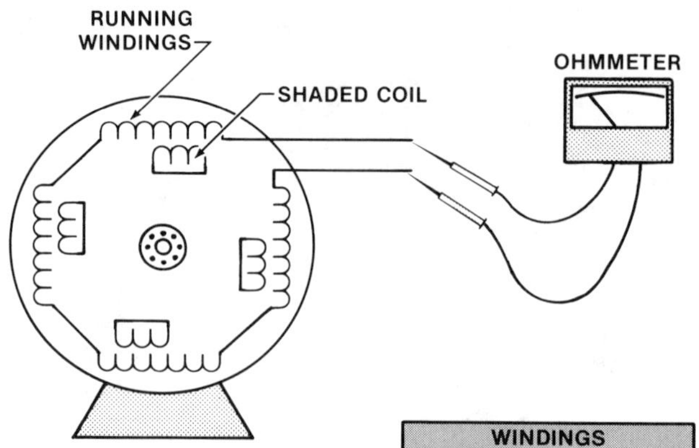

WINDINGS	
Ohmmeter has reading	Good
Ohmmeter has no reading	Bad

Figure 2-13. An ohmmeter is used to measure resistance of the windings in a shaded-pole motor.

When power is applied to the running windings, a magnetic field is set up between the poles and the rotor. The shaded coil cuts through a portion of the magnetic field, creating an out-of-phase condition with the flux lines. A two-phase magnetic field is created, and the phase shifting provides the torque needed to start and rotate the driven load.

Shaded-pole motors use a high-resistance rotor to provide additional starting torque. The high-resistance rotor causes the motor to have a higher slip and poor speed regulation. Shaded-pole motors are available for 115-volt, 230-volt, and dual-voltage operation.

Testing Windings. The running windings can be measured with an ohmmeter. If the windings are good, a reading can be taken. The ohmmeter will show no continuity if there is a broken winding. A test light may also be used for determining continuity. See Figure 2-13.

Reversing Rotation. A shaded-pole motor can be reversed using one of the following methods.

(1) The rotor can be placed in the motor housing in the opposite direction.

(2) Two sets of field windings can be used with each shaded coil.

(3) A switch can open or close the circuit to the correct windings for the direction of rotation. The rotor will always rotate toward the shaded coil. See Figure 2-14.

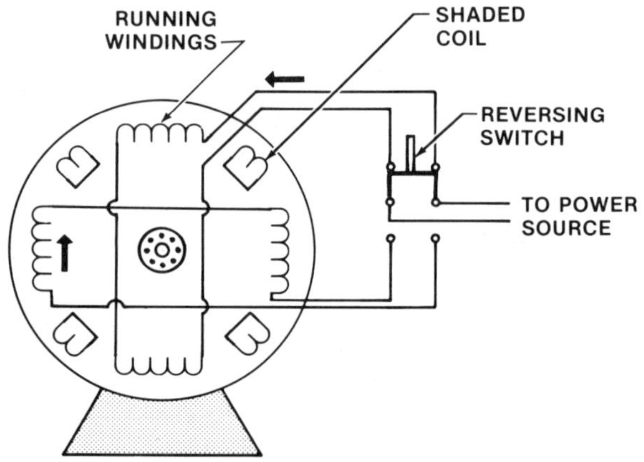

Figure 2-14. Shaded-pole motors may be reversed by using two sets of field (running) windings.

Changing Speeds. A transformer can be tapped to provide different speeds for a shaded-pole motor. For example, a transformer may be constructed to provide three speeds: low (L), medium (M), and high (H). Tapping all the windings produces low speed. Tapping half the windings produces medium speed. Tapping none of the windings produces high speed. See Figure 2-15.

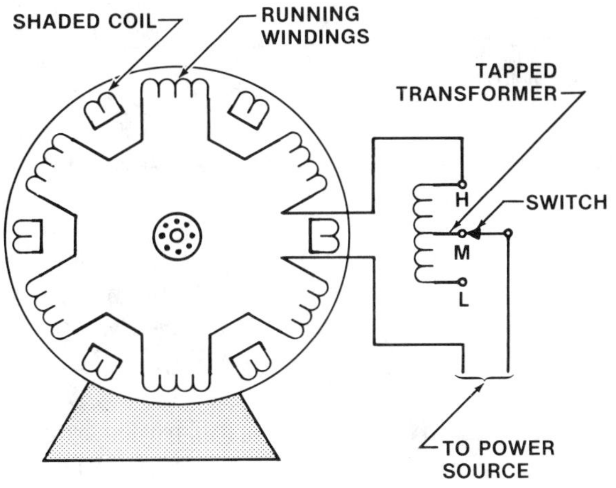

Figure 2-15. Tapped transformers are used to provide different speeds for shaded-pole motors.

Two speeds can be obtained by providing multiple (six) windings. The windings are located in separate slots in the same manner as in a single-speed motor. Five of the windings are wound with the same size wire, while the sixth winding is wound with smaller wire with more turns. The low-speed operation of the motor uses all six windings in the circuit. The high-speed operation has only five of the windings in the circuit. See Figure 2-16.

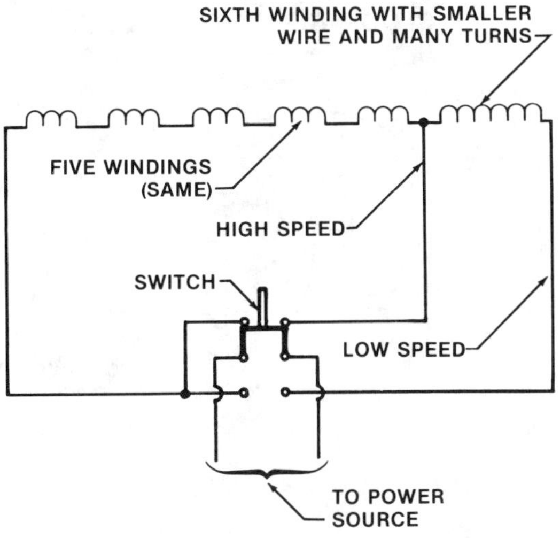

Figure 2-16. Six windings provide two motor speeds.

Universal Motors

Universal motors are small, series-wound motors that operate on AC or DC voltage. The motor will perform in the same manner whether it is used on an AC or a DC system. Universal motors are usually designed and built in sizes of $^3/_4$ HP or less. Fractional horsepower universal motors range from $^1/_{150}$ HP or less.

Universal motors are equipped with field windings and an armature with brushes and a commutator. The commutator keeps the armature turning through the magnetic field of the field windings. It also changes the flow of current in relation to the field windings and armature so there is a push-and-pull action. This push-and-pull action is created by the north and south poles of the field windings and armature. See Figure 2-17.

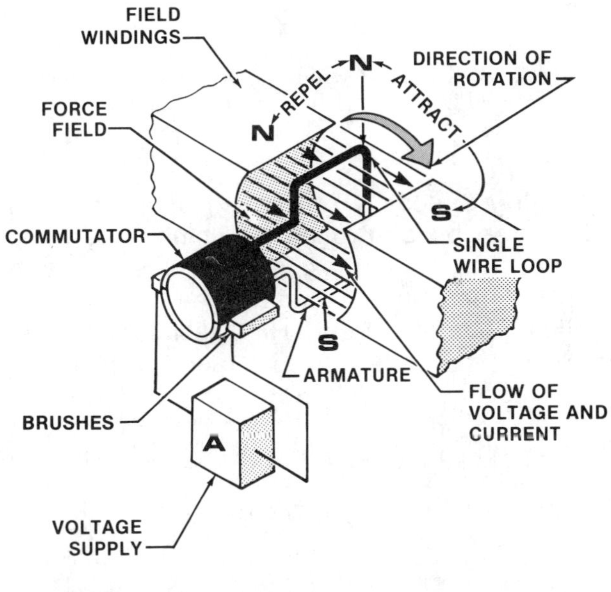

Figure 2-17. A universal motor has field windings, a commutator, brushes, and a single loop of wire representing the armature.

The north pole of the field windings pulls the south pole of the armature (loop) into the main strength of the magnetic field (field force). The commutator and brushes reverse the current flow through the armature, creating a north pole in the loop. The north pole of the field winding then repels the north pole of the armature. This push-and-pull action rotates the armature through the magnetic field of the field windings, establishing motor operation.

When the universal motor operates on AC, the current is constantly changing direction in the field windings. Both the armature and field windings have their current reversed simultaneously. Therefore, the motor operates similar to an inductive motor. The field windings of a universal motor are connected in series with the brushes and armature. See Figure 2-18.

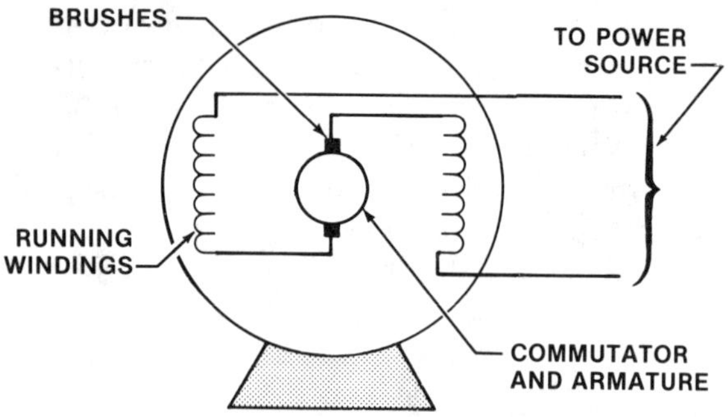

Figure 2-18. The field windings of a universal motor are connected in series with the brushes and armature.

Changing Speeds. The speed of a universal motor can be controlled by using a variable resistor. The resistor is connected in series with one of the motor winding leads, creating resistance in the circuit. High resistance gives lower motor speeds. Low resistance gives higher motor speeds. See Figure 2-19.

Three speeds can be obtained by tapping one of the field windings. All of the tapped winding is used for slow speed. Half of the winding is used for medium speed. The tapped winding is bypassed for fast speed. See Figure 2-20.

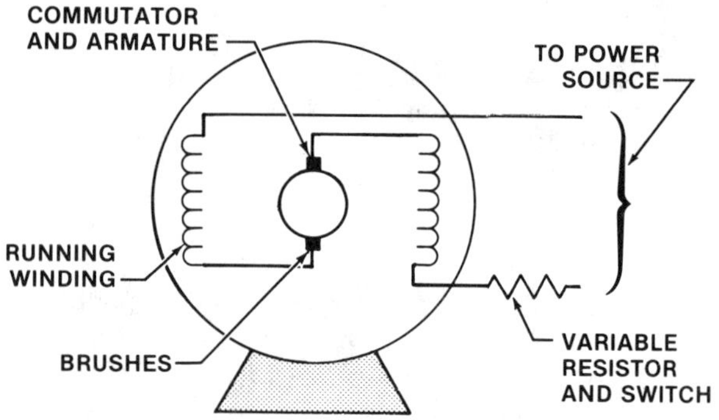

Figure 2-19. A variable resistor produces high resistance in the circuit for low motor speeds and low resistance in the circuit for high motor speeds.

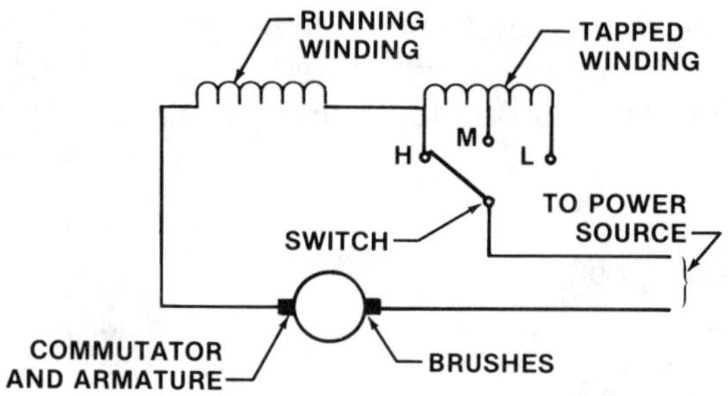

Figure 2-20. A tapped winding is used to provide three speeds for a universal motor.

Reversing Rotation. Universal motors can be reversed by changing the flow of current through the armature. This may be accomplished through the use of a reversing switch or by interchanging the leads on the terminals. See Figure 2-21.

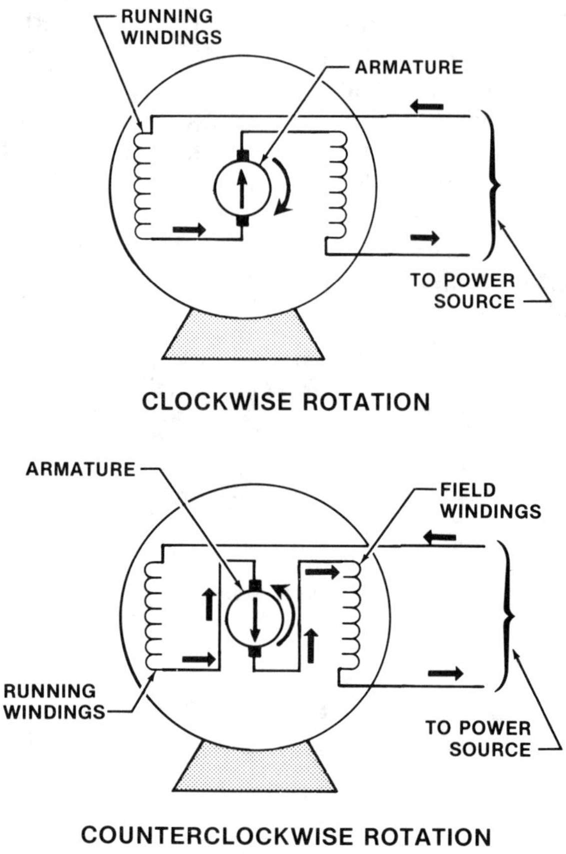

Figure 2-21. The direction of rotation can be reversed by changing the flow of current through the armature.

Testing Windings. The continuity of the field windings can be measured by using an ohmmeter or a continuity tester. If the ohmmeter does not give a reading, the following checks should be made.

(1) Check brushes.
(2) Check for broken leads.
(3) Check for dirty commutator.
(4) Check spring tension.

If the brushes are not properly set against the commutator, the motor will have poor speed control or will not run. The ohmmeter needle will peg if the winding has continuity from point to point. If the ohmmeter needle is not stable or does not show continuity, the winding is broken or has a loose connection.

THREE-PHASE MOTORS

Three-phase motors operate on three-phase systems. The windings can be wound for dual-voltage connections. The motor windings are connected in parallel for 240-volt operation and in series for 480-volt operation. The most commonly used three-phase induction motors are:

(1) squirrel-cage
(2) wound-rotor
(3) synchronous

Squirrel-cage Motors

Squirrel-cage induction motors are equipped with a stator (field poles) and rotor. See Figure 2-22. The field poles on the stator create a magnetic field from the north to south poles. This magnetic field induces voltage and current into the rotor. The rotor is designed with either low or high resistance for starting the driven load. When a rotor has low resistance, the motor has low starting torque with high inrush starting current. A motor with a high-resistance rotor has high starting torque and low inrush starting current. The speed regulation of a motor with a high-resistance rotor is not as accurate as the speed regulation of a motor with a low-resistance rotor.

Three-phase motors have three separate windings per pole on the stator. The three-pole windings have currents that are spaced 120° apart electrically on the stator. Voltages flowing through the windings are 120° out-of-phase from each other. See Figure 2-23.

The currents rotate around the stator windings, and this establishes a rotating magnetic field. The direction that the magnetic field rotates determines the direction that the rotor

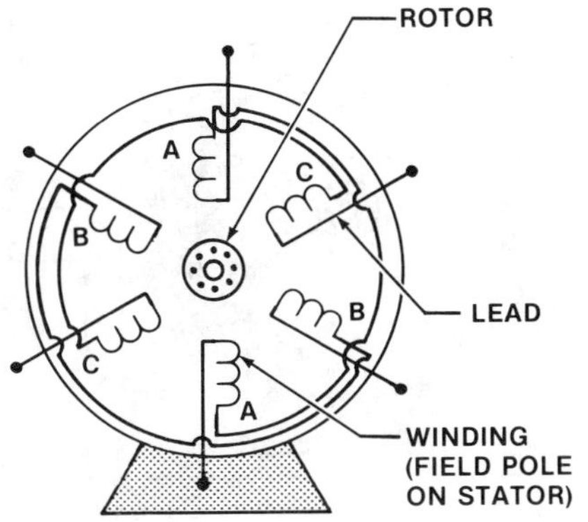

Figure 2-22. Squirrel-cage motors have a stator (field poles) and a rotor.

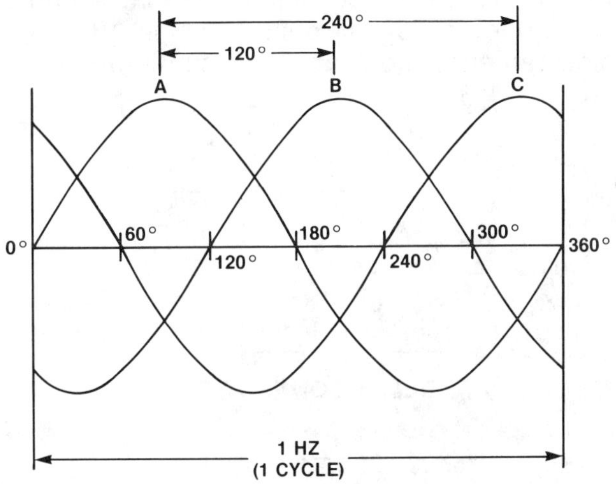

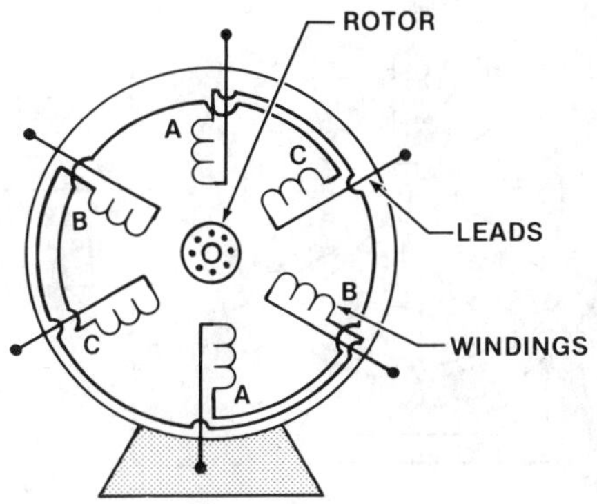

Figure 2-23. Voltage from B to A is 120° behind A. Voltage from C to A is 240° behind A.

will turn. Three-phase motors do not require an additional winding to start as single-phase motors do. Three-phase motors do not need the additional components that single-phase motors must have to start. Therefore, less maintenance is required for the operation of three-phase motors.

Connection of Windings. The windings of squirrel-cage induction motors can be connected to operate either wye or delta. The delta-connected windings are closed and form a triangle configuration. Wye-connected windings form a Y configuration. The windings of a motor can be designed with either six or nine leads to connect the windings to the three-phase supply lines.

Each winding of a three-phase induction motor has its leads marked with numbers for easy connection. A six-lead motor with the internal windings identified for connecting the motor for delta operation is shown in Figure 2-24. The winding leads are connected so that A and B close one end of the delta (triangle). B and C close an end, and C and A close the other end to form a closed-delta connection of the motor windings.

Squirrel-cage induction motors are also available with nine leads to connect the internal windings for delta operation. Six internal windings are connected together to form a closed delta. Three windings are marked 1-4-9, 2-5-7, and 3-6-8. The windings are connected to operate as a closed-delta system. The windings can be wound for single- or dual-voltage operation. See Figure 2-25 for a nine-lead closed-delta connection of the motor windings.

The windings of most squirrel-cage induction motors are wye-connected. One lead of each winding is connected to form the wye or star connection. The three remaining leads are connected to the three-phase supply lines L_1, L_2, and L_3. Figure 2-26 shows the connections for a six-lead motor for wye operation.

A nine-lead wye-connected motor has three leads of its windings connected to form a wye with three remaining leads (7-8-9). The three remaining windings are numbers 1-4, 2-5, and 3-6. The windings may be connected to operate on low or high voltage. For low-voltage operation, the windings are connected in parallel. For high-

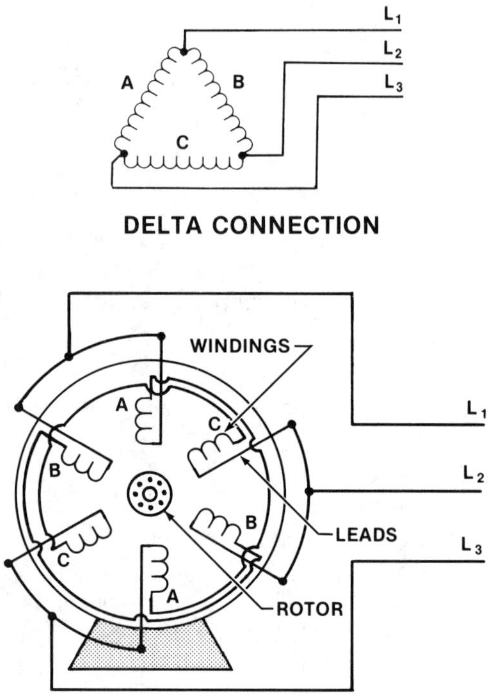

Figure 2-24. Connections of the internal windings of a six-lead squirrel-cage induction motor for delta operation.

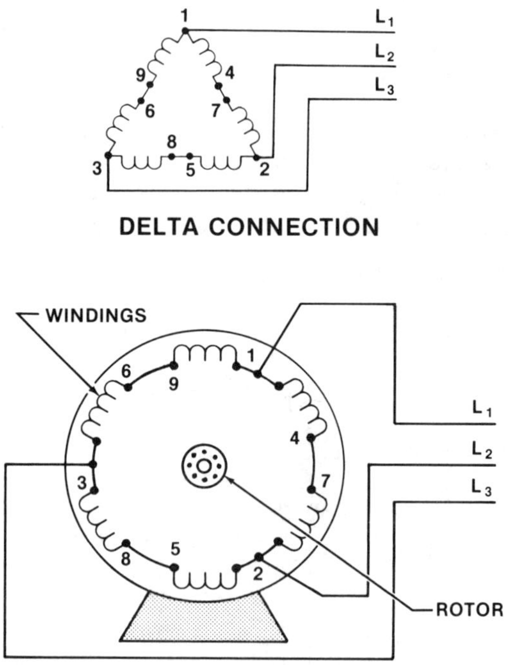

Figure 2-25. Connections of the internal windings of a nine-lead squirrel-cage induction motor for delta operation.

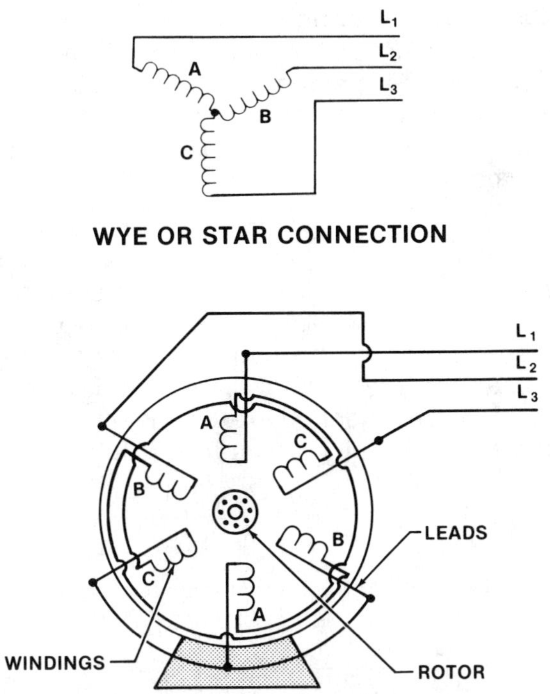

Figure 2-26. Connections of the internal windings of a six-lead squirrel-cage induction motor for wye operation.

voltage operation, the windings are connected in series. This method applies for wye- or delta-connected windings. See Figure 2-27.

Testing Leads. The leads of a squirrel-cage induction motor can be tested by using an ohmmeter or test light. This is accomplished by testing the continuity of the windings from L_1, L_2, and L_3. Measure the resistance from L_1 to L_2, L_1 to L_3, and L_2 to L_3. If there is a reading from these windings, the motor windings are usually good, and the problem is elsewhere in the circuit. See Figure 2-28.

Reversing Rotation. The rotation of any three-phase squirrel-cage induction motor can be reversed by interchanging any two of the three-phase leads. Using windings A, B, and C as a reference, the C winding will follow the B winding rather than the A winding. By reversing the polarity through the windings, the rotating magnetic field will rotate in the opposite direction, carrying the rotor with it. See Figure 2-29.

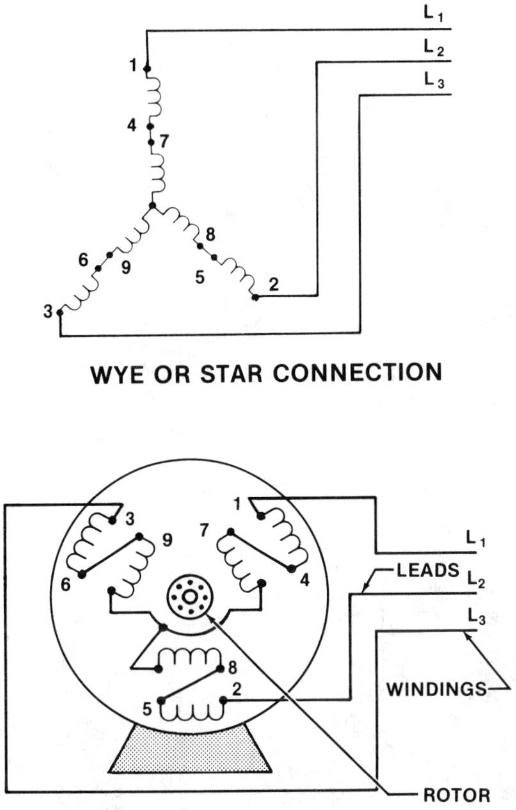

Figure 2-27. Connections of the internal windings of a nine-lead squirrel-cage induction motor for wye operation.

Wound-rotor Motors

Wound-rotor motors are similar in design to squirrel-cage induction motors; however, there is a difference in construction and operation. The stator in a wound-rotor motor is equipped with field poles to produce the magnetic field which induces voltage and current into the windings of the rotor. The rotor of a wound-rotor motor has insulated windings that are connected together to make up the rotor. These windings are not permanently short-circuited, and resistance from a resistor bank can be inserted into the rotor for starting and speed control.

Wound-rotor motors have two sets of leads. One set is the main leads to the motor windings (field poles) and the other set is the secondary leads to the rotor. One end of the secondary leads connects to the rotor through the slip rings, and the other end of the leads connects

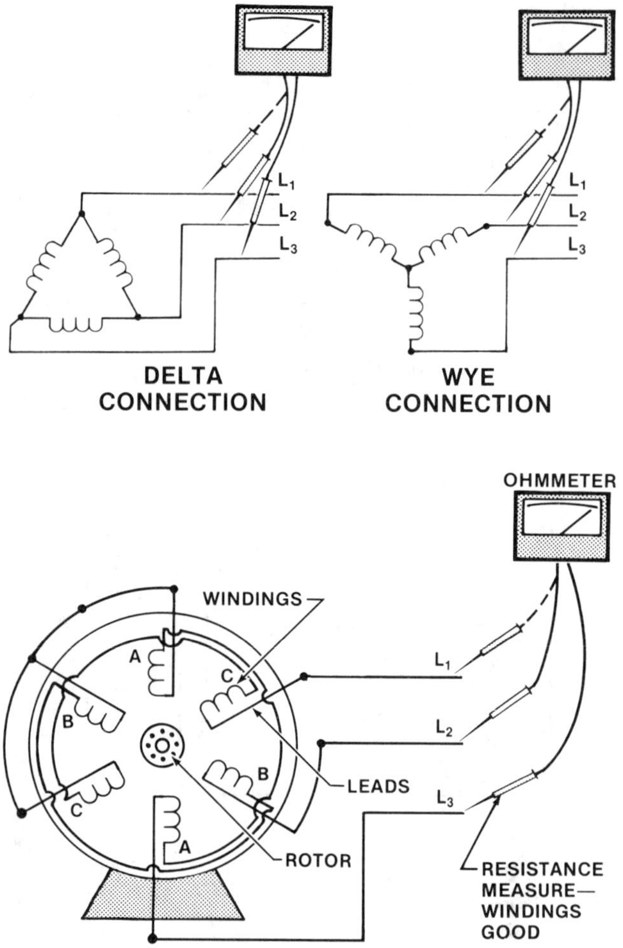

Figure 2-28. The continuity of the internal windings of a squirrel-cage induction motor can be tested with an ohmmeter or test light.

through a controller to a set of resistors. See Figure 2-30.

Starting. The starting principle of the wound-rotor motor is identical to the squirrel-cage induction motor except that the starting current can be varied by inserting resistance into the wound rotor. Usually all the resistance of the resistor bank is inserted into the rotor when starting and accelerating the motor up to its running speed.

If the resistance in the resistor bank is used only for starting and not for speed control also, the leads of the resistor can be sized at 85% or less of the secondary current of the motor. This reduction is permitted in Table 430-23(c) because the resistor conductors from the drum controller carries current during the starting period only. Reduced sizing of the resistor leads is determined by the percentages listed in Table 430-23(c). Each percentage used is selected according to the starting duty classification of each motor.

Speed Control. If the resistor leads are used for speed control as well as lowering the starting current, 35% to 110% of the secondary current can be used to size the leads per Table 430-23(c). For example, the resistor leads can be sized at 35% of the secondary current of the motor if the motor is used for light starting duty and no

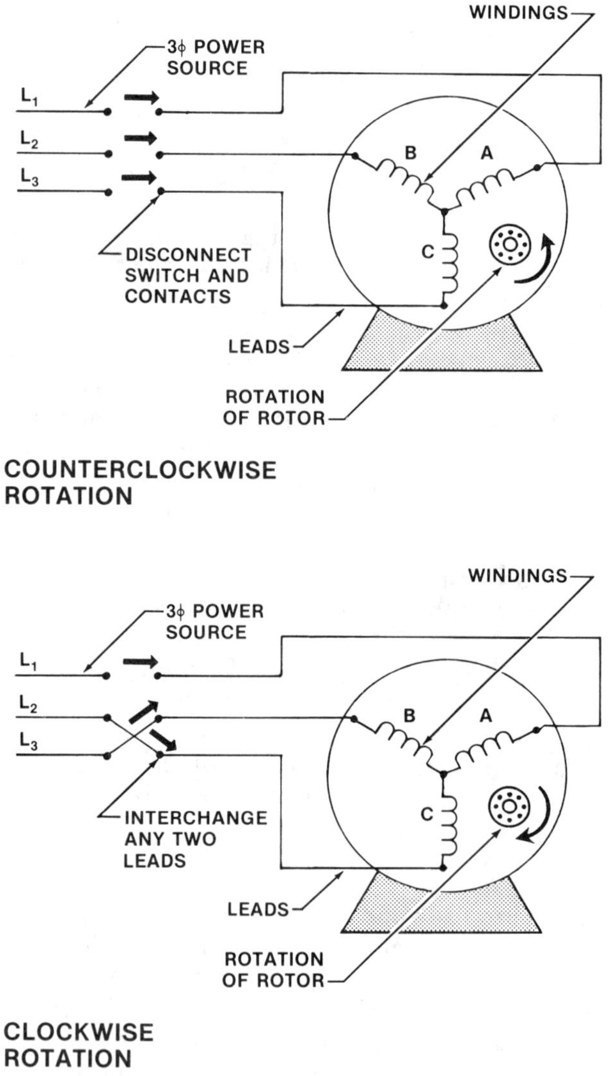

Figure 2-29. The rotation of a three-phase squirrel-cage induction motor is reversed by interchanging any two leads.

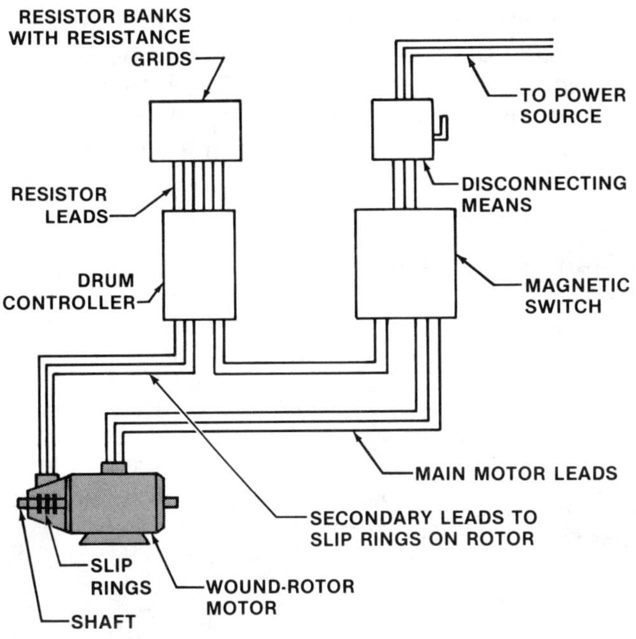

Figure 2-30. Wound-rotor motors have two sets of leads that connect the main motor leads and secondary leads for starting and running the motor.

speed control. If the motor is used for heavy intermittent duty starting and no speed control, the resistor leads are sized at 85% of the secondary current. The secondary current can be found on the motor's nameplate or from the manufacturer.

If the resistor leads are used for controlling the speed of the motor on a continuous basis, the leads must be sized at 110% of the secondary current of the motor. Use 110% when the resistor leads are used for both starting the motor and controlling the motor's speed.

Wound-rotor motors provide two to three times the amount of starting torque of a squirrel-cage motor of the same horsepower rating. By changing the amount of resistance to the rotor through the slip rings, the starting torque of the wound-rotor motor can be varied to accommodate the driven load. *NOTE:* An induction motor requires the rotor to be designed class A, B, C, or D to provide different amounts of torque, and the torque is fixed for each motor design.

Sizing and Selecting Conductors. Two main sets of leads (conductors) must be sized to wire a wound-rotor motor. These are the conductors that supply the main field poles of the motor and the secondary leads to the slip rings and rotor. The leads from the resistor bank to the drum controller must be sized to add or subtract resistance from the rotor. These leads are in addition to the two main sets of conductors to the wound-rotor motor.

Figure 2-31 shows the procedure for selecting the main conductors to the motor windings, the secondary conductors to the rotor, and the resistor conductors between the controller and resistor bank.

If the resistor bank in Figure 2-32 is used for starting and the driven load is rated for heavy intermittent duty, the secondary current is multiplied by 85% instead of 110%, per Table 430-23(c).

Testing. Slip rings can be checked for breaks or openings with a mirror. The slip rings must be set against the rotor. The speed of the motor cannot be regulated if the slip rings are not making contact with the rotor. If the slip rings are not broken or open, check the windings of the resistor bank.

Reversing Rotation. The direction of rotation can be reversed by interchanging any two of the

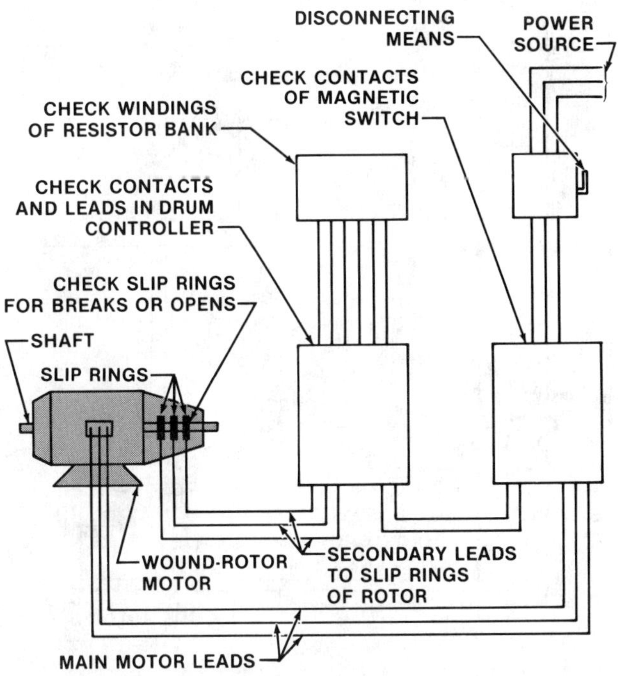

Figure 2-31. Main conductors are selected for the motor windings and the secondary conductors for the rotor.

three-phase conductors supplying the windings of the wound-rotor motor. Refer to Figure 2-29.

Synchronous Motors

Synchronous motors are similar to three-phase squirrel-cage induction motors in that they have stators, but synchronous-motor rotors are excited by direct current. Circuits are run from a control panel to the stator and rotor. A DC generator is either mounted on the end of the motor or remotely installed. See Figure 2-33.

The AC power supply of a synchronous motor is fed to the field pole windings of the motor. The DC power supply is fed into the rotor through the slip rings. North and south poles are set up in the rotor by the DC power supply. There is a rotor pole present for every stator pole. The AC power sets north and south poles on the stator poles. A stator north pole interlocks with a rotor south pole and the rotor moves in the same direction as the stator's magnetic field, creating synchronous speed. There is no percentage of slip as there is in a squirrel-cage induction motor. The rotor of an induction motor turns behind the strength of the magnetic field. The rotor of a synchronous motor turns with the strength of the magnetic field produced by the AC power supply. See Figure 2-34.

Starting. A synchronous motor cannot be started by itself. It is started by using another

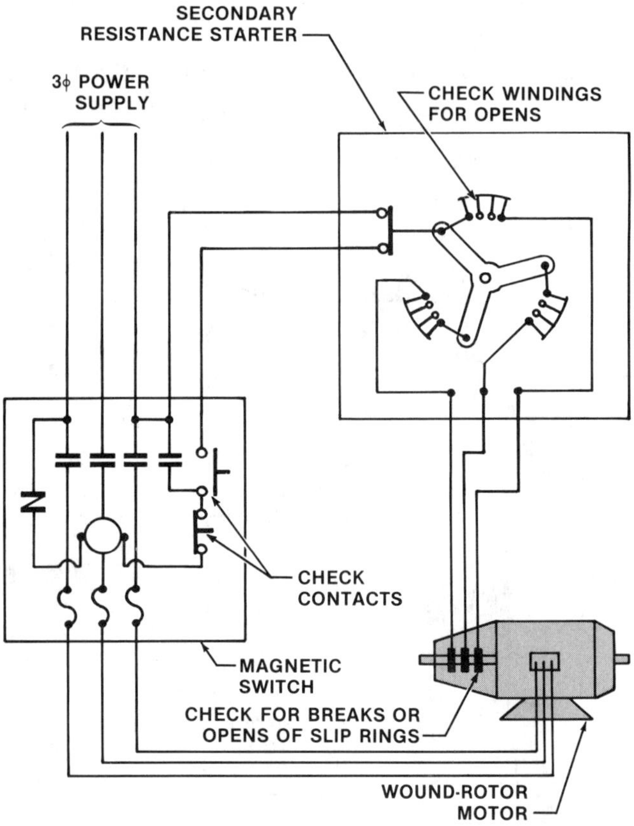

Figure 2-32. The resistance in a resistor bank is used to start and run a motor. It can also be used to control speed.

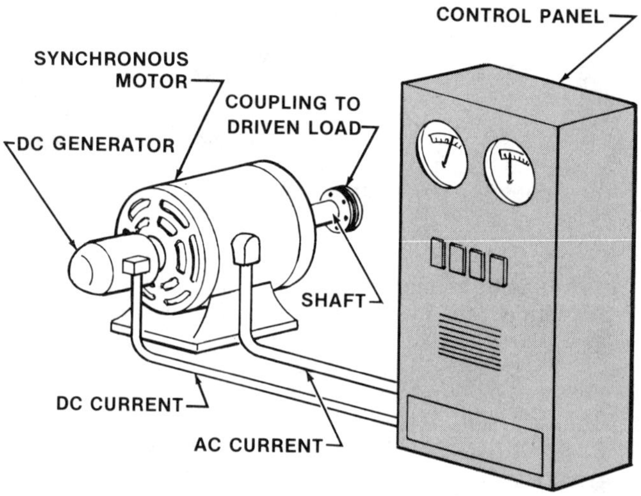

Figure 2-33. Rotors on synchronous motors are excited by direct current.

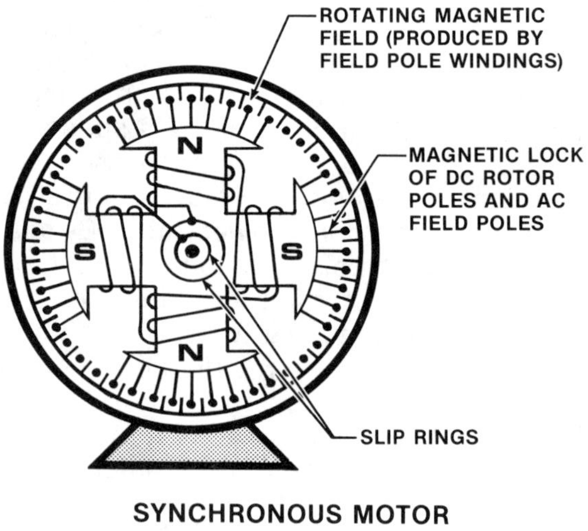

Figure 2-34. In a synchronous motor, the north pole of the stator locks with the south pole of the rotor, which moves along with the field pole's magnetic field at synchronous speed.

motor or providing a damper winding over the regular wound rotor. Synchronous motors cannot be started by themselves because the torque is zero when voltage is applied. An auxiliary motor can be used to bring the synchronous motor almost up to full speed. The auxiliary motor is then disconnected from the line and the synchronous motor accelerates to its synchronous running speed on its own power.

Damper windings can also be used to start synchronous motors. They are placed over the rotor windings to induce self-starting and bring the rotor up to running speed. The DC power is removed on starting but inserted when the motor approaches operating speed. With the DC winding excited, the rotor locks into step with the rotating magnetic field. See Figure 2-35.

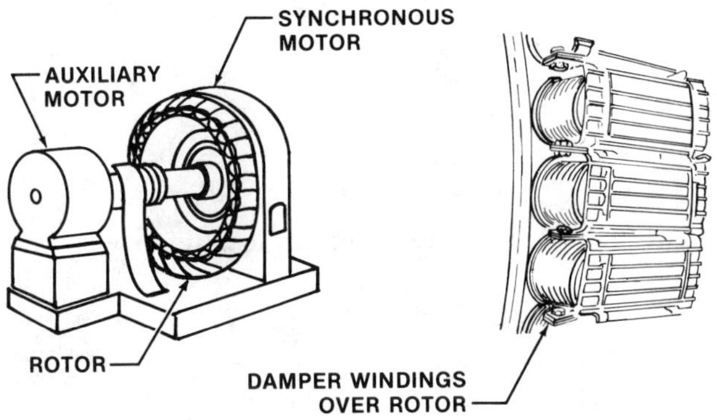

Figure 2-35. Synchronous motors can be started by using an auxiliary motor or by placing damper windings over the rotor.

Small synchronous motors can be started without an auxiliary motor or damper winding. They may be started manually or with special armature windings. The rotors of small synchronous motors are small and light and require very little torque for starting. These rotors are made of solid bars that are slotted. Small brass (or copper) bars are placed in the slots of each pole. An end ring is used to connect these bars and effectively short-circuit them. The rotor reacts like a squirrel-cage induction motor and starts without the aid of another motor or damper winding.

Synchronous motors are sometimes used in industrial applications where a number of induction motors create a poor power factor. Induction motors and transformers have currents that lag the voltage. The greater the lag is, the poorer the power factor will be. By applying a greater amount of direct current to the field than required, the current will lead the voltage, which improves the power factor. The synchronous motor acts as a capacitor and operates as a synchronous capacitor in that it supplies capacitance to the circuit.

Testing. If the rotor of a synchronous motor will not lock into step with the rotating magnetic field, the exciter must be checked. To check the exciter output turn the rotor by hand to measure the output. Check the DC source for a blown fuse or open control device. Check the Variac windings for opens or breaks.

Windings on the AC field poles can be checked with an ohmmeter. If resistance is measured from the leads, the windings are considered good. See Figure 2-36.

REPULSION MOTORS—SINGLE PHASE

Repulsion motors have been replaced by capacitor-start motors. Repulsion motors have more parts and require more maintenance than capacitor-start motors. The stator of a repulsion motor is wound with windings only. The rotor is an armature with brushes and a commutator. Repulsion motors are classified as follows:

(1) standard repulsion motors
(2) repulsion-start induction motors
(3) repulsion-induction motors

Standard Repulsion Motors

Standard repulsion motors operate on single-phase AC power. These motors are equipped with a short-circuiter and brush-lifting mechanism. The repulsion motor operates by the principle of magnetic repulsion. The motor has high starting torque and if lightly loaded it will run at high speeds. The direction of rotation can be reversed by shifting the brushes to the opposite side of the neutral plane. See Figure 2-37.

Repulsion-start Induction Motors

Repulsion-start induction motors start in the same manner as standard repulsion motors. However, repulsion-start induction motors have

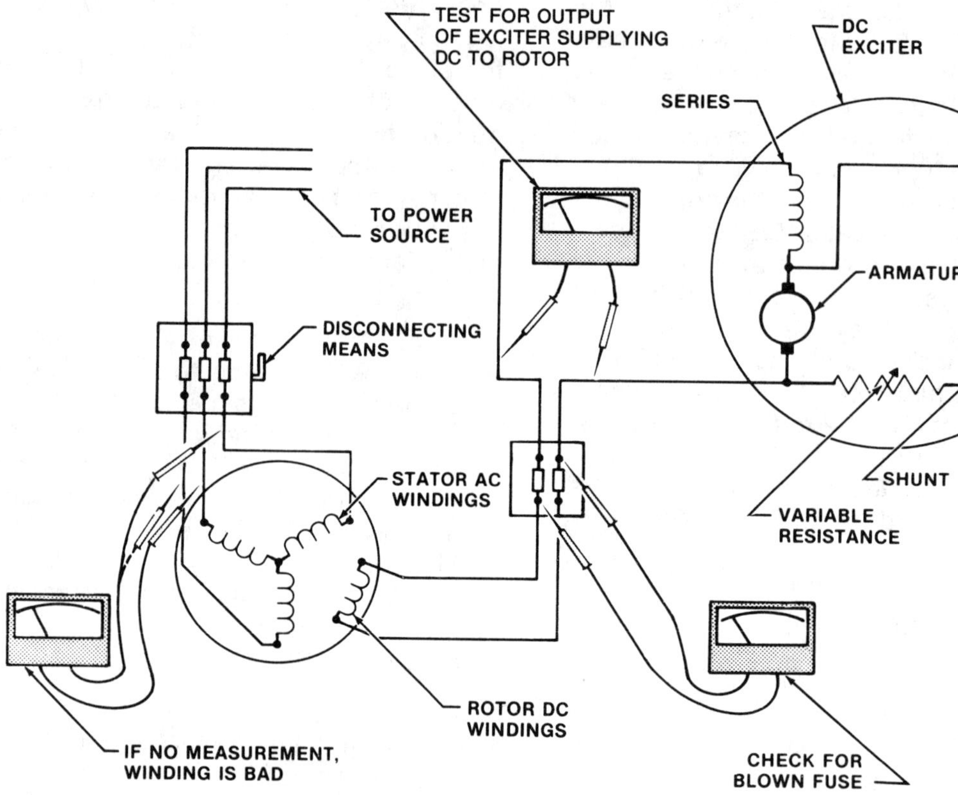

Figure 2-36. The DC exciter must be checked if the rotor of a synchronous motor will not lock into step.

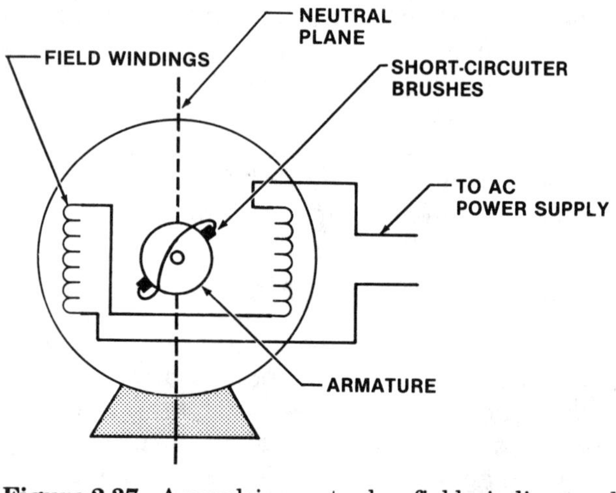

Figure 2-37. A repulsion motor has field windings and a wound rotor with brushes and a commutator.

a centrifugal mechanism that converts the motor operation to a single-phase induction motor when the armature has accelerated near its operating speed. The centrifugal device is a short-circuiter that cuts off the current flow through the brushes. The armature becomes a rotor and rotates through the rotating magnetic field of the field pole windings. See Figure 2-38.

Repulsion-induction Motors

Repulsion-induction motors are equipped with a rotor that combines a squirrel-cage rotor construction and a drum winding with a commutator and short-circuiter brushes. The drum winding produces the starting torque by magnetic repulsion to accelerate the driven load. The squirrel-cage winding provides additional torque to the torque produced by repulsion.

Both windings are in operation as the rotor turns through the rotating magnetic field produced by the field pole windings. This type of repulsion motor does not have a short-circuiter or brush-lifting mechanism. The repulsion-induction motor can be reversed by changing the position of the brushes in relation to the neutral plane. See Figure 2-39.

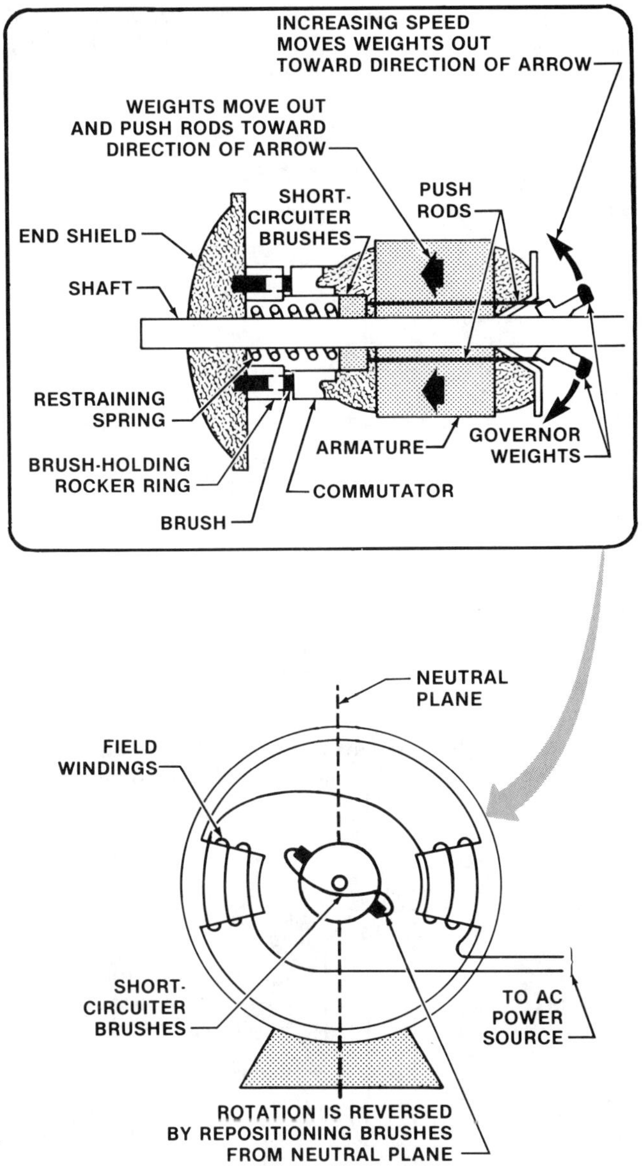

Figure 2-38. A centrifugal force operates the short-circuiter and brush-lifting mechanisms of the repulsion-start induction motor to develop a normal induction motor.

Reversing Rotation

The direction of rotation of repulsion motors can be reversed by shifting the position of the brushes from the neutral plane. The repulsion motor always rotates in the direction of the brushes. If the brushes are shifted to the right of the neutral plane, the rotor will rotate clockwise. If the brushes are shifted to the left of the neutral plane, the rotor will rotate counterclockwise. See Figure 2-40.

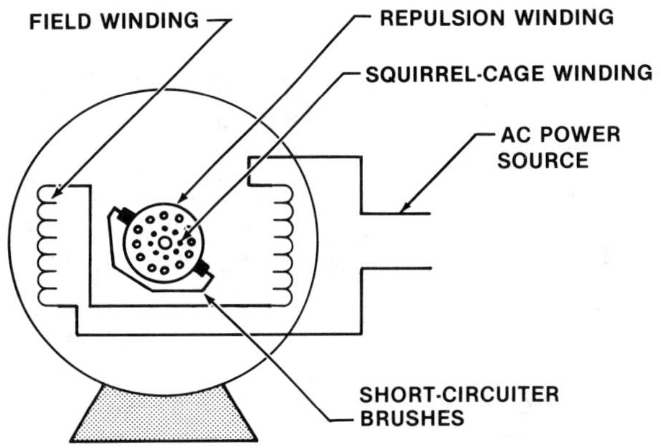

Figure 2-39. The repulsion-induction motor has a squirrel-cage rotor with a wound armature and a commutator with short-circuiter brushes.

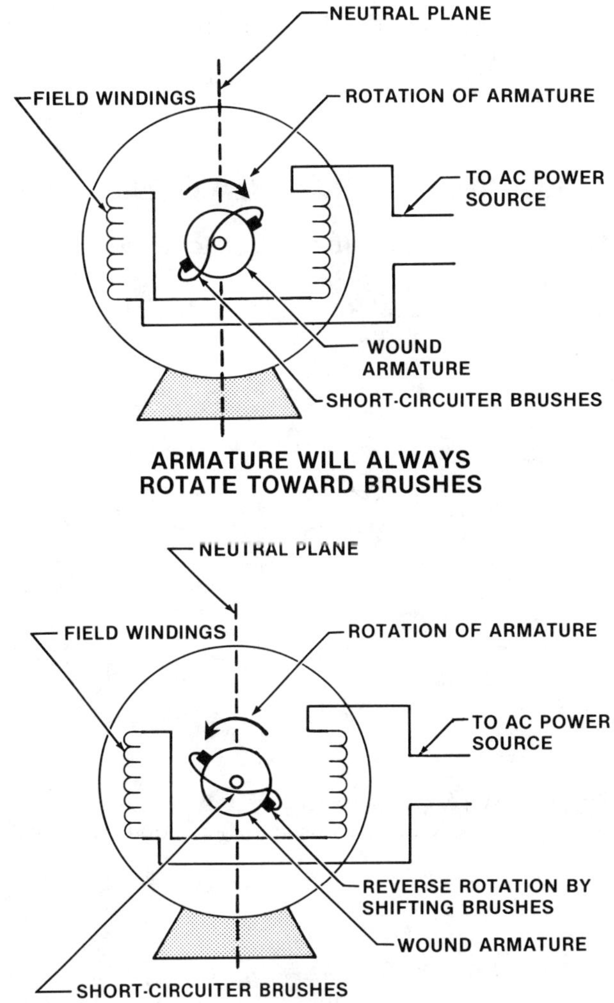

Figure 2-40. The armature of a repulsion motor always rotates toward the position of the short-circuiter brushes from the neutral plane.

Testing

AC field pole windings of repulsion motors can be checked for continuity with an ohmmeter. The brushes can also be checked with an ohmmeter to verify that they are making good contact. Dirty commutators should be cleaned with fine sandpaper so the brushes will make good contact. If the governor mechanism is sticking, replace the defective parts or repair. Incorrect spring tension will cause the motor to operate improperly. Check the spring tension and replace or reset spring tension if necessary.

CHECKING OPERATING AMPS

The operating amps of a motor shall not exceed the nameplate rating of the motor for the motor to have normal operating life. The amp rating is checked with an ammeter for an accurate measurement. See Figure 2-41.

CHECKING OPERATING VOLTAGE

Two measurements are required to determine the operating voltage of a motor. The first check is to measure the system voltage to the motor. This voltage reading can be 10% above or 10% below the operating voltage. See Figure 2-42. The second check is to measure the unbalanced voltage from each phase to ground and average the measurements to obtain a percentage. The percentage shall not exceed 1%. See Figure 2-43.

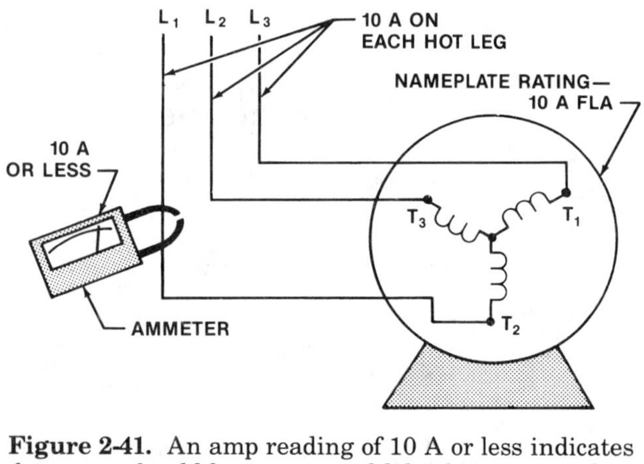

Figure 2-41. An amp reading of 10 A or less indicates the motor should have a normal life of 20 years or less (NEMA 1).

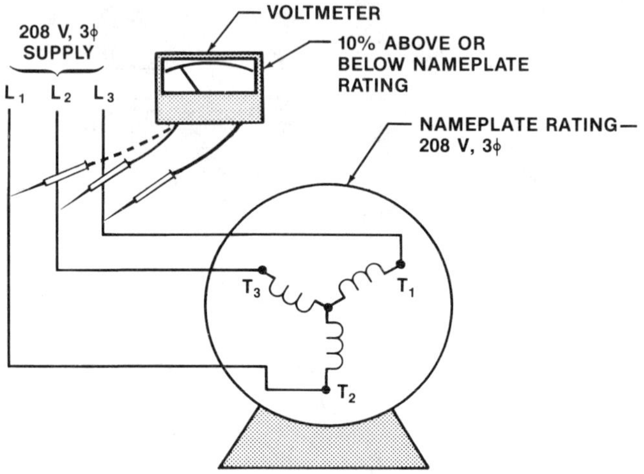

Figure 2-42. The operating voltage should read within 10% above or below the operating voltage listed on the nameplate of the motor.

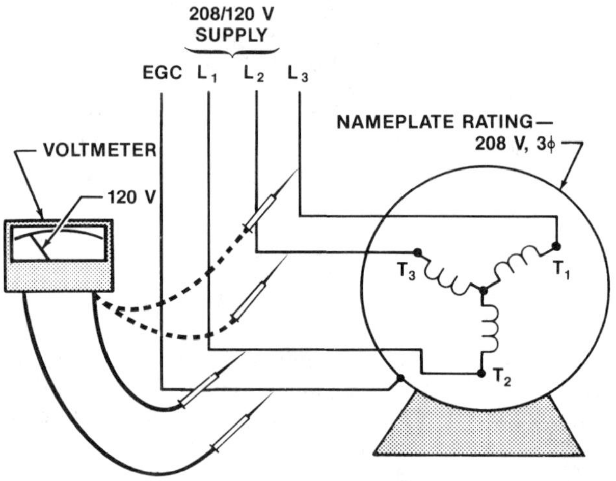

Figure 2-43. Unbalanced voltage (if measured) shall not exceed 1%, or the running current of the nameplate shall be derated to compensate for the percentage that exceeds 1%.

REVIEW—CHAPTER 2

Name ______________________ Date ______________

True-False

T F 1. Split-phase motors can start without the aid of an additional winding.

T F 2. Split-phase motors can be wound for dual voltage with either a 120-volt or 240-volt supply.

T F 3. Split-phase motors must have an auxiliary winding in order to produce the torque needed to start the driven load.

T F 4. The auxiliary winding of a split-phase motor is placed 40° electrically from the main winding.

T F 5. A centrifugal switch may be used to open the contacts and disconnect the starting winding when the rotor reaches 50% of its running speed.

T F 6. Starting windings have a lower resistance than running windings.

T F 7. Running windings have larger wire than starting windings.

T F 8. The running winding is located on the top of the starting winding.

T F 9. An induction motor connected for the rotor to rotate counterclockwise turns to the left, as viewed from the front of the motor.

T F 10. When checking the resistance of the starting winding with an ohmmeter, the resistance measured will be higher than the resistance of the running winding.

T F 11. The direction of rotation of the rotor in a single-phase induction motor can be reversed by interchanging the wiring in the panel that supplies the circuit.

T F 12. Split-phase motors are usually protected by an additional switch to prevent the windings from overheating.

T F 13. It is not necessary to keep the inlets and outlets of the motor free from dirt, dust, or lint so cool air can be pulled across the field-pole windings.

T F 14. A capacitor-start, split-phase motor has a greater starting torque than a regular split-phase motor.

T F 15. The direction of rotation of a capacitor-start motor cannot be reversed by changing the flow of current through the capacitor.

Completion

__________ 1. The line current of a split-phase motor is __________ than that for a capacitor-start motor.

__________ 2. When a starting capacitor and a running capacitor are used to start and run a motor, the capacitance of both capacitors are ______________ together to start and run the motor.

_____ 3. A centrifugal switch is used to disconnect the _____ capacitor on a start-and-run capacitor motor when the rotor accelerates to its running speed.

_____ 4. A permanent-split capacitor motor has the capacitor _____ in the motor circuit while the motor is in operation.

_____ 5. A permanent-split capacitor motor has _____ starting torque.

_____ 6. Shaded-pole motors use a shaded coil to provide _____ torque.

_____ 7. The shaded coil cuts through a portion of the _____ created by the field poles.

_____ 8. The shaded coil provides a two-phase magnetic field that creates a phase _____, providing the torque needed to start the motor.

_____ 9. A(n) _____ resistance rotor is used with a shaded-pole motor.

_____ 10. If _____ can be measured through the running windings of a shaded-pole motor, the windings are good.

_____ 11. If the rotor of a shaded-pole motor is placed in the _____ direction in the motor housing, the rotor will rotate in the reverse direction.

_____ 12. Varying speeds can be obtained in a shaded-pole motor by using a(n) _____ transformer.

_____ 13. Two individual speeds can be provided in a shaded-pole motor by using five windings with the same size wire and a sixth winding with _____ wire with many turns.

_____ 14. Universal motors are designed for use on either _____ or _____ electrical systems.
_____

_____ 15. Universal motors are equipped with field windings, an armature with _____, and a commutator.

Multiple Choice

_____ 1. Universal motors commonly are available in sizes of _____ horsepower or less.
- A. 1/8
- B. 1/2
- C. 3/4
- D. 1

_____ 2. The speed of universal motors can be varied by using a _____.
- A. switch
- B. voltage relay
- C. current relay
- D. resistor

__________ **3.** Three-phase induction motors can be wound with ________ leads.
A. two
B. three
C. four
D. six

__________ **4.** Three-phase induction motors are available with ________ leads.
A. nine
B. ten
C. twelve
D. fifteen

__________ **5.** The windings of three-phase induction motors are located approximately ______° out of phase from each other.
A. 90
B. 120
C. 180
D. 220

__________ **6.** The direction of rotation of the rotor for three-phase motors can be reversed by interchanging any _______ leads.
A. two
B. three
C. four
D. six

__________ **7.** A wound-rotor motor has _______ set(s) of leads.
A. one
B. two
C. three
D. four

__________ **8.** If resistor bank leads are used for starting only (heavy intermittent duty), the leads can be sized at _______% or less.
A. 60
B. 70
C. 85
D. 90

__________ **9.** If resistor bank leads are used for speed control as well as starting, the leads must be sized at _______%.
A. 100
B. 110
C. 125
D. 135

__________ **10.** The main conductors supplying motor windings must be sized at _______%.
A. 110
B. 115
C. 125
D. 150

Problems

1. What size THWN copper conductors are required to supply the field poles of a three-phase, 230-volt, 20-horsepower wound-rotor motor? The motor has a secondary current rating of 42 amps, is started on full voltage, and is rated code letter F.

2. What size THWN copper conductors are required for the resistor bank conductors used to start and control the speed of a wound-rotor motor with a secondary current of 42 amps?

3. What size THWN copper conductors are required for the secondary conductors of a wound-rotor motor with a 42-amp secondary current rating?

4. Connect the leads of the six-pole motor for delta operation.

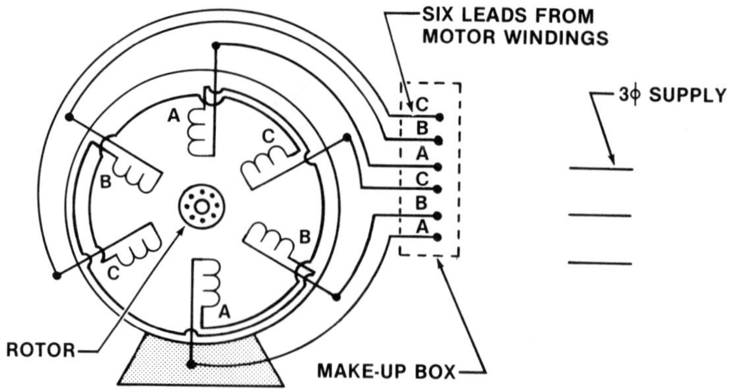

5. Connect the leads of the six-pole motor for wye operation.

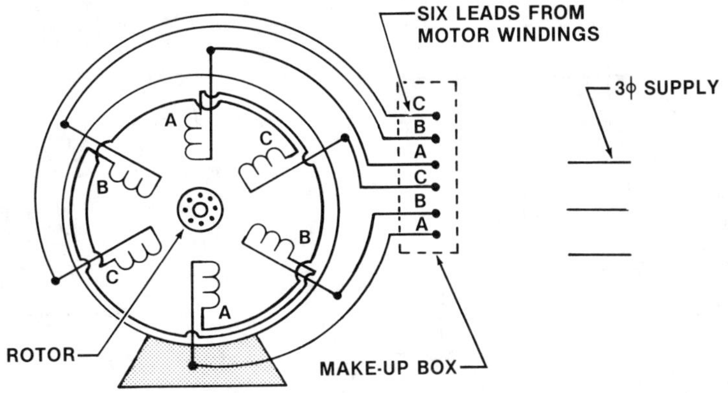

Identification

Identify the components of the single-phase, capacitor-start motor.

1. ____________________
2. ____________________
3. ____________________
4. ____________________
5. ____________________
6. ____________________

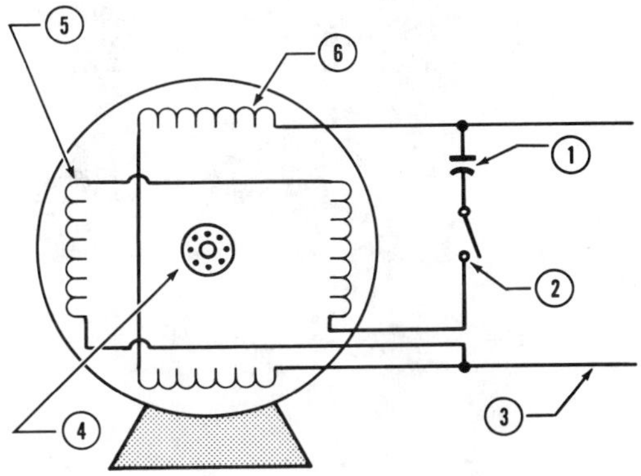

Identify the components of the universal motor.

7. ____________________
8. ____________________
9. ____________________
10. ____________________

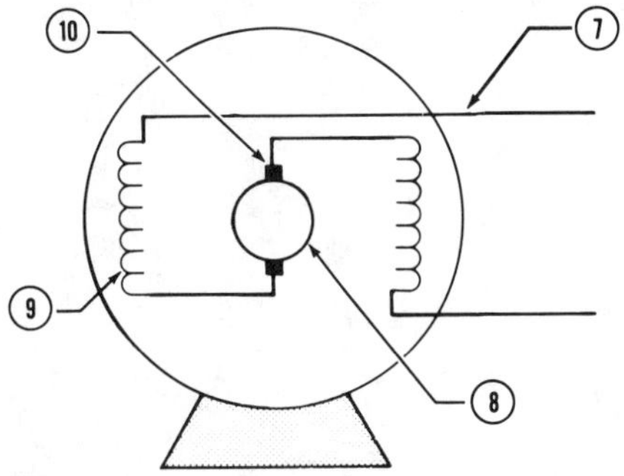

Identify the components of the shaded-pole motor.

11. ______________________________

12. ______________________________

13. ______________________________

14. ______________________________

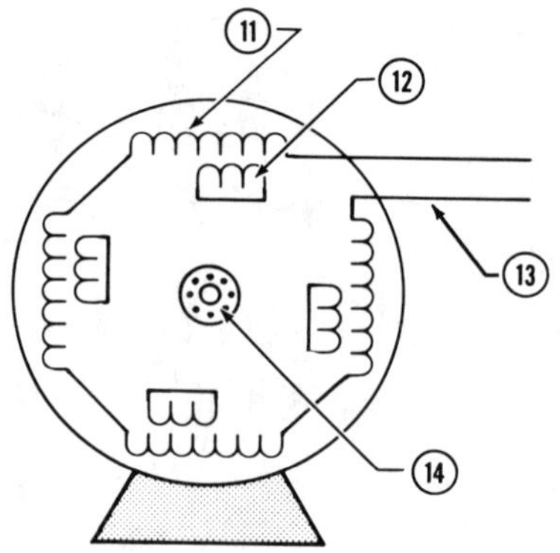

Identify the components of the repulsion motor.

15. ______________________________

16. ______________________________

17. ______________________________

18. ______________________________

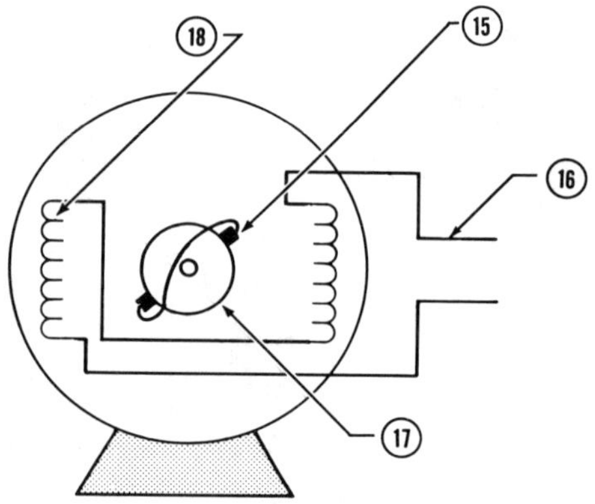

MOTOR COMPONENTS

Chapter 3

Motor circuits are designed to start and run a motor safely. Protection must also be provided to protect motor windings when motors are driving loads. Overcurrent protection devices are used to protect the motor windings. They are selected according to the type of motor used and the amount of starting current required to start the motor. The percentages used to size the overcurrent protection devices are based on the type, starting method, and code letter. The starting method chosen is based on the amount of current required to be reduced. The starting methods are accomplished by using the windings of the motor or external components in motor starters.

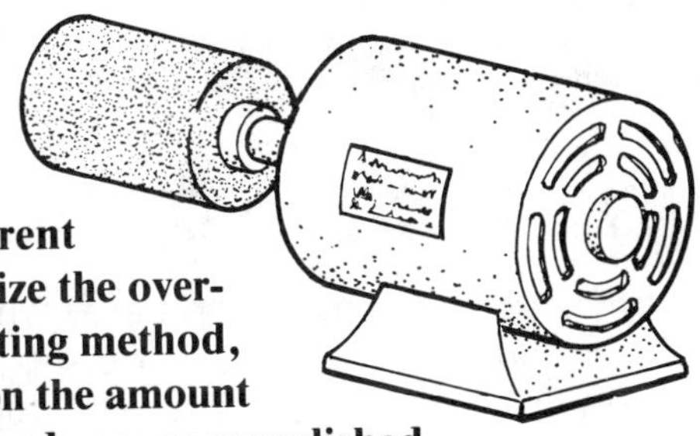

TYPES OF MOTORS—TABLE 430-152

The different types of motors are
(1) single-phase AC squirrel-cage
(2) three-phase AC squirrel-cage
(3) wound-rotor
(4) synchronous
(5) DC

Table 430-152 lists the elements necessary to design a complete motor system. The table is used to size the overcurrent protection device to allow a motor to start and accelerate to its running speed. To use Table 430-152, the type of motor, number of phases, starting method, and code letter must be known. This information is given on the motor's nameplate. By applying this information to Table 430-152, the percentage figure used to size the overcurrent protection device is obtained. For example, a three-phase AC squirrel-cage motor with full-voltage starting and code letter F is to be started with a circuit breaker. Code letter F shows 250% of the motor's full-load current selected from Table 430-152.

Single-phase AC Squirrel-cage Motors

A squirrel-cage motor is an *induction motor*. The field windings act as the primary, and the rotor acts as the secondary. An induction motor operates on the same principles as a transformer's primary and secondary windings. *Split-phase (single-phase) induction motors* require a starting winding on the stator. This winding has a higher resistance than the running winding. The difference in resistance creates an angular-phase displacement between the two windings. The split-phase motor derives its name from the angular-phase displacement.

This angular-phase displacement is approximately 18° to 30° in time. Splitting the phases provides enough starting torque (twist or force) to start the motor. When the rotor starts turning (spinning) and comes up to running speed, which is approximately 75% to 80% of synchronous speed, a centrifugal switch placed in the circuit of the starting winding opens the starting winding. The motor will then operate on the running winding. See Figure 3-1.

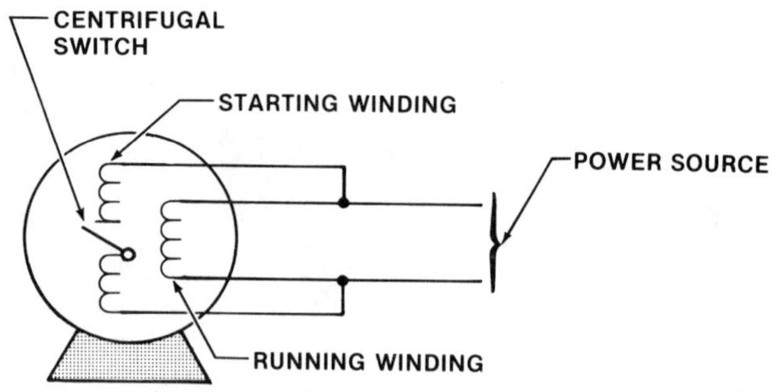

Table 430-152

Figure 3-1. Split-phase (single-phase) squirrel-cage induction motors operate on 120-volt or 240-volt systems. After the starting winding has started the motor, a centrifugal switch opens the starting winding.

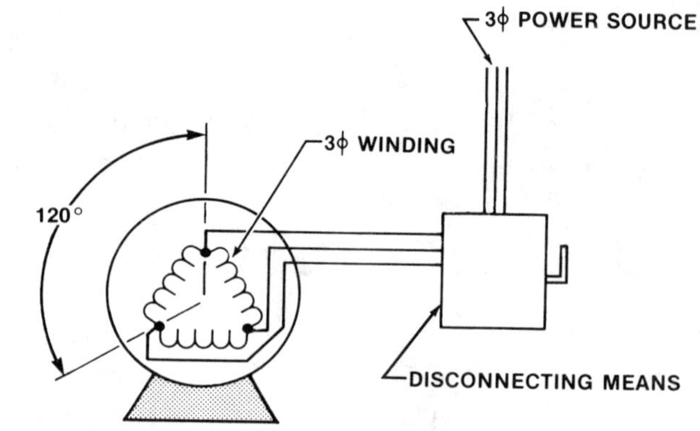

Table 430-152

Figure 3-2. Three-phase squirrel-cage induction motors operate only on three-phase systems. The generated magnetic fields are 120 ° out-of-phase with each other.

Three-phase AC Squirrel-cage Motors

A three-phase induction motor operates with the generated magnetic fields 120° out-of-phase with each other. It has good starting torque and can start by itself. This type of motor is similar in principle to three people riding a bicycle, each making a downward stroke on a pedal at a different time. At least one rider is always making this stroke (delivering a power stroke). As one rider leaves the peak (power stroke), another rider will peak, and so on, delivering a smooth and steady power stroke.

Three-phase motors deliver consistent output since there is always a peak phase of the current. In a single-phase motor, there is a loss of power when the alternating current reverses its direction of flow. In a three-phase motor, when the alternating current of one phase reverses itself, the current of another of the three-phases will peak, creating a smooth and continuous source of power. See Figure 3-2.

Wound-rotor Motors

A wound-rotor motor is a three-phase motor. It operates in the same manner as a squirrel-cage induction motor. The wound-rotor motor, however, has two sets of leads extending from the controller and a bank of resistors to slip rings that are connected to the rotor. When resistance in the rotor circuit varies, the speed of the motor also varies. The greater the resistance in the rotor is, the slower the motor will run, and vice versa. The resistor banks may be separate from the motor, or the resistances may be incorporated in the controller.

Synchronous Motors

A synchronous motor is designed to run at a specified speed. The two types of synchronous motors are *nonexcited* and *direct-current excited*. A wide range of sizes for various applications is available.

A nonexcited synchronous motor has no direct current applied to the rotor. A direct-current excited synchronous motor requires a DC source to excite its field. The DC current of the rotor field interacts with the stator AC current to produce the torque required to turn the rotor at a synchronous speed. See Figure 3-3. A synchronous motor can be used to cause the current to lead or lag the voltage, which will correct poor power factor.

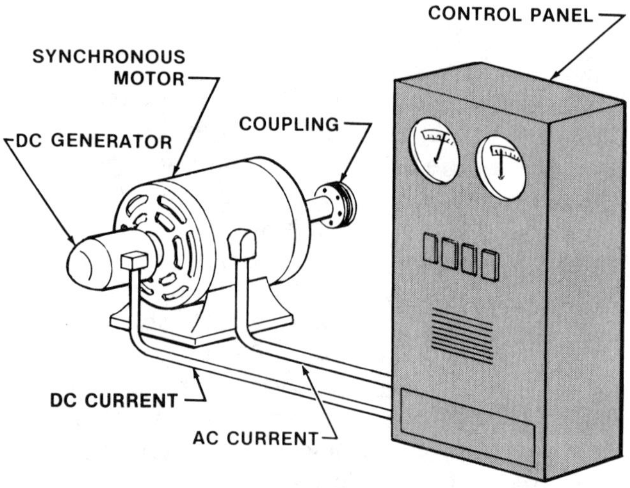

Table 430-152

Figure 3-3. Three-phase synchronous motors are supplied by alternating and direct current to obtain synchronous speed.

DC Motors

A DC motor operates on direct current only. It is designed in two main parts: the stator (field), which is the stationary frame of the motor, and the rotor (armature), which is mounted on the drive shaft. The speed of a DC motor may be adjusted by applying direct current to the rotor. The windings of the rotor and field poles can be connected for either *series, shunt,* or *compound* operation. See Figure 3-4.

Series DC Motors. Series DC motors provide very high starting torque of 300% to 375% of the full-load torque. This motor is used to drive loads that require high torque and low speed regulation. Speed varies with the load requirements. Series DC motors are ideal for traction work, such as that done by a hoist where the speed must vary with the load. The armature and fields are connected in series.

Shunt DC Motors. Shunt DC motors provide medium starting torque of 125% to 200% of the full-load torque. This motor is used to drive loads that require constant or adjustable speeds and loads that do not require high starting torque. DC shunt motors are used to drive such loads as printing presses, woodworking machines, and paper-making machines.

Compound DC Motors. Compound DC motors provide high torque of 180% to 260% of the full-load torque. The motor has fairly constant speed. The compound DC motor is equipped with a series and shunt winding. The shunt winding is connected in parallel with the armature, and the series winding is connected in series with the armature. The operation of the motor has the characteristics of both series and shunt motors. Compound DC motors are used to drive such loads as punch presses, crushers, and reciprocating compressors.

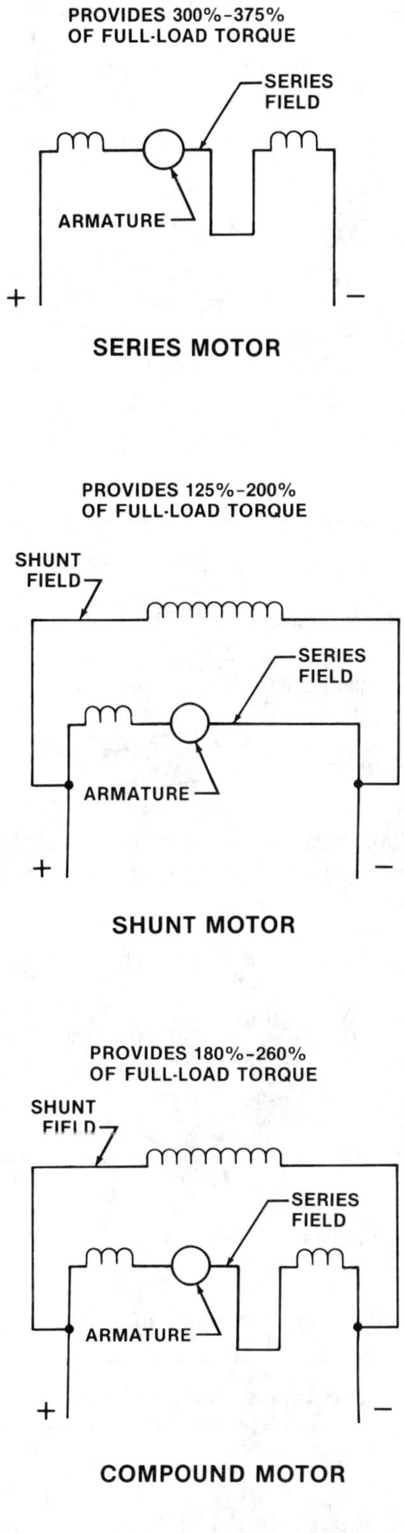

Figure 3-4. Starting torque of a DC motor is determined by the number and connections of the field in relation to the armature.

MOTOR TORQUE

The *torque* of a motor is its ability to accelerate and drive a piece of equipment. Torque (the turning or twisting force of the motor) is measured in foot-pounds or pound-feet. It is determined by dividing the horsepower times 5252 by the rpm of the motor. *NOTE:* The figure 5252 is found by dividing 33,000 foot-pounds per minute by 6.2831853, which is found by multiplying π (3.14159265) by 2.

$$\frac{33{,}000}{6.2831853} = 5252$$

See Figure 3-5.

The starting torque of a motor varies with the classification of the motor. NEMA classifies these motors as design B, C, or D. These standardized motors are the most used motors in the electrical industry. NEMA also classifies motors with design E, F, or G.

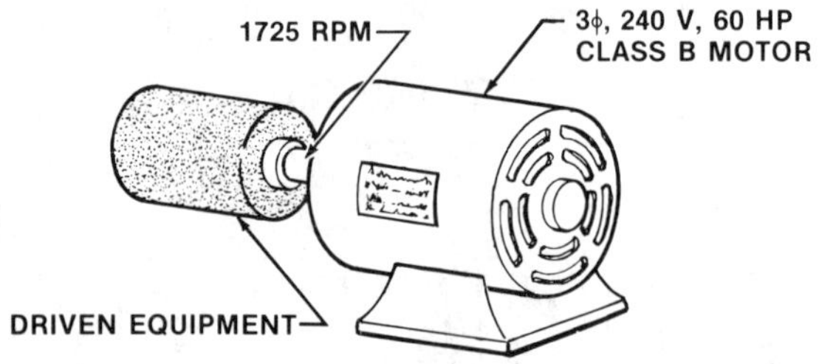

PROBLEM: What is the full-load torque of a three-phase, 240-volt, 60-horsepower motor turning at 1725 rpm?

Step 1: The motor turns at 1725 rpm.

Step 2: Apply formula.

$$torque = \frac{hp \times 5252}{rpm}$$

$$torque = \frac{60 \times 5252}{1725}$$

$$torque = \frac{315{,}120}{1725}$$

Answer: **full-load torque = 182.7 ft lb**

Figure 3-5. The full-load torque of a motor is found by multiplying the horsepower rating times 5252 and dividing by rpm's.

Each class of motor has a different rotor design that offers a different value of starting torque. The NEMA classification of design B, C, or D motors produces a different value of torque, speed, current, and slip to start and drive the various types of loads. The design selected depends on the starting torque of the driven load and the running torque required to drive the load.

A Class B design is the most widely used motor in the electrical industry. For example, a Class B motor will increase the starting torque of an induction motor by 150% of the full-load torque. Most designers only increase the starting torque rating by 125% when using a Class B motor. See Figure 3-6.

A Class C motor increases the starting torque of a squirrel-cage induction motor by about 225% of the full-load torque. However, designers often subtract 25% to keep from overloading the starting torque of the motor.

Example: What is the full-load torque and starting torque of a Class C, 50-horsepower induction motor operating at 1725 rpm?

$$torque = \frac{hp \times 5252}{rpm}$$

$$trq = \frac{50 \times 5252}{1725}$$

$$trq = \frac{262{,}600}{1725}$$

full-load torque = 152 ft lb

Increase full-load torque by 200%:

152 ft lb × 200% = 304 ft lb

starting torque = 304 ft lb

A Class D motor increases the starting torque of a squirrel-cage motor by about 275% of full-load torque. (Again, 25% is deducted to keep from overloading the starting torque of the motor.)

Example: What is the full-load torque and starting torque of a Class D, 40-horsepower squirrel-cage induction motor operating at 1725 rpm?

$$torque = \frac{hp \times 5252}{\text{rpm}}$$

$$trq = \frac{40 \times 5252}{1725}$$

$$trq = \frac{210{,}080}{1725}$$

full-load torque = 121.8 ft lb

Increase full-load torque by 250%:

121.8 ft lb × 250% = 304.5 ft lb

starting torque = 304.5 ft lb

Motor	Percentage
B	125%
C	200%
D	250%

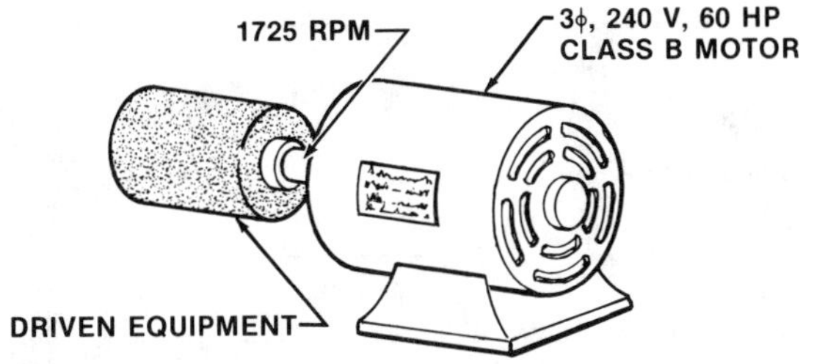

PROBLEM: What is the starting torque of a three-phase, 240-volt, 60-horsepower motor turning at 1725 rpm?

Step 1: The motor turns at 1725 rpm.

Step 2: Apply formula.

$$torque = \frac{hp \times 5252}{rpm}$$

$$torque = \frac{60 \times 5252}{1725}$$

$$torque = \frac{315{,}120}{1725}$$

$$torque = 182.7 \text{ ft lb}$$

*For Class B motor increase 125%.

182.7 × 125% = 228 ft lb

Answer: **starting torque = 228 ft lb**

Figure 3-6. The starting torque of a motor is found by multiplying the full-load torque by the proper motor design (B, C, or D) percentage.

Note that the rpm of the motor determines the full-load torque. A motor turning at 1200 rpm produces more torque than a motor turning at 1800 rpm. If a two-speed, 25-horsepower motor operates either at 1200 or 1800 rpm, the full-load torque of each speed would be:

$$trq = \frac{hp \times 5252}{rpm}$$

$$trq = \frac{25 \times 5252}{1200} = \textbf{109 ft lb}$$

$$trq = \frac{25 \times 5252}{1800} = \textbf{72.9 ft lb}$$

A resistor or a reactor-reduced starting method can be used to reduce the inrush starting current (locked-rotor current) of a motor. Either method reduces the starting current to 65%. If the starting current is reduced, the starting torque must be reduced to 42%. The 42% reduction of starting torque is obtained by multiplying the starting current by itself (65% × 65% = 42%). Care must be taken when selecting a reduced starting method to ensure that enough foot-pounds are provided to accelerate the load. See Figure 3-7.

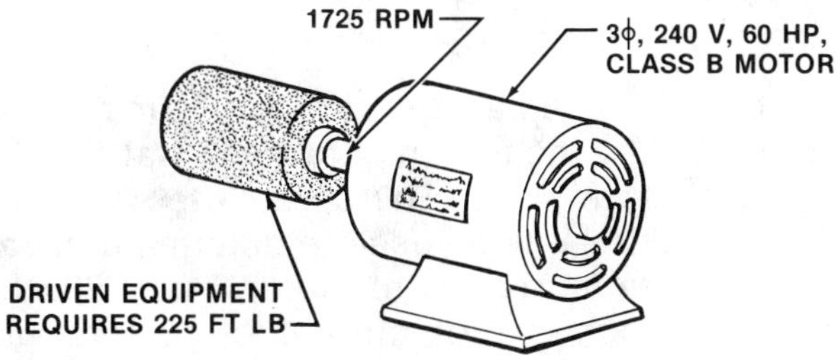

PROBLEM: Will a resistor starting method provide enough starting torque to accelerate the driven equipment?

Step 1: The motor turns at 1725 rpm.

Step 2: Apply formula.

$$torque = \frac{hp \times 5252}{rpm}$$

$$torque = \frac{60 \times 5252}{1725}$$

$$torque = 182.7 \text{ ft lb}$$

Class B = 182.7 ft lb × 125% = 228 ft lb

Step 3: Reduce starting torque to 42%.

228 ft lb × 42% = 95.8 ft lb

Answer: **No. Resistor starting reduces the starting torque from 228 ft lb to 95.8 ft lb.**

Figure 3-7. Resistor or reactor starting reduces the starting torque of a motor.

CODE LETTERS— TABLE 430-7(b) AND TABLE 430-152

Code letters must be put on motors by the manufacturer. These code letters are used for figuring and selecting the size of the running overload protection or short-circuit protection. The *locked-rotor current* (*lrc*) of a motor is the current the motor pulls when the rotor of the motor is stalled or starting. Overcurrent protection devices must be set above the locked-rotor current of the motor to prevent the overcurrent protection device from opening the circuit to the motor.

Finding Locked-rotor Current Using Code Letter

The nameplate on a three-phase, 230-volt, 25-horsepower motor has the code letter G and specifies a temperature rise of 40°C. Table 430-7(b) lists the kVA (kilovolt-amps) per horsepower with the locked-rotor current for the code letter G. To find locked-rotor current, apply values to the following formula. *NOTE*: The figure 1.732 is the square root of 3.

$$lrc = \frac{k \times VA \times hp}{V \times \sqrt{3}} = A$$

Using the figure 6.29 (for code letter G) from Table 430-7(b), calculate the locked-rotor current of a motor as follows:

$$lrc = \frac{1000 \times 6.29 \times 25 \text{ hp}}{\underbrace{230 \text{ V} \times 1.732}} = \mathbf{395\ A}$$

NOTE: USE 398 V FOR CALCULATION

The overcurrent protection device must be sized to hold 395 amps for a period long enough to allow the motor to start and accelerate up to its running speed. For example, a motor that has a locked-rotor current of 395 amps can be started under normal conditions by a 150-amp circuit breaker (cb). A 150-amp circuit breaker will hold three times its rating for about 35 seconds (3 × 150 A cb = 450 A). Four hundred fifty (450) amps will hold 395 amps of locked-rotor current long enough for the motor to start.

A time-delay fuse holds five times its rating for 10 seconds. For example, an 80-amp time-delay fuse holds 400 amps (5 × 80 A = 400 A) for 10 seconds, which is long enough for the motor to start. The running current of the motor cannot exceed 80% of the 80-amp time-delay fuses (80 A × 80% = 64 A). The motor cannot pull more than 64 amps in the run operation when using 80-amp time-delay fuses. See Figure 3-8. Sizing conductors at 125% of FLC, per 430-6(a), gives the derating factor.

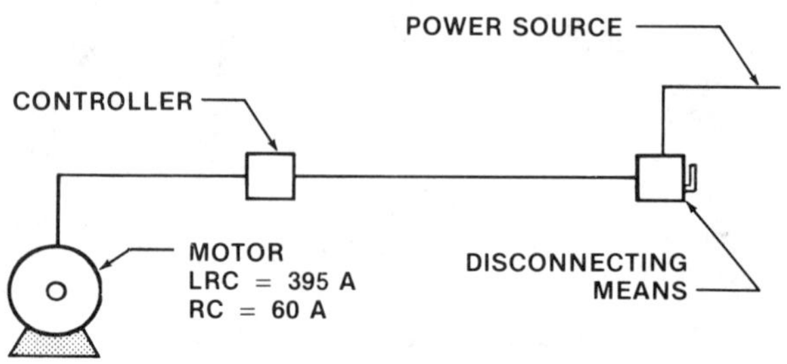

PROBLEM: What size overcurrent protection device is required using time-delay fuses? (Use minimum size fuses.)

Step 1: *lrc* = 395 A
rc = 60 A

Step 2: 80-A tdf × 5 = 400 A

400 A will hold 395 A for 10 seconds.

Step 3: 60-A running current

80-A tdf × 80% = 64 A

60 A is below 64 A.

Answer: **80-A TDF**

NOTE: The 60-A running current is below 80% of the time-delay fuses and 400 A (5 × 80 A) holds 395 A for 10 seconds.

Figure 3-8. Selecting time-delay fuses to start and run a motor.

Torque causes a motor to start from rest and is created by the action of current that flows in the stator windings. The code letters listed in Table 430-7(b) can be used to determine the actual starting current of a motor. Table 430-151 lists the locked-rotor current of motors with code letters A through H for motors starting and running under normal conditions. Table 430-152 lists code letters to select the percentages for determining the proper size overcurrent protection device. Four types of devices are listed in four columns for starting motors. Note that code letter A stands alone. Code letters B through E are grouped together and code letters F through

V are in a group. These three groups of code letters are separated because motors are designed for many conditions of use.

Code letter A has the lowest percentages permitted for selecting overcurrent protection devices. Code letters B to E have percentages higher than code letter A. Code letters F to V have the highest percentages of all the code letters for selecting overcurrent protection devices to allow motors to start. Percentages are also listed in Table 430-152 for motors without code letters. Because motors are built and used in various locations under many types of uses, these code letters are grouped with percentages to compensate for such conditions.

Figure 3-9 illustrates a step-by-step procedure for determining the locked-rotor current based upon the motor's code letter.

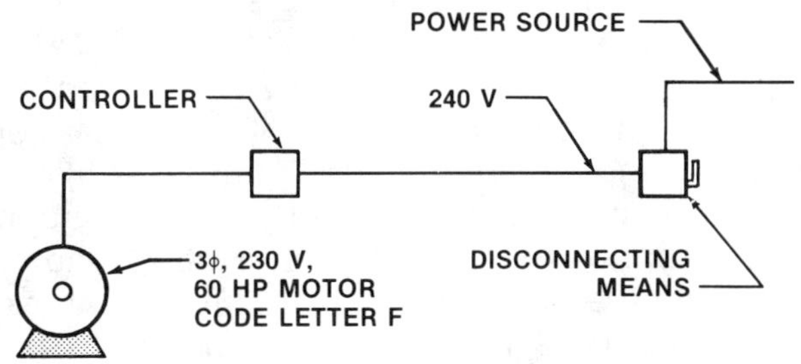

PROBLEM: What is the locked-rotor current for a three-phase, 230-volt, 60-horsepower motor with code letter F?

Step 1: The code letter of the motor is F.

Step 2: *Table 430-7(b).*
5.59 kVA for code letter F

Step 3: Apply formula.

$$Irc = \frac{k \times VA \times hp}{V \times \sqrt{3}} = A$$

Step 4: $Irc = \dfrac{1000 \times 5.59 \times 60\ \text{hp}}{240\ \text{V} \times 1.732} = 806\ \text{A}$

Answer: **Irc = 806 A per code letter F**

Figure 3-9. Finding the locked-rotor (inrush starting) current using the code letter of the motor.

FINDING LOCKED-ROTOR CURRENT USING HORSEPOWER RATING—TABLE 430-151

Table 430-151 may be used to find the locked-rotor current of a motor. The locked-rotor currents in this table are calculated from the full-load current ratings in either Table 430-148 or Table 430-150 and are based on the voltage and horsepower rating of the motor.

The locked-rotor current equals six times the full-load current rating of the motor listed in Table 430-150. See Figure 3-10. However, when the code letter is not known, electricians and engineers, to be safe, will select the locked-rotor current from Table 430-151 for sizing the overcurrent protection device above the locked-rotor current of the motor to prevent the circuit from opening when the motor is started. In most cases, four times the full-load current of the motor equals the locked-rotor current for the motor (4 × 154 A = 616 A).

The locked-rotor current values listed in Table 430-151 for motors based on the voltage and horsepower can be used safely for motors marked with code letters A through H.

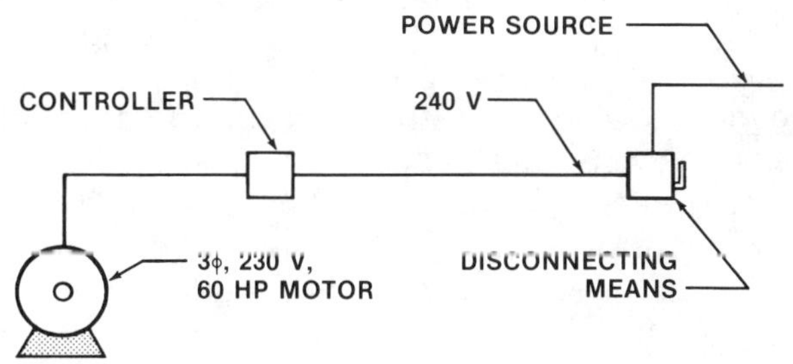

Table 430-150
Table 430-151

PROBLEM: Determine the locked-rotor current required for this motor.

Step 1: *Table 430-150, Column 2 and Footnote.*
Irc rating = 154 A × 6 = 924 A

Step 2: *Table 430-151.*

Answer: **Irc = 924 A**

Figure 3-10. Finding the locked-rotor current using the horsepower rating of the motor.

Motors marked with code letters above H through V will have higher locked-rotor currents, which require overcurrent protection devices with greater trip settings. These higher settings prevent the overcurrent protection devices from opening due to the higher inrush starting current (locked-rotor current) of the motor. Note that this higher current can require larger electrical equipment such as disconnects for fuses and panels for circuit breakers.

STARTING METHODS—TABLE 430-152

The starting method of a motor must be considered when sizing the overcurrent protection device for the motor circuit. The seven starting methods are

*(1) full-voltage starting
*(2) resistor starting
*(3) reactor starting
*(4) autotransformer starting
(5) solid-state starting
(6) part-winding starting
(7) wye-delta starting

*Listed in Table 430-152

Full-voltage Starting

Full-voltage starting applies the supply voltage directly to the motor. See Figure 3-11. If the supply voltage from the power company is three-phase and 480 volts, then full-voltage starting will apply 480 volts directly to the motor's windings. The motor is connected across the line in one step by closing a single main switch, such as a disconnecting switch or circuit breaker. A START and STOP pushbutton station may be used to control a coil in a magnetic starter that bridges the line to the load terminals of the motor.

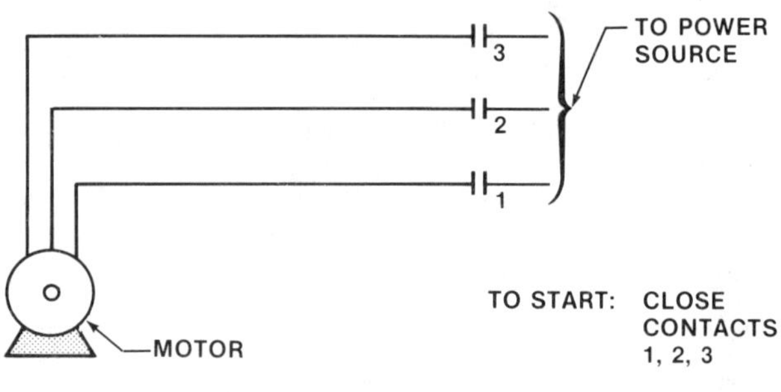

Figure 3-11. Full-voltage starting applies full supply voltage and starting current directly to the motor.

Full-voltage starting has the basic advantage of producing high torque. The starting torque per ampere of line current is the highest of all starting methods. All motors started on full voltage have an inrush starting current that varies from 3½ to 10 times the normal full-load running value. The power system must be capable of delivering its starting current without a voltage dip. The driven equipment must be designed to withstand such heavy currents. Full-voltage starting is selected whenever possible because of its low cost. Full-voltage starting also requires less maintenance than other motor-starting methods.

Full-voltage starting is used on small integral motors or on high-voltage systems where the starting current of the motor is low due to the high voltage. For example, if a three-phase, 240-volt, 60-horsepower motor is started with full voltage applied across the line to its windings, the locked-rotor starting current is 924 amps per Table 430-151. The locked-rotor starting current of a motor can be less than 924 amps per Table 430-151 by using the code letter of the motor. However, the locked-rotor current can be greater by using the code letter per Table 430-7(b).

The components that make up a motor circuit system based on the code letter of the motor must be carefully selected. Code letters above H can require a larger disconnect due to the larger size fuses needed to allow the motor to start and come up to its running speed.

Resistor Starting

Resistor starting is accomplished by placing a resistor in series with each phase of a motor. Reduced voltage and current are obtained when utilizing this starting method. Resistor starting reduces the torque efficiency of a motor to a value less than that of full-voltage starting. See Figure 3-12.

Resistor starting is the least expensive of all reduced starting methods. It provides high starting torque with smooth starting and acceleration up to the motor's running speed.

Resistor starting does not provide as high a starting torque as autotransformer starting and does not reduce the starting current to the value of autotransformer starting. Note that the starting current is not limited to the value of autotransformer starting. Resistor starting reduces the normal inrush starting current to about 65% lrc. The normal starting torque of the

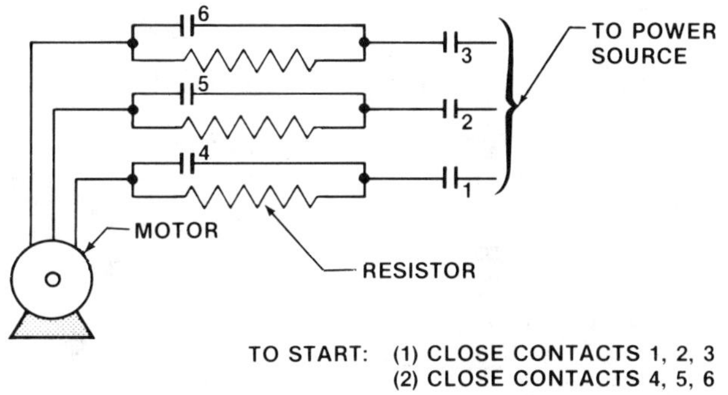

Figure 3-12. Resistor starting applies a percentage of reduced supply voltage and starting current to the motor.

motor will be reduced to about 42% of the starting torque. See Figure 3-13. Care must be taken to ensure that the amount of reduced torque will start the motor and driven load. The amount of starting torque required to start a driven load and accelerate to running speed must be considered when selecting a reduced starting method. Refer to Figures 3-5 and 3-6 for step-by-step procedures for determining the full-load torque and starting torque of a motor.

Starting Current	Starting Torque
Reduced 65% of Normal	Reduced 42% of Normal

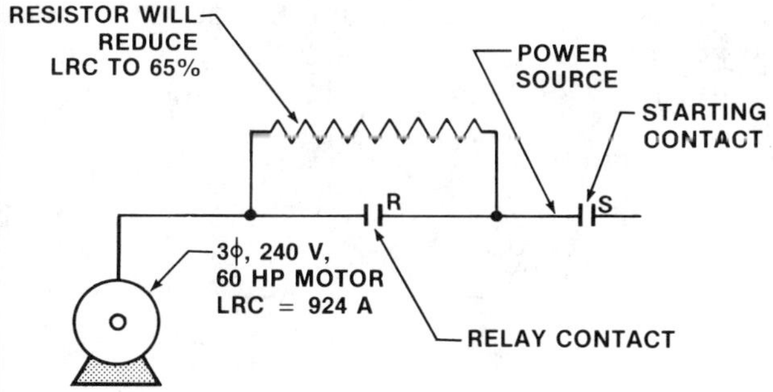

PROBLEM: What will be the reduced starting current by applying resistor starting?

Step 1: *Table 430-151.*
The *lrc* of the motor is 924 A.

Step 2: 924 A × 65% = 600.6 A

Answer: **Starting current will be reduced to 600.6 A.**

Figure 3-13. Reducing the inrush starting current by applying resistor starting.

Resistor starting is used for motors operating on low-voltage systems that are usually 600 volts or less. Note that the starting torque of the motor with applied resistor starting is reduced 42%. This reduced starting torque may not be high enough to accelerate the motor up to its running speed. A resistor starting method with two steps of acceleration is selected when this situation occurs. See Figure 3-14.

Starting Current	Starting Torque
Reduced 65% of Normal	Reduced 42% of Normal

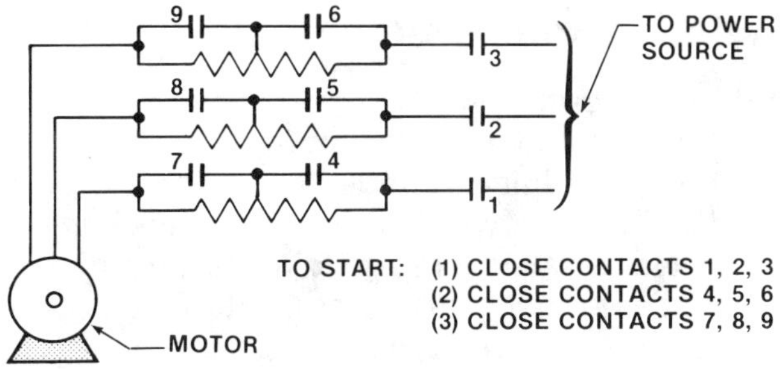

Figure 3-14. Two steps of resistance are added to reduce starting voltage and current during acceleration.

Reactor Starting

Reactor starting is accomplished by placing a reactor in series with each phase of the motor to achieve reduced-voltage starting. See Figure 3-15. Reactor starting does not create the insula-

Starting Current	Starting Torque
Reduced 65% of Normal	Reduced 42% of Normal

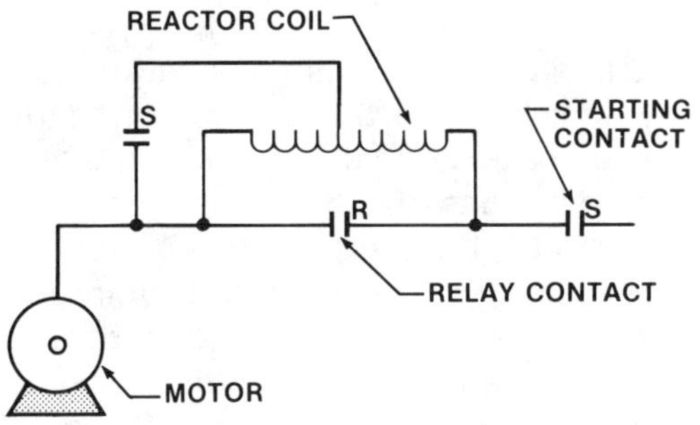

Table 430-152

Figure 3-15. Reactor starting applies a percentage of reduced voltage and starting current to the motor.

tion problems that occur when resistors are used to reduce voltage and current in starting a motor. Reactor starting is used mainly for motors with high-voltage systems. As with resistor starting, the torque efficiency of reactor starting is less than that of full-voltage starting.

Reactor starting reduces the inrush starting current of a motor to about 65%. It reduces the starting torque to about 42% of the normal starting torque. The starting current and starting torque of a motor will be reduced about the same by applying either resistor or reactor starting. However, reactor starting could affect the power system's power factor by producing a lagging current. Reactor starting is usually used on high-voltage systems over 600 volts.

Autotransformer Starting

Autotransformer starting is accomplished by providing taps to start the motor at 65% to 80% of the applied line voltage. When the motor is above 50 horsepower, a tap of 50% at the applied line voltage must be provided to start the motor. Autotransformer starting consists of an autotransformer with step-down taps and a switching device to start the motor. Once started, the switching device switches the autotransformer out of the circuit, and the motor is connected directly to the line. Autotransformer starting is the most popular reduced starting method because it has the same effect (torque efficiency) as full-voltage starting.

Torque efficiency is determined by dividing the starting torque of the motor by the locked-rotor current. The autotransformer is switched into the motor circuit by contacts that connect the desired tap to reduce the inrush starting current. When the motor accelerates up to its running speed, the autotransformer is switched or transferred out of the motor circuit. See Figure 3-16.

Autotransformer starting provides taps to reduce the inrush starting current of a motor to 50%, 65%, or 80% of the applied line voltage. These percentages of voltage taps produce from 25% to 64% of the full-load starting torque and from 25% to 64% of the locked-rotor inrush current. When 240 volts are supplied to a three-phase, 60-horsepower motor, the locked-rotor starting current will be 924 amps per Table 430-151.

The starting current and starting torque are determined by squaring the percentage of the

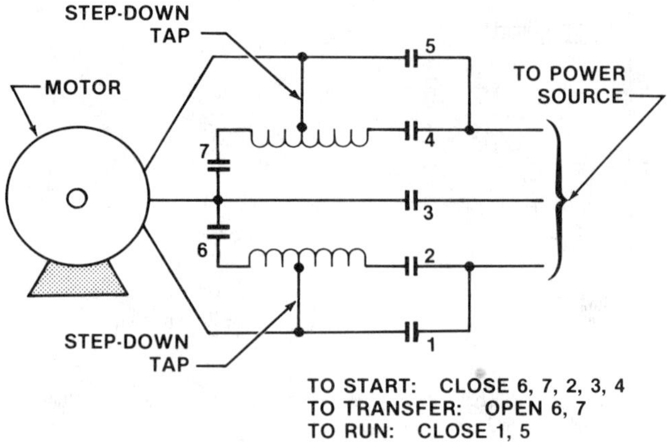

Table 430-152

Figure 3-16. Autotransformer starting applies a percentage of reduced voltage and starting current to the motor.

tap (50% × 50% = 25%). The starting current and starting torque will be reduced to 25% by applying an autotransformer with a 50% tap. The chart in Figure 3-17 can be used to estimate the reduced starting torque and inrush current in a motor circuit using an autotransformer-reduced starting method. If an autotransformer with a 50% tap is used to reduce the starting current of a motor, the 50% tap will reduce the starting current to 25%. However, the starting torque will be reduced to 25%.

Care must be taken in selecting the percentage of tap required to reduce the starting current. The tap selected must not reduce the starting torque to a point where the motor cannot start the driven load and accelerate up to its run-

Autotransformer Tap	Starting Current	Starting Torque
50%	25%	25%
65%	42%	42%
80%	64%	64%

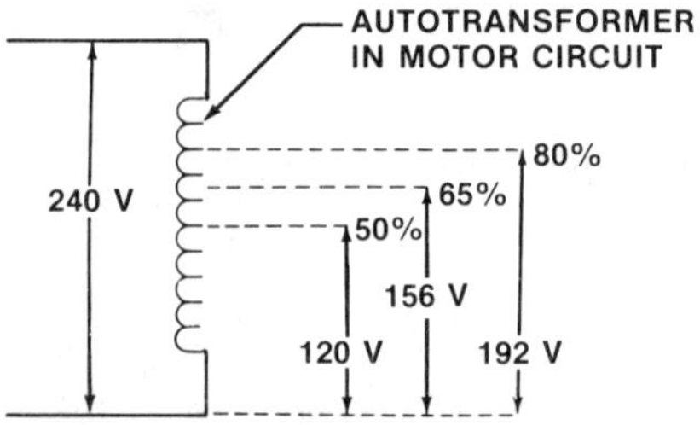

50% TAP	240 V × 50% = 120 V
65% TAP	240 V × 65% = 156 V
80% TAP	240 V × 80% = 192 V

Figure 3-17. Standard values may be used to determine reduced voltage for autotransformer starting.

ning speed. For example, a three-phase, 60-horsepower motor has a locked-rotor current of 924 amps per Table 430-151. The code letter on the nameplate of the motor will alter the locked-rotor current (starting current) per Table 430-7(b). An autotransformer with a 50% tap reduces the starting current and torque to 25% of the original value.

If the starting current is 924 amps, a 50% tap on an autotransformer reduces the starting current to 231 amps for the transformation (line to winding) current. The current on the line wire conductors will be reduced to 231 amps (924 A × 25% = 231 A). However, the current to the motor windings will be reduced to about 462 amps (924 A × 50% = 462 A). See Figure 3-18.

An autotransformer with a 65% tap reduces the starting current and torque to 42% (65% × 65% = 42%) of the original value of the motor. The starting current of a motor with 924 amps is reduced to 388 amps (924 A × 42% = 388 A). The winding current, line current, and transformation current can be determined by multiplying the locked-rotor current of 924 amps by the 65% tap on the autotransformer. That is, 924 A × 65% = 600.6 A. The line current is determined by multiplying 600.6 amps by 65% (600.6 A × 65% = 390 A). The transformation current for the tapped winding of the autotransformer can be determined by 600.6 A − 390 A = 210.6 A.

An 80% tap can also be used on an autotransformer to reduce the starting current (lrc) of a motor. An 80% tap reduces the winding current from 924 amps to 739 amps. The line wire current is reduced to 591 amps (739 A × 80% = 591 A). The transformation current will be 148 amps (739 A − 591 A = 148 A).

Note that the starting torque of the motor is reduced to 25% for a 50% tap. The starting torque is reduced to 42% for a 65% tap and to 64% for an 80% tap. For example, if a motor has a starting torque of 228 foot-pounds, a 65% tap will reduce the starting torque to 95.8 foot-pounds (228 ft lb × 42% = 95.8 ft lb).

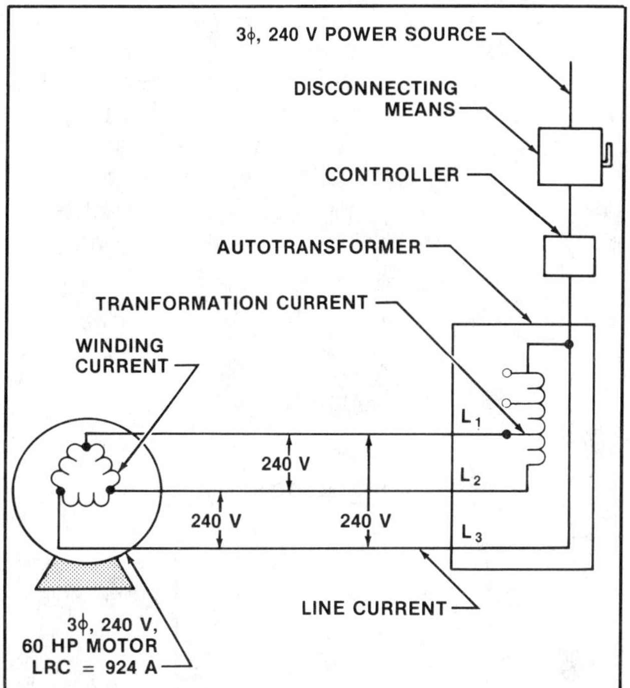

PROBLEM: The line (starting) current of a three-phase, 230-volt, 60-horsepower motor is 924 amps. Using a 50% tap, find the winding current, line current, and transformation current.

Step 1: *lrc* = 924 A

winding current = 924 A × 50%

winding current = **462 A**

Step 2: *line current* = 462 A × 50%

line current = **231 A**

Step 3: *transformation current* = 462 A − 231 A

transformation current = **231 A**

Answer: **winding current** = **462 A**
line current = **231 A**
transformation current = **231 A**

Figure 3-18. Autotransformer starting with a 50% tap has 50% winding and line wire current and 25% transformation current.

Solid State Starting

Solid state reduced starters use *silicon controlled rectifiers (SCRs)* to start and accelerate motors up to their running speeds. Voltage and current flow are controlled and directed through the SCRs by a gate (terminal). Voltage and current through the SCRs are switched ON and OFF by applying a low voltage signal to the gate.

Current only flows in one direction through an SCR. When AC power is applied to the circuit, a signal is sent to the gate, allowing current to flow. The SCRs will turn OFF for each half cycle. By using an AC waveform, the flow of current can be traced as the gate switches the flow of current through the SCR. See Figure 3-19.

Position one (P_1) sends the gate a signal that turns the SCR ON and allows the current to flow. At P_2, no gate signal is present and the current changes its direction of flow. P_4 sends a signal to the gate and current flows for the last half of the cycle until P_5 turns the signal to the gate OFF. This process is repeated as long as the AC power supply is connected to the current.

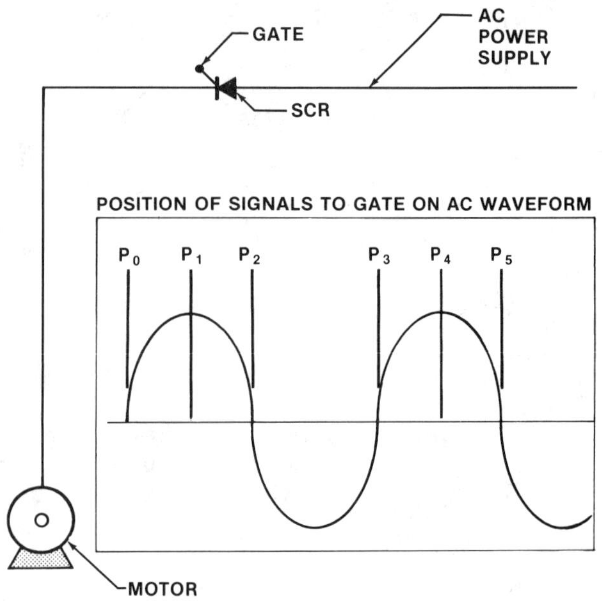

Figure 3-19. P_1 sends a signal to the gate and current flows. At P_2 the flow of current changes direction. P_4 again signals the gate and current flows for the last half of the cycle.

Two SCRs are connected in parallel in each phase leg to the motor in opposite directions. Signals are sent to the gate and the voltage supply to the motor is controlled. SCRs change the current flow through the motor windings each half cycle and the circuit detects alternating current. The starting current and voltage can be adjusted to the desired level by controlling the signals to the gate.

When the power source is connected, the starting contacts close and the circuit is energized to the motor windings. Acceleration of the motor is controlled by switching the signals to the gates and controlling the SCRs. Different voltage and current levels are obtained by turning ON the SCRs with gate signals. When the motor comes up to its running speed, the running contacts close and connect the motor windings directly to the supply line. The SCRs are disconnected from the line by the opening of the starting contacts and the motor runs at full voltage. Figure 3-20 illustrates the operation of a solid state starter with the contacts closing and opening to start and accelerate motor loads.

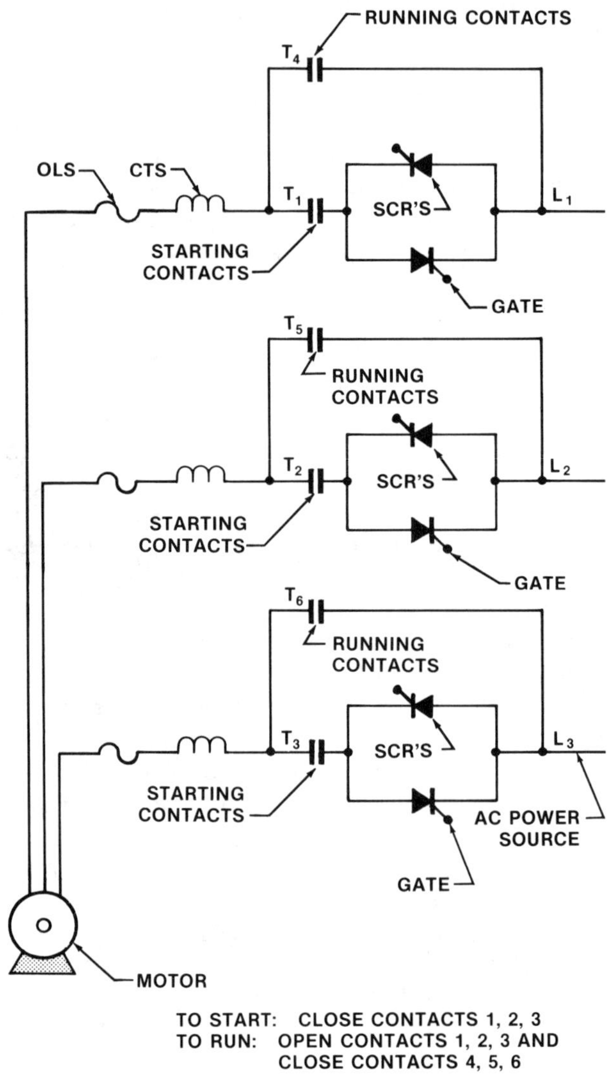

Figure 3-20. Contacts 1, 2, and 3 close to start the motor. When the motor reaches its running speed, contacts 4, 5, and 6 close and contacts 1, 2, and 3 open to run the motor.

Solid state starters reduce the starting current of motors to approximately 100% to 400% of the motor's full-load current rating. For example, a three-phase, 480-volt, 450-horsepower motor has a full-load current rating of 572 amps. What is the reduced starting current where the voltage is reduced 50% and provides 200% reduced starting current? Multiply 572 amps by 200% to find the reduced starting current of 1144 amps. The starting current is approximately 3432 amps (572 A × 6 = 3432 A) without the use of the solid state starter. See Figure 3-21 for a step-by-step procedure for reducing the starting current using a solid state starter.

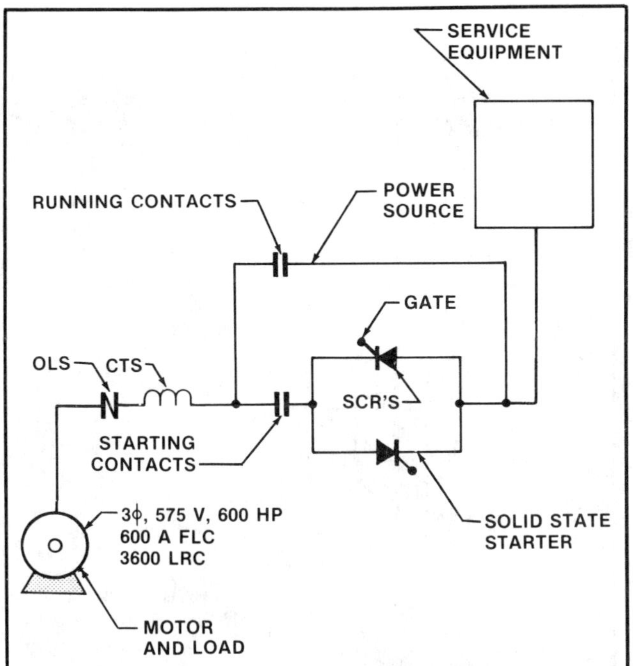

PROBLEM: What is the reduced starting current using a solid state starter with a reduction of 250% of the full-load current rating?

Step1: Apply formula.

starting current = *flc* × %

Step 2: *starting current* = 600 A × 250%

starting current = 1500 A

Answer: **Starting current is 1500 A using solid state starting.**

Figure 3-21. The starting current of a three-phase, 600-horsepower motor may be reduced by using solid state starting. *NOTE:* Starting torque is reduced to 30% of normal starting torque.

Overloads in a solid state starter can be set to sense any amount of overload current above the running current of the motor; therefore, closer protection can be provided for the windings due to all types of overload conditions. Special motors can be protected because the overloads in the solid state starter disconnect the motor from the power supply where overloads existed. After the overload condition is corrected, the power supply is again connected to the solid state starter and motor.

Starting torque varies with the percentage of starting current during the starting period. If the starting current is reduced to 400% the starting torque will be reduced to approximately 60% of the motor's starting torque. At 300% the starting torque is reduced to approximately 40%, and at 200%, the starting torque is reduced to approximately 30% of the motor's starting torque. See Figure 3-22 for a step-by-step procedure for calculating the starting torque using a solid state reduced starting method.

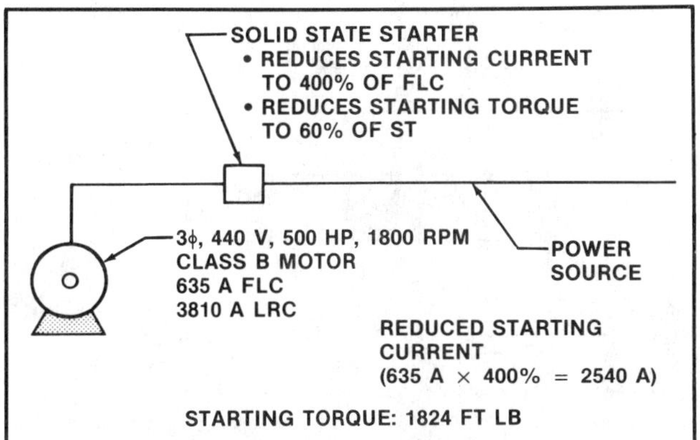

PROBLEM: What will be the reduced starting torque for this motor using solid state starting?

Step 1: Apply formula.

reduced starting torque = *starting torque* × %

Step 2: *reduced starting torque* = 1824 ft lb × 60%

reduced starting torque = 1094 ft lb

Answer: **Solid state starting will reduce the starting torque to 1094 ft lb.**

Figure 3-22. A solid state starter may be used to reduce the starting current to 400% of full-load current and the starting torque to 60% of the starting torque.

Part-winding Starting

Part-winding starting is used mostly on a weak power system to reduce the voltage and prevent a voltage disturbance. Using part-winding as the starting method no voltage dip will occur during starting and acceleration of the motor. Part-winding refers to a motor that has two separate parallel windings and consists of two basic starter units. Each starter unit is selected for half the horsepower rating of the motor. One winding is connected to the supply voltage when the motor is started. The second winding is connected at a predetermined time with a preset time delay.

Part-winding starting has the disadvantage of utilizing only half of the motor's copper during the starting operation. Part-winding starting reduces the inrush starting current and starting torque to 65% of normal. See Figure 3-23.

Starting Current	Starting Torque
Reduced 65% of Normal	Reduced 65% of Normal

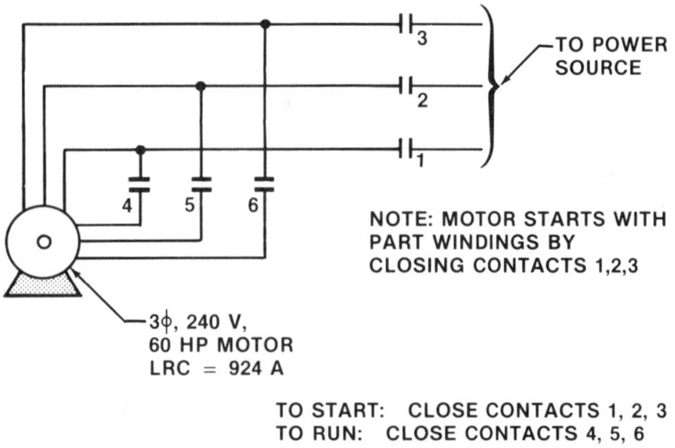

Figure 3-23. Part-winding starting reduces starting current and starting torque to 65% of normal.

Overcurrent Protection. Section 430-3 lists the requirements for sizing the overcurrent protection device for a part-winding motor. The protective device must be selected by 430-52, which refers to Table 430-152. Since only half of the motor's horsepower is used for starting, the overcurrent protection device is selected on half the percentages listed in Table 430-152. For example, if a three-phase, 240-volt, 60-horsepower part-winding motor has a full-load current rating of 154 amps, 175% is applied per Table 430-152 for selecting delay fuses. Section 430-3 requires half of the motor's full-load current rating to be used for sizing the fuses. The full-load current is multiplied by 175% (154 A × 175% = 269.5 A). Half of 269.5 amps is 134.8 amps. Section 240-6 lists the next lower rating as 125-amp fuses. See Figure 3-24 for the step-by-step procedure for selecting the overcurrent protection devices for part-winding motors. *NOTE:* The next lower size fuse is selected per 240-6.

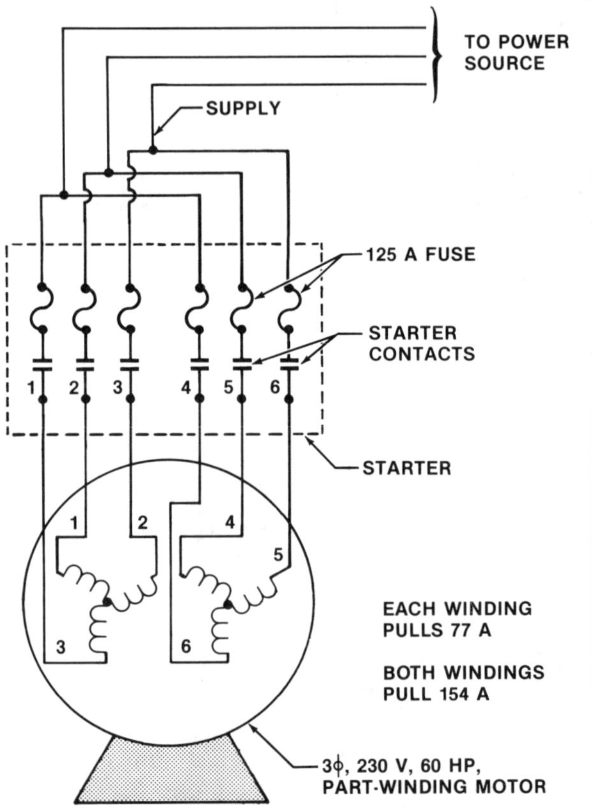

PROBLEM: What size time-delay fuses are required for a three-phase, 230-volt, 60-horsepower part-winding motor?

Step 1: *Motor's flc*
60 hp = 154 A

Step 2: *Table 430-152.*
tdf = 175%

Step 3: 154 A × 175% = 269.5 A

430-3.
one-half of 269.5 A = 134.75 A

Step 4: *240-6.*
125-A tdf

Answer: **(1) 125-A fuses for the starting winding**
(2) 125-A fuses for the second winding

Figure 3-24. The overcurrent protection device for a part-winding motor is determined by multiplying full-load current by 175% and dividing by 2.

The exception to 430-3 permits a single device for both windings if the device selected at half the rating will start the motor. See Figure 3-25. The exception to 430-3 also allows time-delay fuses to be used as a single device for both windings if sized at 150% or less of the motor's full-load current rating. For example, a three-phase, 230-volt, 60-horsepower motor with part-winding starting has a full-load current of 154 amps. The size of time-delay fuses required is determined by multiplying 154 amps by 150% per 430-3, Exception (154 A × 150% = 231 A). Section 240-6 shows a 225-amp fuse as the next lower rating.

Wye-delta Starting

The wye-delta starting method requires a special wound six-lead motor. The standard nine-lead motor cannot be used to apply wye-delta starting to bring a motor up to its running speed. The motor is started on the wye winding and switched to the delta winding in the run operation. Wye windings produce a lower current due to the lower voltages. Line current in the wye winding is equal to the phase current and not 1.732 times the phase current. Wye windings have 139 volts impressed across each winding instead of the 240 volts in delta windings. This value of voltages is derived by using a three-phase, 240-volt supply to the motor.

The phase voltage and line voltage are not the same value in wye-connected windings. The line voltage is equal to 1.732 times the phase voltage. For example, if the phase voltage is 120 volts, then the line voltage is 208 volts (120 V × 1.732 = 208 V). However, the phase current and the line current are the same value. If the phase current is 10 amps, the line current is 10 amps. See Figure 3-26. Phase voltage in wye-connected windings can also be determined by multiplying the phase voltage of 240 V × 58% (240 V × 58% = 139 V).

Delta-connected Windings. The phase voltage and line voltage are always the same value in

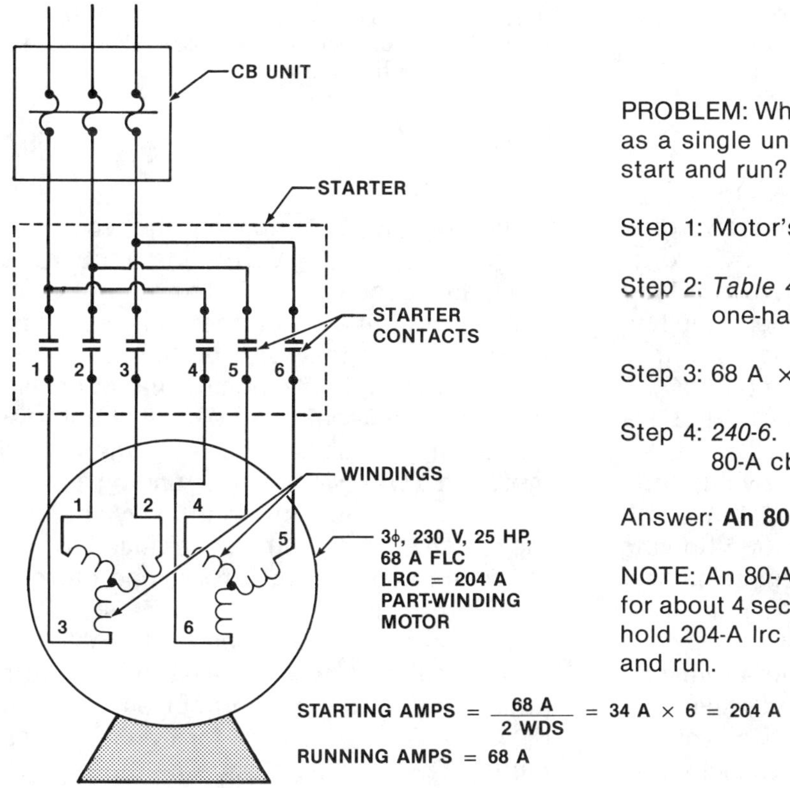

PROBLEM: What size circuit breakers are required as a single unit to allow a part-winding motor to start and run?

Step 1: Motor's flc rating is 68 A.

Step 2: *Table 430-152*; *430-3*, *Ex.*
one-half of 250% = 125%

Step 3: 68 A × 125% = 85 A

Step 4: *240-6*.
80-A cb

Answer: **An 80-A three-pole unit is required.**

NOTE: An 80-A cb will hold three times its rating for about 4 seconds (80 A × 3 = 240 A); 240 A will hold 204-A lrc long enough for the motor to start and run.

Figure 3-25. Single unit overcurrent protection may be utilized for part-winding starting by applying 430-3, Exception.

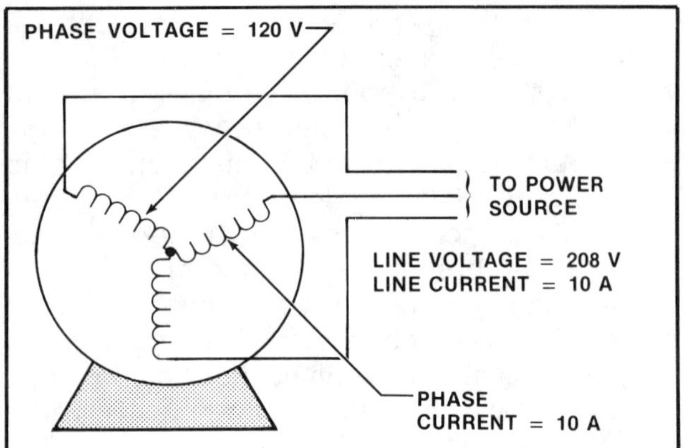

PROBLEM: What is the line voltage and line current in relation to the phase voltage and phase current for wye-connected windings?

Step 1: *phase voltage* = 120 V

120 V × 1.732 = 208 V

Answer: **line voltage = 208 V**

Step 2: *phase current* = 10 A

Answer: **line current = 10 A**

Figure 3-26. Phase current and line current are the same value in wye-connected windings. Line voltage is 1.732 times phase voltage.

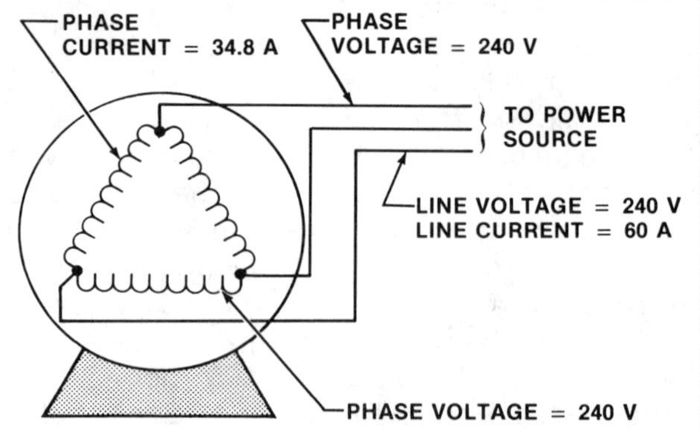

PROBLEM: What is the phase voltage and current in relation to the line voltage and current for delta-connected windings?

Step 1: *line voltage* = 240 V

Answer: **phase voltage = 240 V**

NOTE: Phase and line voltage remain constant.

Step 2: *line current* = 60 A

phase current = 60 A × 58% = 34.8 A

Answer: **phase current = 34.8 A**

Figure 3-27. The phase and line voltage of delta-connected windings remain constant. The phase current is 58% of the line current.

delta-connected windings. If the phase voltage is 240 volts, the line voltage is 240 volts. However, the phase current and line current are always different values. For example, if the line current is 60 amps, then the phase current is equal to 60 amps times 58% (34.8 amps). *NOTE*: The reciprocal of the square root of 3 is 58% (1 ÷ 1.732 = .58). See Figure 3-27.

The phase current has a different value than the line current because the current flows into two phase windings from the line. Multiplying the line current by 58% gives the phase current for delta-connected windings.

Wye-delta-connected Windings. In a wye-connected winding, the line current and phase current are the same current, but the phase voltage is 58% of the line voltage. In a delta-connected winding the voltage value does not change, but the current value in each winding is 58% of the line current.

By starting a motor on the wye winding, voltage and starting current are reduced. When the motor accelerates to its running speed, the delta windings are automatically connected to the circuit for the run operation. *NOTE*: Wye windings are cut out when the motor accelerates to running speed. Figure 3-28 illustrates the values of voltage and current when starting a motor on a wye winding and running it on a delta.

By starting the motor on the wye connection, the inrush starting current is reduced to one-third the value of the locked-rotor line current. The locked-rotor line current, per Table 430-151, is 924 amps for a three-phase, 240-volt, 60-horsepower motor. A wye-delta starting method reduces the inrush starting current to 305 amps (924 A × 33% = 305 A) for the 60-horsepower motor. The starting torque of the motor is reduced to 33% of the normal starting torque. See Figure 3-29. Note that the starting torque of a motor starting on a wye connection and running

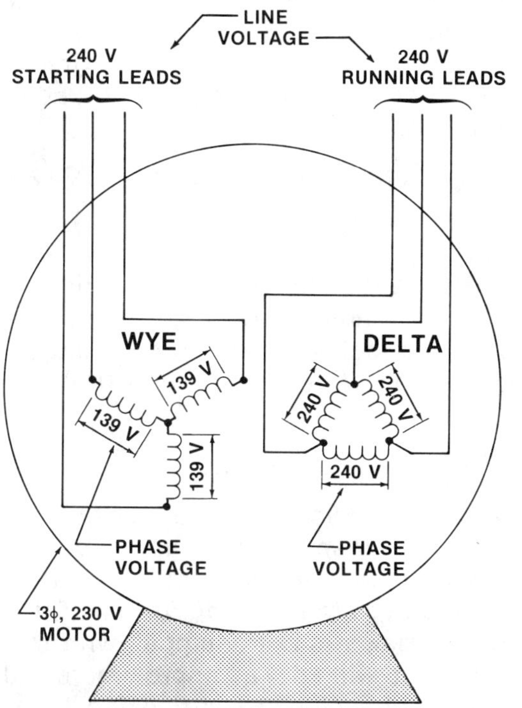

PROBLEM: What is the voltage in the wye and delta windings of this motor applying the wye-delta starting method?

Step 1: Delta winding voltage is 240 V.

Step 2: $\frac{1}{1.732} = 58\%$

240 V × 58% = 139 V

Answer: **wye winding voltage = 139 V**
delta winding voltage = 240 V

Figure 3-28. Wye-delta motors start on wye windings and run on delta windings. Phase voltage is 58% of line voltage in the wye windings.

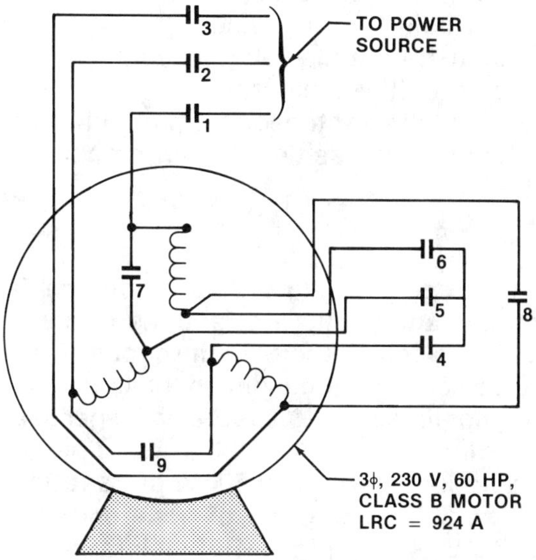

TO START: CLOSE CONTACTS 1, 2, 3, 4, 5, 6
TO RUN: OPEN CONTACTS 4, 5, 6
CLOSE CONTACTS 7, 8, 9

PROBLEM: What is the reduced starting current and torque of a wye-delta starting?

Step 1: *Table 430-151.*
lrc = 924 A

Step 2: 924 A × 33% = 305 A

Answer: **reduced starting current = 305 A**

NOTE: Refer to Figure 3-6 to determine starting torque.

Answer: *starting torque* = 228 ft lb × 33%
starting torque = 75 ft lb

Figure 3-29. Starting current and torque of a wye-delta connected motor is 33% of normal current and torque.

on a delta connection is reduced to one-third. For example, a wye-delta motor with a starting torque of 228 foot-pounds is reduced to 75 foot-pounds (228 ft lb × 33% = 75 ft lb).

TYPES OF OVERCURRENT DEVICES—TABLE 430-152, COLUMNS 1, 2, 3, 4

Overcurrent protection devices must be sized to protect the branch-circuit conductors. Also, the motor must start without the circuit opening as a result of the momentary inrush current of the motor. Section 430-6(a) refers to Table 430-148 for selecting the full-load current ratings for single-phase motors. Table 430-150 must be used for three-phase motors.

Section 240-3(f) refers to 430-52, which refers to Table 430-152 for selecting overcurrent protection devices for motor circuits.

The selection of the overcurrent protection

device may be made using Table 430-152, which lists four columns of devices:

Column 1: nontime-delay fuses
Column 2: time-delay fuses
Column 3: instantaneous trip circuit breaker
Column 4: inverse time circuit breaker

Column 1

The first column lists ordinary nontime-delay fuses. Instantaneous trip features in nontime-delay fuses detect shorts in the circuit. Thermal features sense heat buildup in the circuit. Once blown, nontime-delay fuses are no longer usable. An ordinary fuse (contains no time-delay characteristics) will hold 500% of its rating for approximately one-fourth of a second. *NOTE:* Some NTDF will hold 500% of their rating for up to two seconds. See Figure 3-30. The motor must start and reach its running speed in one-fourth of a second when nontime-delay fuses are used for overcurrent protection.

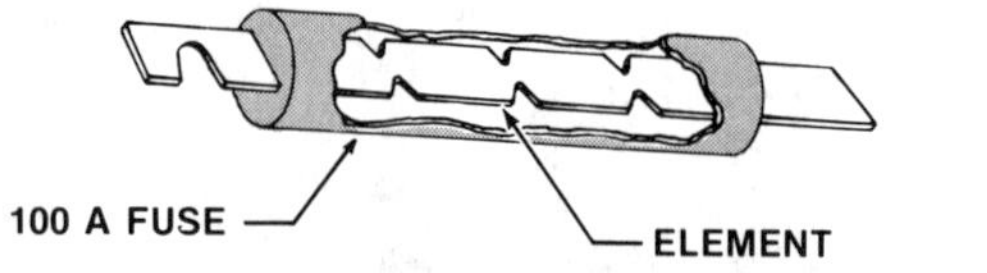

100 A FUSE × 5 = 500 A FOR ¼ TO 2 SECONDS

NONTIME-DELAY FUSE

Table 430-152

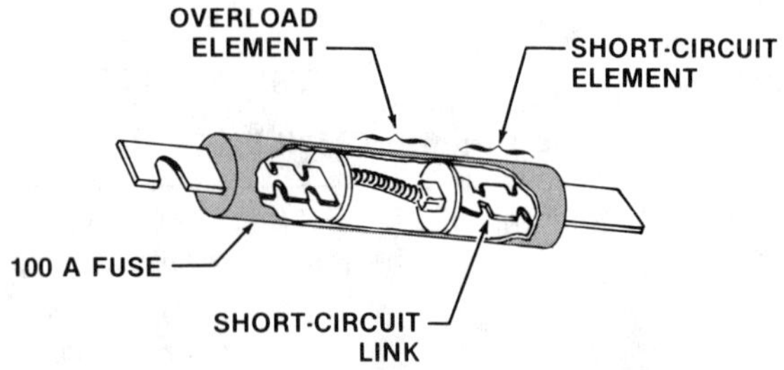

100 A FUSE × 5 = 500 A FOR 10 SECONDS

TIME-DELAY FUSE

Table 430-152

Figure 3-30. A time-delay fuse will hold five times its rating for 10 seconds. A nontime-delay fuse will hold five times its rating for ¼ to 2 seconds.

Column 2

The second column lists time-delay fuses with thermal and instantaneous trip features (dual elements). Their time-delay action allows the motor to start as a result of the starting current of the motor, which is usually six times the full-load current rating of the motor. Time-delay fuses will hold 500% the ampacity rating for 10 seconds. For example:

$$100\text{-A rated fuses} \times 500\% = 500\text{ A}$$

This delay of time will allow almost any motor to start without opening the circuit. Refer to Figure 3-30. Note that with normal conditions of use, 100-amp time-delay fuses will start any motor with a locked-rotor current rating of 500 amps or less (100-A tdf × 5 = 500 A).

Column 3

The third column lists instantaneous trip circuit breakers. They respond to instantaneous (immediate) values of current from a short circuit ground fault or stalled rotor current. This type of circuit breaker is not equipped with thermal protection. It will trip at approximately three times its rating on the low setting and at approximately ten times its rating on the high setting. Some instantaneous trip circuit breakers have adjustable trip settings. See Figure 3-31.

INVERSE TIME CB TRIP SETTINGS			
Size (amps)	Volts	Percent of Load Held	Time (seconds)
100	240	300%	4
100	480	300%	9
110–225	240/480	300%	35
400–500	240/480	300%	50
600	240/480	300%	40

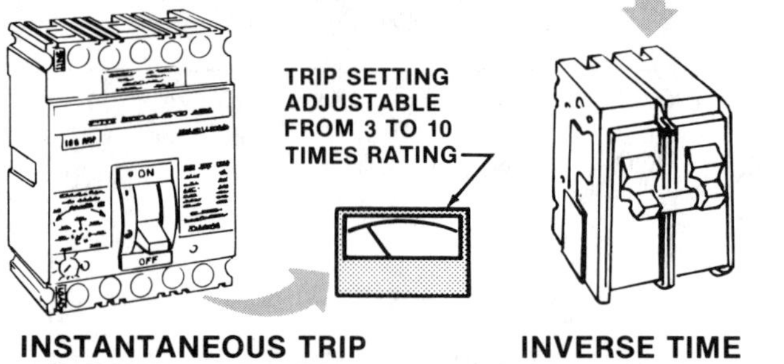

INSTANTANEOUS TRIP CIRCUIT BREAKER **INVERSE TIME CIRCUIT BREAKER**

Table 430-152

Figure 3-31. Instantaneous trip circuit breakers are adjustable. Inverse time circuit breakers are preset.

Trip settings of instantaneous trip circuit breakers are adjustable to hold the inrush starting current of a motor. They are used where time-delay fuses set at five times their ratings or circuit breakers at three times their rating will not hold the starting current of a motor.

Care must be exercised not to adjust the trip setting above 1300% per 430-52(a). Settings above 700% per Table 430-152 cannot be used if the motor will start and run up to speed. Where instantaneous trip circuit breakers are used, overload protection must be provided to protect the motor. A stuck bearing or a blanket of lint covering the inlets and outlets of the motor's enclosure can cause the motor to overheat and damage the windings.

An instantaneous trip circuit breaker only trips open due to a short circuit, either phase-to-phase or phase-to-ground. This type of circuit breaker will never trip from a slow heat buildup due to motor windings overheating. Overload current protection must be provided for this type of overload.

Column 4

The fourth column lists inverse time circuit breakers. They have both thermal and instantaneous trip features. The thermal action of this circuit breaker responds to heat. If a motor's ventilation inlets and outlets are not adequate to dissipate heat from the windings of the motor, the heat will be detected by the thermal action of the circuit breaker. If a short should occur, the magnetic action of the circuit breaker will detect the instantaneous values of current and trip the circuit breaker. This is the most common type of circuit breaker used in the building trades for residential, commercial, and heavy construction.

A 100-amp or less circuit breaker holds a 300% overload for four seconds on 240 volts. A 300% overload holds for nine seconds on 480 volts. A circuit breaker of over 100 amps to 225 amps holds 300% overload on 240 volts or 480 volts for 35 seconds. A 400-amp to 500-amp circuit breaker holds a 300% overload for 50 seconds, and a 600-amp circuit breaker holds a 300% overload for 40 seconds. Refer to Figure 3-31.

NOTE: The preceding trip settings are rule of thumb. For example, if a motor has a locked-rotor current of 250 amps, a 100-amp circuit breaker (3 × 100-A cb = 300 A) allows the motor to start. However, the motor must start and accelerate up to its running speed within four seconds for 240-volt systems or nine seconds for 480-volt systems.

RATINGS FOR INDIVIDUAL MOTOR CIRCUITS—430-52

When the percentages in Table 430-152 do not allow the motor to start because of unusually large starting currents, larger percentages may be applied.

(1) Ratings for nontime-delay fuses not exceeding 600 amps may be increased to 400% for the full-load current of the motor but must not exceed 400%.

(2) Ratings for time-delay fuses may be increased to 225% of the full-load current of the motor but must not exceed 225%.

(3) Ratings for instantaneous trip circuit breakers may be increased to 1300% of the full-load current of the motor, but must not exceed 1300%.

(4) Ratings for inverse-time circuit breakers may be increased to 400% of the full-load current of the motor up to 100 amps, but must not exceed 400%. An increase of 300% is permitted for a full-load current greater than 100 amps.

SIZING OVERCURRENT PROTECTION DEVICES—TABLE 430-152

Table 430-152 is used to size overcurrent protection devices above the locked-rotor current of motors allowing them to start and run up to their running speeds without tripping open the overcurrent protection device of the circuit. The percentages shown in Table 430-152 for each protection device listed will usually start and run motors. Where unusually heavy currents are required to start and run motors, one of the exceptions to 430-52(a) must be applied.

Nontime-delay Fuses (ntdf)

Figure 3-32 shows the step-by step procedure for sizing the overcurrent protection device for a motor using a nontime-delay fuse. Section 430-52(a), Exception 1 allows use of the next higher size fuse or circuit breaker when the values desired do not correspond to standard size devices adequate to carry the load. If the motor still cannot be accelerated up to running speed by applying 430-52(a), Exception 2 can be utilized.

A nontime-delay fuse holds five times its

rating for about one-fourth of a second. Therefore, a 30-amp nontime-delay fuse holds 150 amps and will permit any motor with a locked-rotor current of less than 150 amps to start and run under normal conditions. Nontime-delay fuses will not provide overload current protection for motors because they must be sized above 125% to allow the motor to start and run. Overload protection must be provided to protect the motor windings from heat buildup that can cause insulation failure. Overload protection may be provided by sizing nontime-delay fuses at 115% to 125% of the motor's nameplate current rating. However, these lower percentages generally will not allow the motor to start.

The nameplate full-load current rating listed on a motor is the actual amperage the motor will pull when driving a load. This current rating can vary with the driven load. Note that 125% times the motor's full-load current must not be exceeded. For example, a motor with a 42-amp full-load current listed on the nameplate requires 50-amp fuses (42 A × 125% = 52 A). Since 52 amps must not be exceeded, the next lower size fuse is 50-amp. *NOTE*: The 225-amp fuses in Figure 3-32 will not provide overload current protection for the 30-horsepower motor. A 100-amp set of fuses could be required (80 A × 125% = 100 A) to provide both overload protection and overcurrent protection. See 430-32(a)(1) for selecting running overload protection.

Applying 430-52(a), Exception 2a. If a motor will not start using 430-52(a), Exception 1, Exception 2a can be utilized by applying the following procedure: Fuses not exceeding 600 amps can be used by multiplying the motor's full-load current by 400%. For example, a three-phase, 230-volt, 30-horsepower motor with an 80-amp full-load current per Table 430-150 requires a 300-amp nontime-delay fuse (80 A × 400% = 320 A). The 300-amp nontime-delay fuse is then chosen per 430-52(a), Ex. 2a because it will start and run the motor as shown in the following procedure.

Step 1: *430-52(a), Ex. 2a.*
ntdf not exceeding 600 A, use 400% of flc

Step 2: *Table 430-150 and Table 430-152.*
30 hp = 80 A (80 A × 400% = 320 A)

Step 3: *240-3(f); 240-6.*
300 A ntdf is the next lower rating

Answer: **300-A nontime-delay fuse**

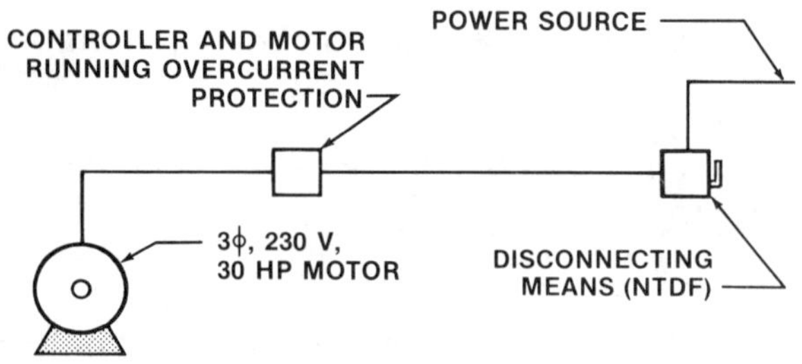

240-6; Table 430-150
Table 430-152

PROBLEM: Find the size overcurrent protection device required for this motor with full-voltage starting and code letter F. (Use lower rating.)

Step 1: *Table 430-150, Col. 2.*
A 230-V, 30-hp motor has an 80-A flc rating.

Step 2: *Table 430-152, Col. 1.*
The maximum allowed is 300%.

80 A × 300% = 240 A

Step 3: *240-6.*

Answer: **The nearest standard size smaller fuse is 225 A.**

NOTE: If the motor will not start, a 250-A fuse may be used per 430-52(a), Ex. 1.

Figure 3-32. Nontime-delay fuses hold five times their rating for approximately ¼ second. This rating is usually above the motor's locked-rotor current.

Time-delay Fuses (tdf)

Time-delay fuses hold five times their rating for approximately 10 seconds, which in most cases is enough time to allow a motor to start. A 50-amp fuse holds 250 amps (50-A tdf × 5 = 250 A) for 10 seconds and will permit any motor with less than 250 amps of starting current to start. For most motors with code letters A through G, 125% of the nameplate current rating of the motor will start and run the motor. Note that 125% must not be exceeded. For example, a motor with a 42-amp full-load current listed on the nameplate requires 50-amp fuses (42 A × 125% = 52.5 A) to provide both overload and overcurrent protection for the motor. Figure 3-33 shows the step-by-step procedure to size a time-delay fuse for a motor.

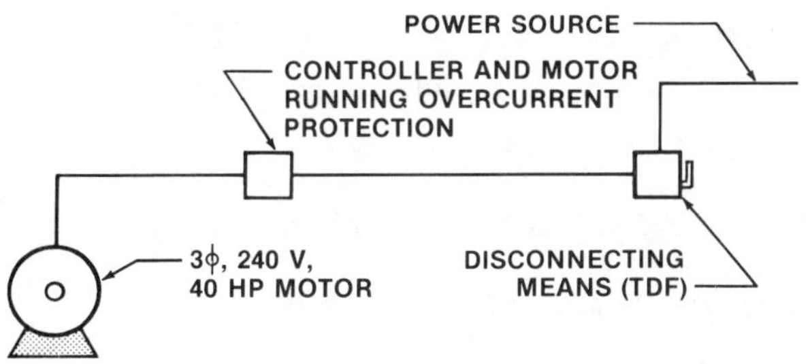

240-6; Table 430-150
Table 430-152

PROBLEM: Find the size of the overcurrent protection device required for this motor with full-voltage starting and code letter G. (Use a lower rating.)

Step 1: *Table 430-150, Col. 3.*
A 40-hp, 240-V motor has a 104-A flc rating.

Step 2: *Table 430-152, Col. 2.*
The maximum allowed is 175%.

104 A × 175% = 182 A

Step 3: *240-6.*

Answer: **The nearest standard size smaller fuse is 175 A.**

NOTE: If the motor will not start, a 200-A fuse may be used per 430-52(a), Ex. 1.

Figure 3-33. Time-delay fuses hold five times their rating for approximately 10 seconds, allowing motors to start and run.

Applying 430-52(a), Exception 2b. According to 430-52, Exception 2b allows the full-load current rating of a motor to be increased to 225% if the motor will not start due to heavy starting current. *NOTE*: When selecting the rating of the overcurrent protection device 225% must not be exceeded. Exception 2b can be applied by multiplying the motor's full-load current by 225%. For example, to find the size time-delay fuse required to start and run a three-phase, 220-volt, 40-horsepower induction motor, the following procedure is used.

Step 1: *430-52(a), Ex. 2b.*
tdf not exceeding 225% can be used

Step 2: *Table 430-150.*
40 hp = 104 A (104 A × 225% = 234 A)

Step 3: *240-3(f); 240-6.*
225 A tdf is the next lower rating

Answer: **225-A tdf**

Instantaneous Trip Circuit Breaker

Instantaneous trip circuit breakers provide no thermal protection for a motor. They will only trip out due to short circuits occurring from phase-to-phase, phase-to-ground, or heavy inrush currents. Instantaneous trip circuit breakers are never used to provide running overload protection to a motor. They are used, however, to start and run a motor. Their trip settings must be adjusted above the locked-rotor current of the motor.

For example, suppose that a shop superintendent wants to provide proper electrical protection and free an operator from resetting a cb that has regularly tripped open due to thermal buildup. An instantaneous trip circuit breaker could be installed ahead of the combination starter and disconnecting means to eliminate the problem. Figure 3-34 shows the step-by-step procedure to determine instantaneous trip circuit breaker setting. The instantaneous trip ratings of an instantaneous trip circuit breaker can be adjusted above the locked-rotor current of a motor to allow the motor to start and come up to its running speed. For example, a motor has a locked-rotor starting current of 650 amps. An instantaneous trip circuit breaker can be set at 700 amps to permit the starting of the motor.

Applying 430-52(a), Exception 1. It a motor will not start due to the inrush starting current, an instantaneous trip circuit breaker can be adjusted up to 1300% of the motor's full-load current rating. *NOTE*: This setting should only be increased until the motor will start. The following procedure is used to find the maximum setting allowed for a three-phase, 440-volt, 60-horsepower motor.

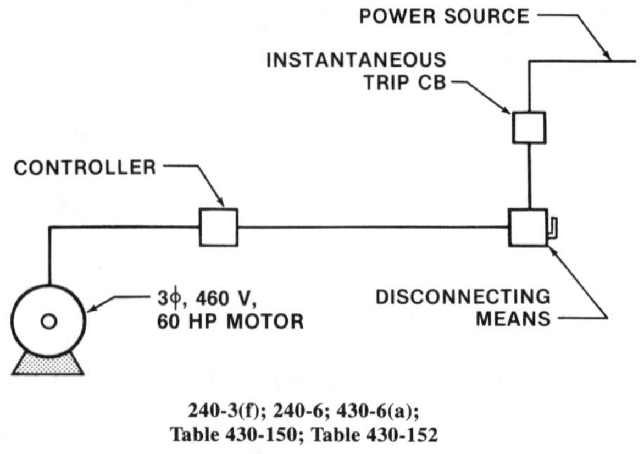

PROBLEM: Find the setting of the instantenous trip circuit breaker required to hold the starting current of this motor, which has no code letter.

Step 1: *Table 430-150, Col. 3.*
A 3ϕ, 460-V, 60-hp motor has a 77-A flc.

Step 2: *Table 430-152, Col. 3.*
The maximum allowed is 700%.

77 A × 700% = 539 A

Step 3: *240-6.* 539 A is maximum setting.

Answer: **539 A is maximum setting.**

NOTE: If the motor will not start, a setting of up to 1001 A may be used per 430-52(a), Ex. 1 (77 A × 1300% = 1001 A).

Figure 3-34. Instantaneous trip circuit breakers can be adjusted above the locked-rotor current of motors to allow motors to start and run. The 1300% maximum setting can be automatically used to start high-efficiency motors where the need has been shown by engineering evaluation.

Step 1: *430-52(a), Ex. 1.*
instantaneous trip cb setting not exceeding 1300%

Step 2: *Table 430-150.*
60 hp = 77 A
77 A × 1300% = 1001 A

Step 3: *240-3(f); 240-6.*
1001 A is a maximum setting

Answer: **1001 A is maximum setting.**

Inverse Time Circuit Breakers

An inverse time circuit breaker holds three times its rating for different time periods based on its frame size and voltage. The rating of an inverse time circuit breaker can be multiplied by 3 and this total amperage will start any motor with less locked-rotor amperage. A circuit breaker with 525 amps (175-A inverse time cb × 3 = 525 A) will start any motor under normal starting conditions with a locked-rotor current less than 525 amps. See Figure 3-35 for step-by-step procedures to size circuit breakers.

Applying 430-52(a), Exception 2c. If the 150-amp circuit breaker permitted by 430-52(a), Exception 1 will not start a motor, Exception 2c can be applied. The following procedure should be followed to apply 430-52(a), Exception 2c.

For example, a three-phase, 460-volt, 50-horsepower motor with a 65-amp full-load current can be protected by a 250-amp circuit breaker (65 A × 400% = 260 A). The next lower standard device is used per 240-6.

Step 1: *430-52(a), Ex. 2c.*
flc for 65 A 400% can be applied

Step 2: *Table 430-150.*
50 hp = 65 A
65 A × 400% = 260 A

Step 3: *240-3(f); 240-6.*
250 A is the next lower rating

Answer: **250-A inverse time cb**

LOCKED-ROTOR CURRENT—TABLE 430-151

Sizing the overcurrent protection device above the locked-rotor current rating of the motor will usually permit the motor to start without opening the circuit. The size of the overcurrent protection device and the running overload protection depends on the load the motor is driving and

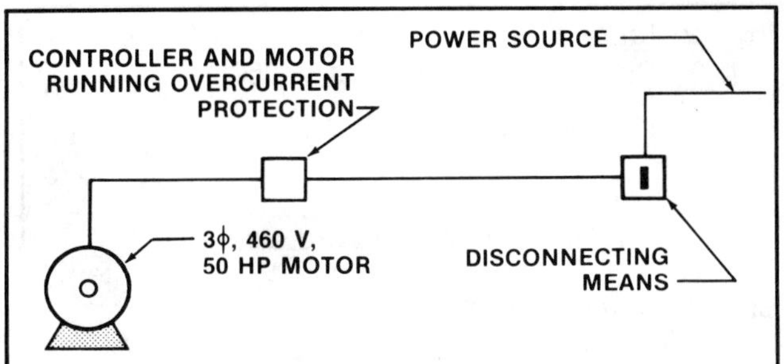

240-6; Table 430-150
Table 430-152

PROBLEM: Find the size of the inverse-time regular circuit breaker required for this motor, which has autotransformer starting and code letter H.

Step 1: *Table 430-150, Col. 3.*
A 3ϕ, 460-V, 50-hp motor has a 65-A flc rating.

Step 2: *Table 430-152, Col. 4.*
The maximum allowed is 200%.

65 A × 200% = 130 A

Step 3: *240-6.*

Answer: **The nearest standard size smaller cb is 125 A.**

NOTE: If the motor will not start, a 150-A cb may be used per 430-52, Ex. 1.

Figure 3-35. Inverse time circuit breakers hold five times their rating for varying frame sizes, allowing motors to start and run.

the environment where the motor and controller are located.

Locked-rotor current values listed in Table 430-151 are based on six times the full-load current ratings of each motor per Table 430-150. The locked-rotor current values of each motor per Table 430-151 are sufficient for code letters A through H per Table 430-7(b).

Overcurrent protection devices can be selected just above the locked-rotor current, which provides closer protection for the motor. Locked-rotor current for a motor can be selected from Table 430-151, which is very conservative, or by the motor's code letter per Table 430-7(b). The locked-rotor current based on the motor's code letter is closer to the actual starting current of the motor than the values selected from Table 430-151.

Selecting Locked-rotor Current Per Table 430-151

The locked-rotor current of a motor can be selected from Table 430-151 based on the horsepower and voltage of the motor. A circuit breaker or fuses can be used to start the motor based on the amount of overload and holding time of the device.

Inverse Time Circuit Breaker. Figure 3-36 shows the step-by-step procedure for sizing an overcurrent protection device above the locked-rotor current of the motor using an inverse time circuit breaker. The circuit breaker is selected to be close to, but a higher value than, the locked-rotor current rating of the motor. Normally circuit breakers are not selected so that the possibility of frequent tripping could occur. However, when personnel are available for monitoring the system, settings closely above locked-rotor current ratings are an efficient and logical design.

NOTE: The overcurrent protection device is usually sized according to Table 430-152, although field application sometimes requires the table's percentages to be adjusted higher or lower. Notice that Table 430-152 permits a 100-amp circuit breaker to allow a motor to start and run. However, a 90-amp circuit breaker will start and run the motor because of its 270-amp holding power (90-A cb × 3 = 270 A).

The locked-rotor current of a motor can be selected from Table 430-7(b) or from Table 430-151. The following example illustrates the use of Table 430-7(b) to find a motor's locked-rotor current.

$$Irc = \frac{k \times VA \times hp}{V \times \sqrt{3}} = A$$

Step 1: *Table 430-7(b).*
code letter F = 5.59 kVA/hp

Step 2: $\frac{1000 \times 5.59 \times 15}{230 \times 1.732} = 210.7 \text{ A}$

Answer: **Irc = 210.7 A**

By calculating the locked-rotor current of the motor using the code letter, an 80-amp circuit breaker (80-A cb × 3 = 240 A) can be used to start and run the motor. An 80-amp circuit breaker will hold 240 amps, which is above the

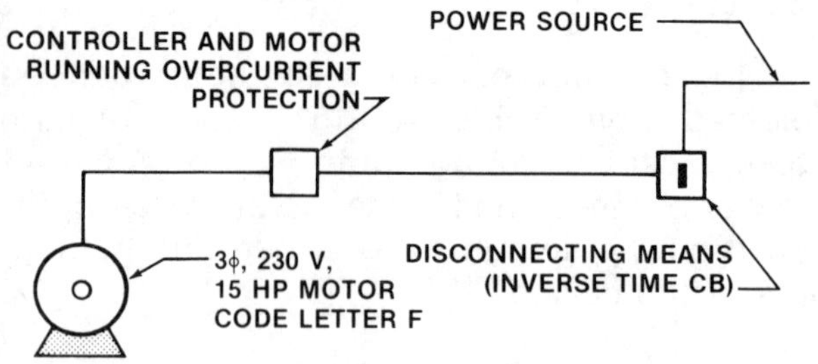

240-6; Table 430-150
Table 430-151; Table 430-152

PROBLEM: Find the size of the required inverse-time circuit breaker above the locked-rotor current of this motor. (Use lower rating.)

Step 1: *Table 430-150, Col. 2.*
A 3ϕ, 230-V, 15-hp motor has a 42-A flc.

Step 2: *Table 430-152, Col. 4.*
The maximum allowed is 250%.

42 A × 250% = 105 A

Step 3: *240-6.*
A 100-A cb is a standard size.

Step 4: *Table 430-151, Col. 3.*
The lrc is 252 A. See Footnote.

42 A × 6 = 252 A

Step 5: Multiply circuit breaker amps by 3.

100 A × 3 = 300 A

The result, 252 A, is above the lrc rating of the motor.

Step 6: *240-6.*
A smaller size circuit breaker may be used. For example, 90 A × 3 = 270 A.

Answer: **90-A cb**

Figure 3-36. Inverse time circuit breakers may be selected by sizing the device slightly above the motor's locked-rotor current.

210.7-amp locked-rotor current. *NOTE*: The amount of time that a circuit breaker will hold is three times its rating, depending on the frame size and setting. (This is a rule of thumb.)

Time-delay Fuse (tdf). Any of the protection devices in Table 430-152, columns 1, 2, 3, or 4 may be sized above the locked-rotor current rating of the motor. Under normal conditions of use, the motor will start as a result of the inrush current of the motor. The same step-by-step procedure shown in Figure 3-36 is followed when sizing overcurrent protection devices above the locked-rotor current ratings in Table 430-152. See Figure 3-37.

NOTE: The 50-horsepower motor will never pull 65 amps under normal loaded conditions. The 80-amp fuses are selected at a value less than 125% and they provide backup protection to the thermal overloads in the magnetic starter (65 A × 125% = 81 A). The next lower standard size overcurrent protection device per 240-6 is an 80-amp time-delay fuse.

NOTE: Designers use 125% times the full-load current of the motor when designing with time-delay fuses. This sizes the fuse above the locked-rotor current rating of the motor. For example, 65-A flc × 125% = 81 A. The next smaller size standard fuse, 80-amp, will comply. By sizing the time-delay fuse at 110% to 125%, backup protection is provided in case the thermal overloads should fail.

Normally, motors are never fully loaded and will not pull their rated full-load current; therefore, smaller overcurrent protection devices can be utilized. If overcurrent protection devices are sized and selected at 125% or less of the motor's nameplate full-load current rating, overload protection is also provided. See 430-6(a).

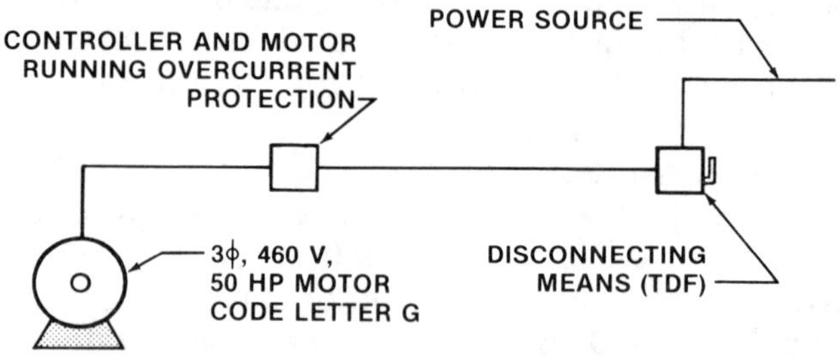

240-6; Table 430-150
Table 430-151; Table 430-152

Figure 3-37. Time-delay fuses may be selected by sizing the device slightly above the motor's locked-rotor current.

SIZING THE OVERCURRENT PROTECTION DEVICE FOR A FEEDER SUPPLYING TWO OR MORE MOTORS—430-62 AND 430-63

To size the overcurrent protection device for a feeder supplying two or more motors, first determine which motor requires the largest overcurrent protection device. The motor's full-load current rating times the percentages from Table 430-152 gives the ampacity to size the device.

To use the next higher rating, 430-52(a), Exception 1 must be applied. Add the rating of this overcurrent protection device to the full-load current ratings of the remaining motors. From the total obtained, select the next lower standard size overcurrent protection device from 240-6. Figure 3-38 shows the step-by-step procedure for sizing the overcurrent protection device for a feeder supplying two or three motors.

There is no exception to 430-62 that will permit the next larger size overcurrent protection device to be used. Therefore, a smaller size must be used. *NOTE*: Conductors are selected per 430-24, which requires 125% of the largest motor's full-load current rating plus the remaining motors of the group. The conductors will be lower in rating than the feeder circuit overcurrent protection device that is permitted per 240-3(f) and Table 430-152. Exceptions 1 and 2 to 430-52(a) allow the larger size overcurrent protection device when the motor will not start and run from the percentages in Table 430-152.

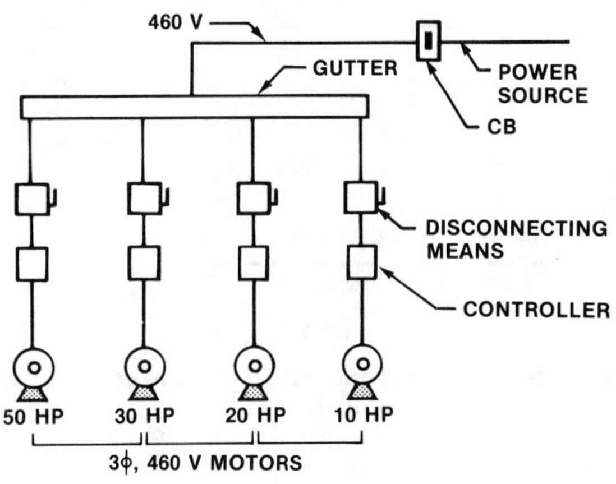

PROBLEM: Find the size circuit breaker required for these motors.

Step 1: *Table 430-150, Col. 3.*
The motor ratings are:

Horsepower	Amps
50	65
30	40
20	27
10	14

Step 2: *Table 430-152, Col. 4.*
The maximum allowed is 250%.

65 A × 250% = 162.5 A

NOTE: 430-52(a), Ex. 1 allows a 150 A cb.

Step 3: *430-62(a); 240-3(f); 240-6.*

150 A cb + 40 A + 27 A + 14 A = 231 A

Answer: **The next lower standard size cb is 225 A.**

Figure 3-38. Overcurrent protection for two or more motors is based on the largest overcurrent protection device of any motor in the group plus the full-load current of the remaining motors.

DESIGNING A SINGLE BRANCH CIRCUIT TO SUPPLY TWO OR MORE MOTORS—430-42, 430-52, AND 430-53

Two or more motors, or one or more motors and other loads, may be connected to one circuit protected by the circuit overcurrent protection device if the requirements of 430-53 are followed.

Motors Not Over 1 Horsepower in Rating—430-53(a)

Two or more motors rated less than 1 horsepower each, and drawing 6 amps or less of full-load current, may be installed without individual overcurrent protection devices. They must be portable, manually started, and within sight of the starter. If these conditions are not met, running overload protection must be provided for each motor per 430-32. The overcurrent protection device must not be larger than 20-amp at 125 volts or less, or 15-amp at 600 volts or less.

Figure 3-39 shows several fractional-horsepower motors on one circuit. The combined amperages must not exceed the branch circuit rating. For example: Two motors (disposal and compactor) rated at 6-amp and 5-amp are connected to a 20-amp branch circuit. Can these two motors

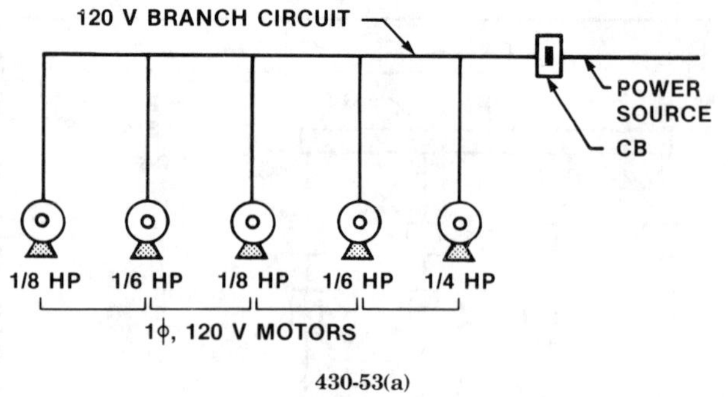

Figure 3-39. Several fractional-horsepower motors may be installed on one branch circuit.

be supplied by the 20-amp branch circuit? Yes, a 20-amp circuit breaker protecting the circuit will start the motors under normal conditions. A 20-amp circuit breaker will hold 60 amps for four seconds (20-A cb × 3 = 60 A). The 6-amp motor should pull no more than 18 to 24 amps when the motor is started.

Smallest Motor Protected—430-53(b)

Two or more motors of different ratings may be connected to a branch circuit as long as the maximum rating of the overcurrent protection device will protect the smallest motor of the group and still allow the largest motor to start. The overcurrent protection device must be set no higher than permitted by Table 430-152 for the smallest motor of the group. Overload protection must be provided per 430-32. See Figure 3-40.

Other Group Installation—430-53(c)

Two or more motors of any size may be connected to a single branch circuit. Fuse or circuit breaker protection is set by the percentages of Table 430-152 for the largest motor of the group. Each motor controller and component must be approved for group installation. Circuit components that must be protected include

(1) overcurrent protection devices
(2) controllers
(3) running overload protection devices

See Figure 3-41.

These components may be factory-installed as an approved assembly or field-installed as separate assemblies. Field-installed components must be factory-listed for such use. *NOTE*: If a fuse or circuit breaker is installed at the point where each motor is tapped to the line, the branch circuit becomes a feeder and any number of motor taps may be made. See Figure 3-42.

Section 430-28(1)(2) must be applied when making taps to supply the components required to serve a motor system. Any size tap can be utilized when applying the 10′ tap rule. For example: Number ten conductors can be tapped from 2/0 conductors as long as the conductors

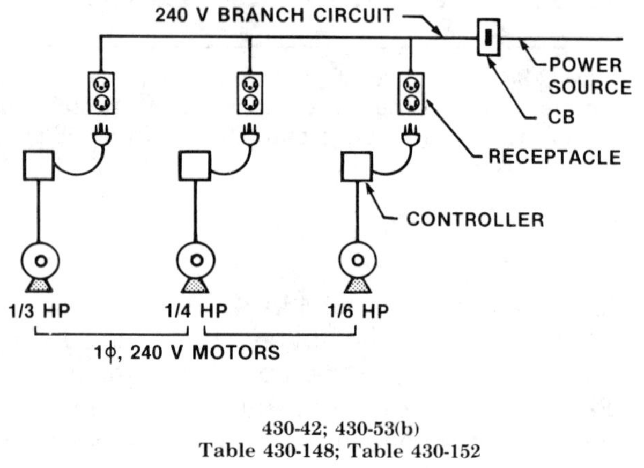

PROBLEM: Find the size circuit breaker required for these motors.

Step 1: *Table 430-148, Col. 2.*
The motor ratings are:

Horsepower	Amps
1/3	3.6
1/4	2.9
1/6	2.2

Step 2: *Table 430-152, Col. 4.*
The maximum allowed is 250%.

2.2 A × 250% = 5.5 A

Answer: *430-53(b).*
15-A cb

Figure 3-40. An overcurrent protection device must protect the smallest motor in the branch circuit while allowing the largest motor to start.

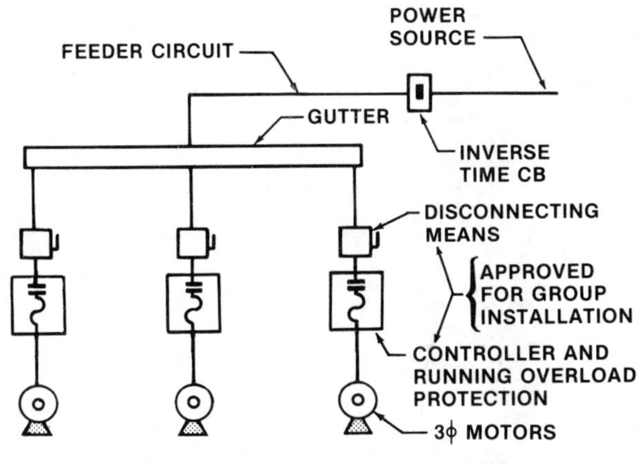

430-53(c)

Figure 3-41. Two or motors may be installed in a group installation.

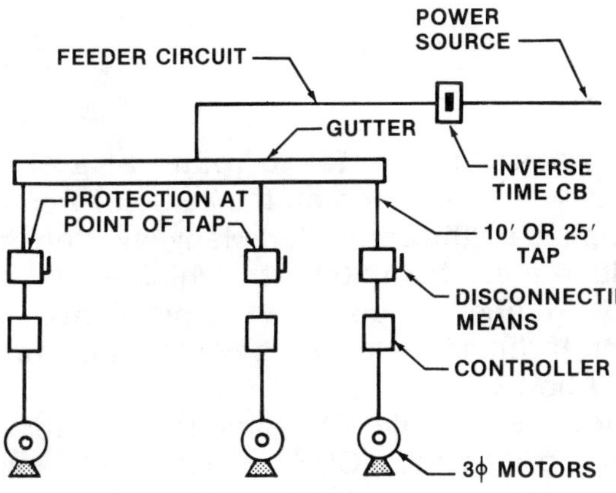

430-28
430-53(d)

Figure 3-42. Any number of motors may be tapped if protection is installed at the point of the tap.

are 10' or less in length and terminated in an approved manner. For example: the tap conductors from a feeder circuit with a 200 A OCPD shall equal at least 20 A ($\frac{200\text{ A OCPD}}{10} = 20$ A). See 240-21(b) and 430-28(1). However, a 25′ tap (any tap over 10′ to 25′) must be one-third of the #2/0 conductors. For #2/0 THWN Cu conductors, Table 310-16 lists a rating of 175 A per conductor. The 25′ tapped conductors, must be at least 58 amps ($\frac{1}{3} \times 175$ A = 58 A) in rating, which requires No.6 THWN copper conductors. See 240-21(c) and 430-28(2). *NOTE:* The feeder circuit conductors can terminate in the lugs of a panelboard instead of the gutter and each motor served by a circuit breaker. This type of installation removes the requirements of applying the tap rules because overcurrent protection is at the point of the tap. See Figure 3-43. For taps over 25′ in length, see 430-28, Exception.

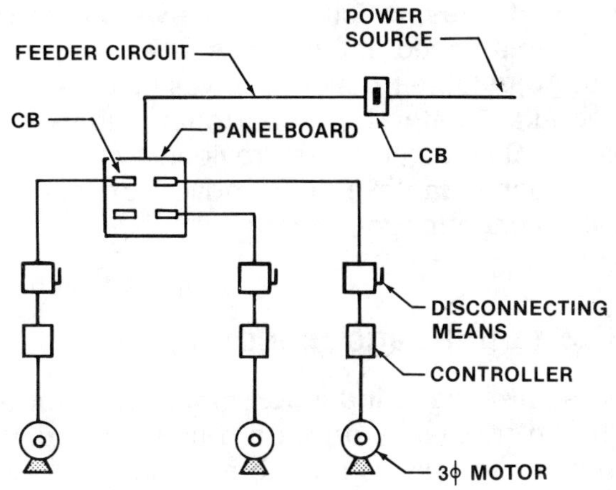

430-28
430-53(d)

Figure 3-43. Feeder circuit conductors may be terminated in a panelboard with a circuit breaker supplying each motor.

ADJUSTABLE FREQUENCY DRIVES

A basic adjustable frequency drive system consists of a *squirrel-cage motor*, an *inverter*, and an *operator's control station* (the control station may be on the inverter cabinet). Figure 3-44 shows a basic adjustable frequency drive system. Adjustable frequency drives are used to

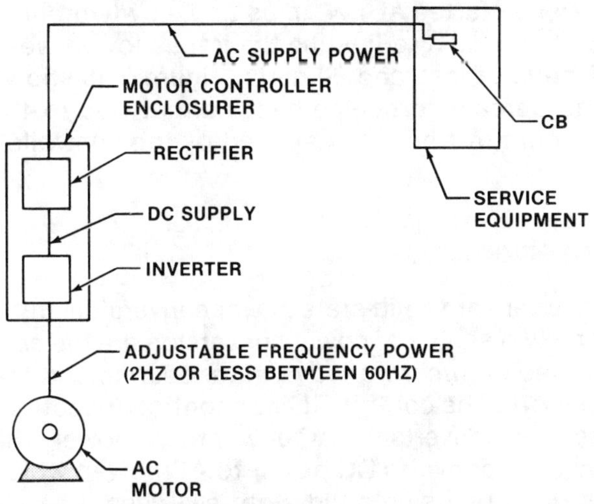

Figure 3-44. Adjustable frequency drives convert AC supply power to DC supply power to start and run AC motors. Rectifiers are used to convert AC to DC and inverters are used to regulate the level of frequency to start and drive the AC induction motor.

control the speed of standard AC squirrel-cage induction motors by varying the frequency of the power supply to the motors.

Adjustable frequency drives are available in a variety of sizes, which permits a wide range of adjustable speed applications at a reasonable cost. Adjustable frequency drives have become a popular method of controlling speeds of motors. Speed control is provided by using the most dependable prime mover of all—the squirrel-cage induction motor.

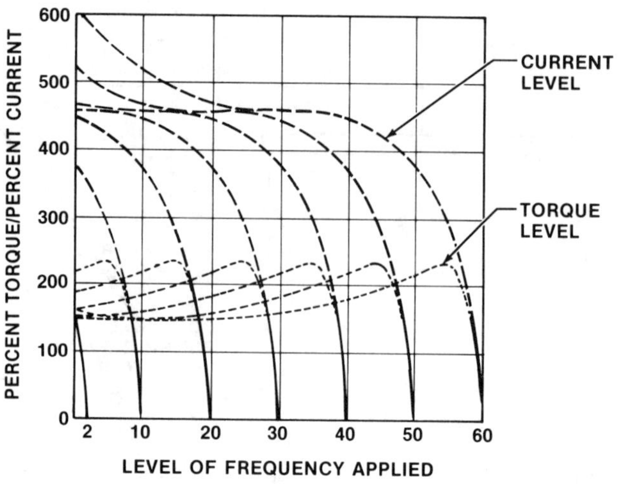

Figure 3-45. The percentage of torque and current vary with the amount of frequency applied.

AC Squirrel-Cage Induction Motor

The squirrel-cage induction motor used for adjustable frequency drives are usually Class B, three-phase, 460-volt motors connected to a three-phase, 480-volt supply. However, any size AC standard induction motor can be fitted in an adjustable frequency drive system.

Standard AC squirrel-cage induction motors equipped with an adjustable frequency drive system usually are started by reducing the applied frequency to a value of 2 hertz or less. Starting the motor at such a low frequency reduces the inrush current to approximately 150% of the rated current of the motor.

For example, a three-phase, 460-volt, 75-horsepower induction motor pulling 96 amps is started by applying 2 hertz or less (per Table 430-150). The inrush current would be 144 amps (96 amps x 150% = 144 amps). The AC motor can be started at 144 amps and slowly brought up to speed by adjusting the frequency between 2 hertz or less and 60 hertz. Figure 3-45 shows the percent torque/percent current due to the amount of frequency applied by the controller.

Inverters

Inverters are solid state power conversion units. The two stages of power conversion are the *controlled* or *uncontrolled rectifier section* and the *inverter*. The controlled or uncontrolled rectifier section converts the AC power to DC power. The inverter converts DC power to AC power, which is done by using solid state switches.

Inverters receive three-phase, 480-volt power at 60 hertz. A full 60 hertz applied to the inverter by the controller causes the AC motor to rotate at its maximum speed. Applying less than 60 hertz causes the motor to rotate at a slower speed, which is based on the value of frequency applied by the controller. Many inverters may produce more than 60 hertz, which would produce a higher output speed (rpm's) from the motor. If this is the case, the amount of torque would be less.

Care must be taken in sizing the inverter and matching it to the AC motor. Generally, an inverter with a certain horsepower can be used with an AC motor having the same horsepower rating. The motor rating in horsepower must equal the amount of current required to drive the load. This is the method used to size the AC motor. The inverter is selected and sized based on this rating.

Where the AC motor is oversized to provide a wider range of speed control, the inverter must be sized for the amount of current required for the motor at the maximum operating torque. If the AC motor is operated at full-load current, the inverter output current rating must equal or be greater than the current listed on the nameplate of the AC motor.

Most inverters have a maximum ambient temperature of 40°C. A few manufacturers have raised their maximum ambient temperature to 50°C. If installing an inverter in an area with a higher ambient temperature, the higher temperature usually can be compensated for by oversizing the inverter by one size. Check with the manufacturer to verify this.

Operator's Control Station

The operator's control station contains start and stop pushbuttons with normally open or normally closed contacts that are used to start and stop the motor circuit. A *speed-setting potentiometer* is used to adjust the rotating speed of the AC motor. A potentiometer is a rheostat or resistor with three terminals having one or more sliding contacts that are adjustable and act as an adjustable voltage divider.

Starting Torque. An AC motor started across the line at full voltage and full frequency has an inrush starting current of approximately 600% of the motor's full load-current rating. The 600% will apply only for AC motors marked with code letter A through H. For code letters J through V, the kilovolt-amperes per horsepower is based on each individual code letter per Table 430-7(b). For example, a three-phase, 460-volt, 75-horsepower induction motor with code letter S has an inrush current (IC) of 1693 amps per Table 430-7(b).

$$IC = \frac{kVA \times hp}{V \times 1.732}$$

$$IC = \frac{17.99 \times 1000 \times 75}{460 \times 1.732}$$

$$IC = 1693 \text{ amps}$$

However, the same AC motor with code letter F can be computed with an inrush current of 576 amps per Table 430-151 (96 amps × 600% = 576 amps).

EDDY-CURRENT DRIVES

An eddy-current drive consists of soft iron bars *(electromagnets)* shaped like a "U" with a coil of wire around the bases. The bars are magnetized by applying DC voltage to the coil of insulated wire. A solid ring of soft iron *(drum assembly)* is added to encircle the poles of the electromagnets. These components develop an *eddy-current clutch.* See Figure 3-46.

Eddy-current drives are used to obtain a wide range of stepless, adjustable speeds from AC power supply lines operating at standard frequencies. Eddy-current drives consist of an *AC squirrel-cage induction motor* and a magnetic *eddy-current clutch.* Eddy-current drives are used in industries where a variety of speed control or regulated torque is required for equipment.

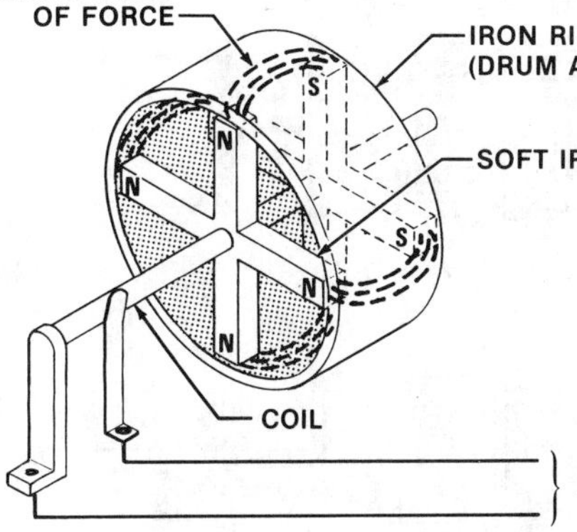

Figure 3-46. A basic eddy-current drive consists of soft iron bars (electromagnets), a coil of insulating wire, and an iron ring (drum assembly).

AC Squirrel-Cage Induction Motor

Eddy-current drives are used to convert AC supply power to rotational power using AC squirrel-cage induction motors. The basic output speeds of standard induction motors are 3600, 1800, and 1200 rpm. Squirrel-cage induction motors have full-load speeds of 2% to 5% less than synchronous speeds. For example, if the synchronous speed of an induction motor is 3600 rpm and the motor is operating at 5% slip, the actual speed would be 3420 rpm.

$$3600 \text{ rpm} \times 5\% = 180 \text{ rpm}$$
$$3600 \text{ rpm} - 180 \text{ rpm} = 3420 \text{ rpm}$$

An eddy-current clutch is added to the AC induction motor in which the motor drives to obtain variable output speeds.

Eddy-current Clutch

The three main components of the eddy-current clutch are the *drum, rotor,* and *rotating coil.* The drum is the input member and is essentially a steel drum that is driven by the standard squirrel-cage induction motor. The rotor is the output member and is free to rotate in the clutch drum. The rotating coil is wound around the rotor

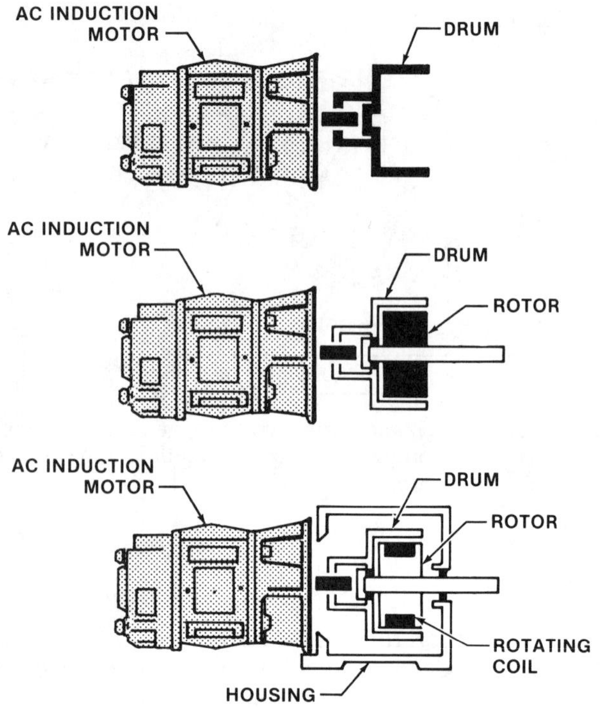

Figure 3-47. A basic eddy-current clutch consists of an AC motor, drum, rotor, and rotating coil.

clutch and is supplied with DC voltage to create a flux pattern through the drum and rotor. See Figure 3-47.

Poles are cast in each section of the rotor, and as the field coil is excited, each section of the rotor develops a north and south pole. Each section of the rotor has a polarity opposite that of the other section. When the field coil is excited, magnetic lines of force will flow through the north poles of the rotor into the drum, through the south poles of the rotor, and return to the field assembly.

Eddy currents are generated in the drum by the motion between the rotating drum and the rotor. Eddy currents are the circulating currents induced in a conducting material when they cut magnetic flux lines. These small currents are created by the voltage through the conducting material. A second magnetic field is created by the generated eddy currents. The interaction between the magnetic field generated by the field coil and the magnetic field generated by the eddy currents causes the rotor to rotate in the same direction as the drum.

When no voltage is applied to the coil, the drum and rotor will rotate freely with no rotation of the output shaft. When voltage is applied to the coil, the output shaft will pick up speed and continue to increase until it is rotating slightly less than the motor. The output shaft never rotates at the same speed as the motor because of the slip generated by the difference in speed between the drum and rotor.

The speed of the output shaft can be varied by adding or subtracting the amount of DC voltage applied to the coil. The output shaft speeds up when the voltage to the coil is increased and slows down when the voltage is decreased. See Figure 3-48 for the components required for an eddy-current drive unit.

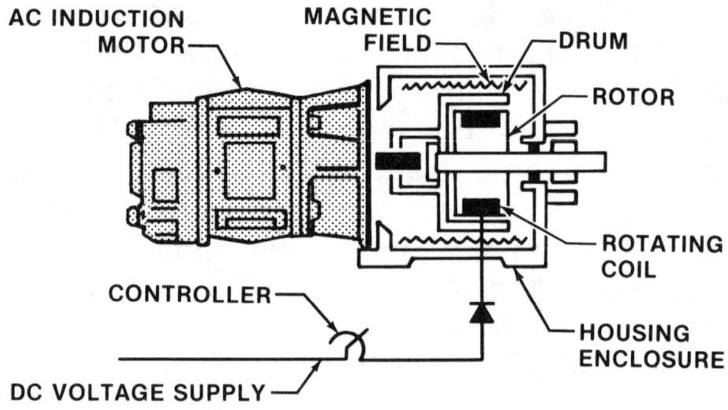

Figure 3-48. The torque of an eddy-current drive can be varied by adjusting the DC excitation voltage level to the coil.

Output Torque. A fixed amount of excitation voltage can be applied to the coil, and the output speed of the clutch will vary as the load is increased or decreased.

The amount of torque transmitted from the AC induction motor to the output shaft can be varied by adjusting the level of excitation to the coil. See Figure 3-49. The more excitation voltage applied to the coil, the greater the magnetic field will be on the drum and the faster the drum will rotate the output shaft.

Speed Control. A tachometer generator is used to provide a signal that is proportional to the output speed of the shaft. This signal is used to adjust the excitation to the coil, which will realign the difference in speed between the preset speed of the controller and the actual speed of the load. The tachometer generator is mounted integrally with the output shaft.

The accuracy of realigning the output speed to the preset speed of the controller is accomplished with the aid of a drive regulator. See Figure 3-50.

Controller

The eddy-current controller varies the excitation voltage to the clutch field coil, which changes the strength of the magnetic field on the rotating drum. The controller can be set to add more DC voltage to the coil, which will speed up the output shaft of the clutch. Less excitation voltage applied to the coil causes the magnetic field to be weaker on the drum. Therefore, the drum slows down and the output shaft will turn slower.

The controller can be set to moderate the amount of current flow and protect the motor from overload. The controller can be programmed to shut the motor off and reverse the motor's rotation and then start again in the original rotation. This programming is necessary where a blockage could occur in the supply line.

The controller is provided with solid state transistorized boards, which are easy to replace to facilitate troubleshooting procedures. The boards can then be repaired and used again. See Figure 3-51.

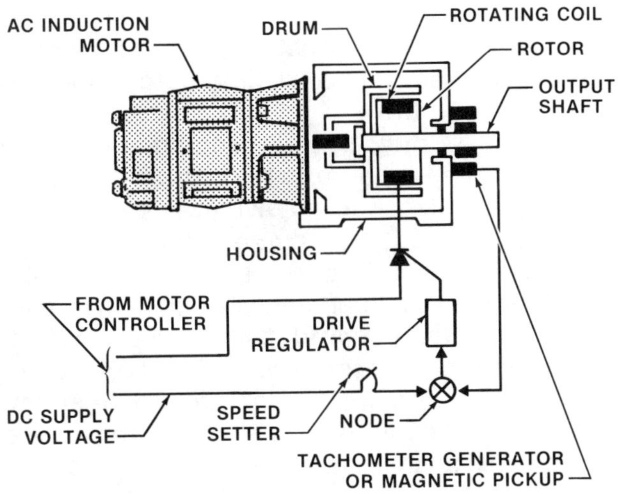

Figure 3-49. A tachometer generator and drive regulator are used to adjust and correct the realignment of the actual output speed of the output shaft to the preset speed where the load is varied.

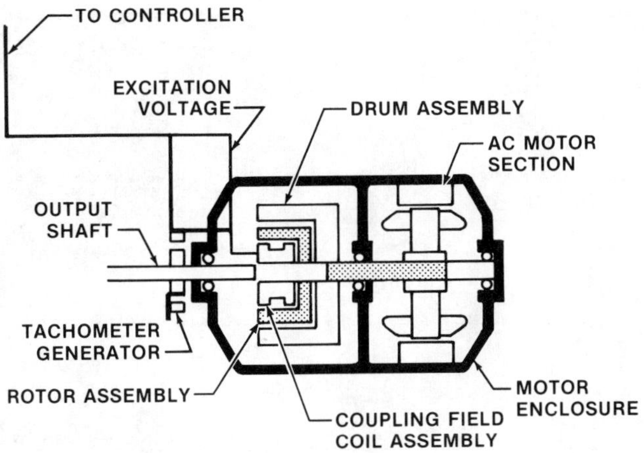

Figure 3-50. An eddy-current drive with its basic operating components.

TROUBLESHOOTING VARIABLE FREQUENCY DRIVES

Maintenance required for the variable frequency drive is minimal. The drive should be installed away from any high heat-producing equipment.

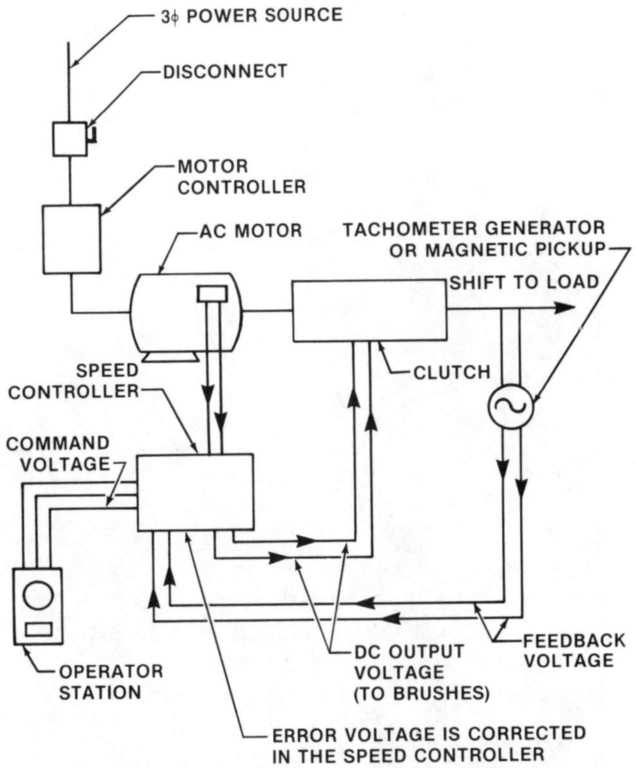

Figure 3-51. An operating station and controller are used to control the amount of excitation voltage required to increase or decrease the rotation speed of the eddy-current drive.

Air flow must remain high around the drive unit with no other equipment or materials restricting air flow. All connections should be tightened every six months.

Most maintenance and troubleshooting may be performed with the following tools: voltmeter, oscilloscope, digital voltmeter, AC ammeter, clamp-on ammeter, and standard hand tools.

Most systems today have a self-diagnostic system. It is imperative to have the manufacturer's manual for a particular unit and to follow the manufacturer's instructions on the use of the diagnostic system. The following is an example of a manufacturer's diagnostic system.

For basic troubleshooting,

(1) check incoming AC power supply;
(2) check all fuses;
(3) check output to motor;
(4) check tach feedback to controller if used;
(5) consult manufacturer's manual for correct troubleshooting procedure.

Figure 3-52 is a chart illustrating the troubleshooting procedures using a readout board to determine the cause of trouble. *NOTE:* Within the adjustable speed drive there are high voltages of both AC and DC present.

Always remember:

(1) Check all AC power sources.
(2) Check the DC bus for voltage.
(3) Allow sufficient time to discharge any power supply capacitors.
(4) The equipment ground and DC bus ground may not be the same potential. Most DC bus grounds float. Do not let the oscilloscope cabinet touch the chassis of the adjustable speed controller.
(5) Incorrectly connecting the leads to the SCRs will damage the SCRs.

TROUBLESHOOTING PROCEDURES

Most eddy-current drive controllers consist of a voltage reference circuit, anti-hertz circuit, DC amplifier circuit, feedback circuit, and various potentiometers for controlling output speed of the magnetic drive.

If a higher drive speed is required, the speed potentiometer can be turned to a higher point, which results in the following:

(1) The DC output voltage of the potentiometer is increased. This is the *command* voltage.

(2) When the command voltage is increased, an *error* voltage is produced between the command voltage and the *feedback* voltage coming from the tachometer generator (or magnetic pickup frequency to voltage converter). The error causes an increased voltage to be applied to

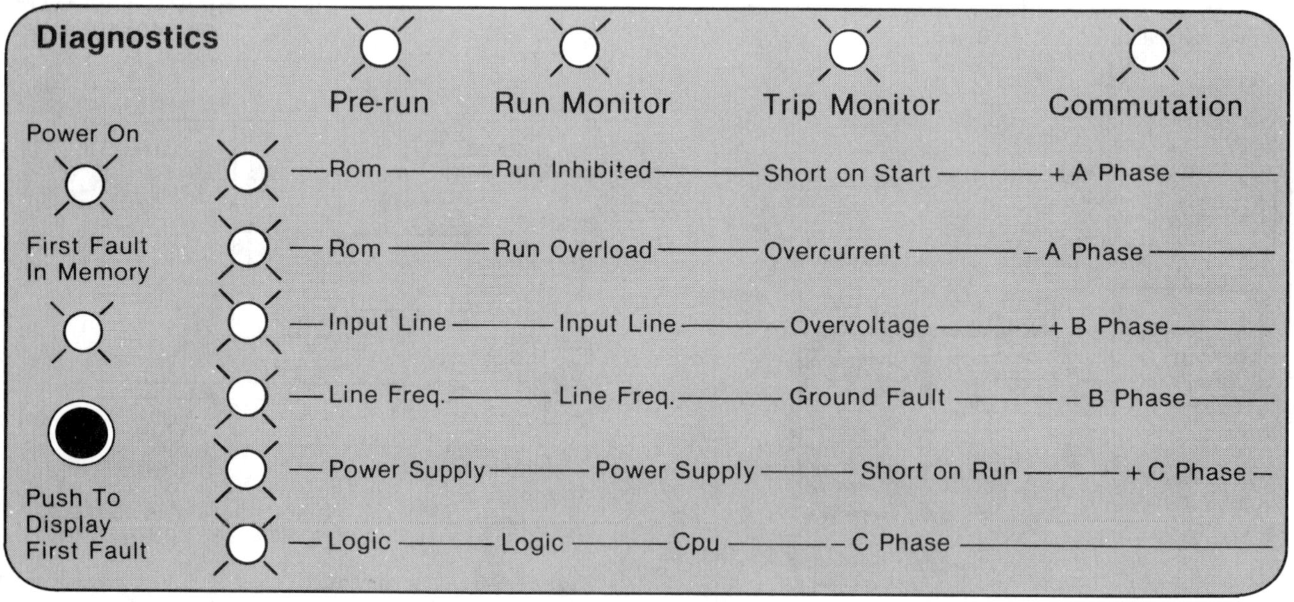

Figure 3-52. Readout boards are used when troubleshooting variable frequency drives.

the *magnetic drive coupling* field through a silicon controlled rectifier (SCR). This increased field excitation causes the magnetic drive coupling to accelerate until the feedback voltage signal again matches the command voltage signal.

NOTE: The difference between the command voltage and the feedback voltage is the error voltage. During operation at a steady speed, the error voltage is practically zero.

See Figure 3-53 for the control circuits and components required to regulate the speed of an eddy-current drive unit. Figure 3-54 shows a list of common problems that occur in eddy-current drives along with common solutions. These can be used as a general troubleshooting guide along with the manufacturer's troubleshooting manual.

TESTING SCR'S

The SCR is a solid state switch having the following characteristics:

(1) It can operate at high voltages and carry high currents.

(2) It can operate at high speeds (millionths of a second).

(3) Very little current is required to drive the SCR.

(4) It is small in size and light weight.

(5) Having no moving parts, it offers long life, low maintenance, and high reliability.

The SCR allows current to move forward at full force when a small amount of current is applied to the gate. The flow of forward current is only limited by the resistance in the circuit. Removing the gate signal does not take the SCR out of the conducting state. Instead, the SCR can be turned off by interrupting the forward current, removing the anode voltage, or reversing the forward current by applying negative voltage to the anode.

It is necessary to understand the operation of the SCR in order to troubleshoot these components. SCRs are used in most adjustable speed drives, AC, DC, and eddy-current drives.

Shorted diodes or SCRs can be easily found by checking across their individual terminals with a volt-ohmmeter. Good rectifier cells have infinite resistance with reverse polarity and read approximately mid-scale with forward polarity. SCRs have resistances of 12K to infinity with any polarity or terminal combination if they are not shorted.

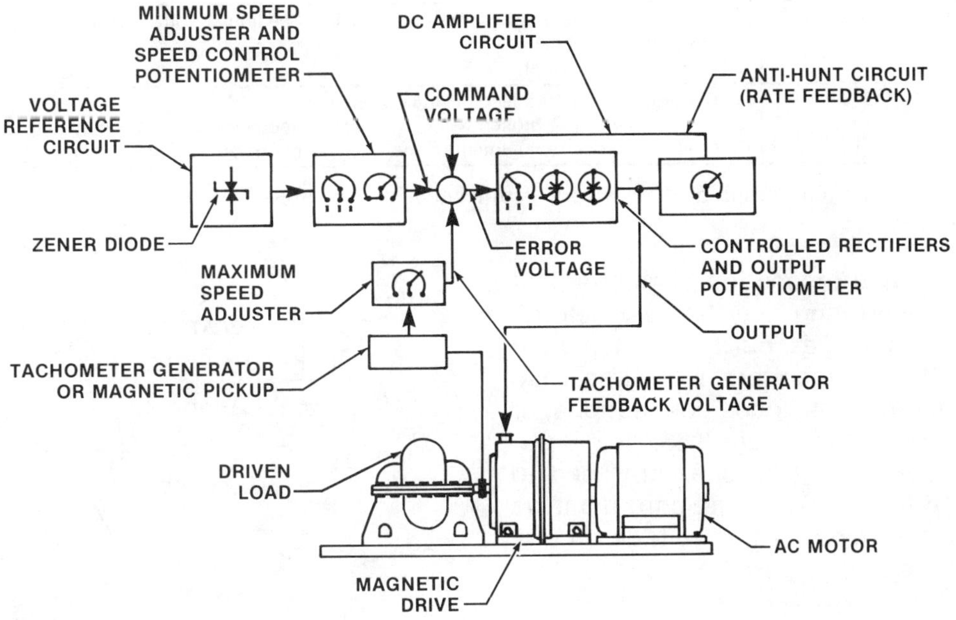

Figure 3-53. An eddy-current drive with control circuits and components is used to regulate the speed of the driven load.

TROUBLESHOOTING EDDY-CURRENT DRIVES	
PROBLEM	**POSSIBLE FAULT**
Motor does not start.	1. Loss of AC power. 2. Defective switch or breaker. 3. Blown fuses. 4. Motor starter not closing. 5. Overload or safety interlock open. 6. Loose or incorrect wiring. 7. Defective AC motor.
Motor runs, but no output.	1. Check controller for input voltage. 2. Loose or incorrect wiring. 3. Open safety interlock. 4. Brushes not making contact. 5. Brake not releasing. 6. Open or defective clutch coil.
Drive stops during operation.	1. Controller malfunction, check controller. 2. Drive is overloaded. 3. Safety interlock is open. 4. Loss of AC power. 5. Loose connection. 6. Open or defective clutch coil. Check brushes first.
Unit overheats.	1. Overload, check motor current. 2. Operating below minimum speed. 3. Air passages blocked on magnetic drive unit. 4. Recirculating cooling air or ambient temperature too high. 5. Brake not releasing or machine binding.
Erratic operation.	1. Controller malfunction, check controller. 2. Velocity feedback malfunction (Tach. Gen. or Mag. Pickup malfunction). 3. Electric noise or radio frequency interference. 4. Loose wiring connection. 5. Contaminated slip rings. 6. Sticking or worn out brushes.
Runs at full speed only.	1. Controller malfunction, check controller. 2. Loss of velocity feedback signal (Tach. Gen. or Mag. Pickup). 3. Mechanical lock up of clutch drum and rotor.
Magnetic drive at stand-still or lower speed than expected with the speed potentiometer set at a higher speed.	1. Possible stall condition or overload on the drive unit. Turn off control. Check driven load for restriction or fault condition. 2. Possible open circuit in the drive unit field circuit. Check brushes and slip rings for continuity. 3. Check output voltage to the drive unit. If no or incorrect voltage check the silicon controller rectifiers (SCR's) or the gate pulse board is not sending pulse to the SCR's. Replace SCR's and pulse board as needed.
Magnetic drive at 100% speed with no control.	1. If voltage to drive unit is correct then check "Feedback" voltage from the tachometer generator (magnetic pickup). 2. Check pulse board and SCR's.
Mag drive has intermittent speed up or slow down.	1. Check brushes and collector rings on drive unit. 2. Check for loose or broken wires on the tachometer generator (or magnetic pickup). 3. Check for proper adjustment of the tach generator or magnetic pickup.

Figure 3-54. Check common problems first when troubleshooting variable frequency drives.

A thorough visual inspection at the earliest point of troubleshooting may be very helpful. Look for such things as burned or damaged resistors, rectifiers, capacitors, and wires. Also inspect for broken wires, loose connections, or open solder joints. See Figure 3-55.

When multiple sets of brushes are used to supply voltage to the slip rings, care must be taken to keep the negative (−) voltage on one slip ring and the positive (+) voltage on the other slip ring.

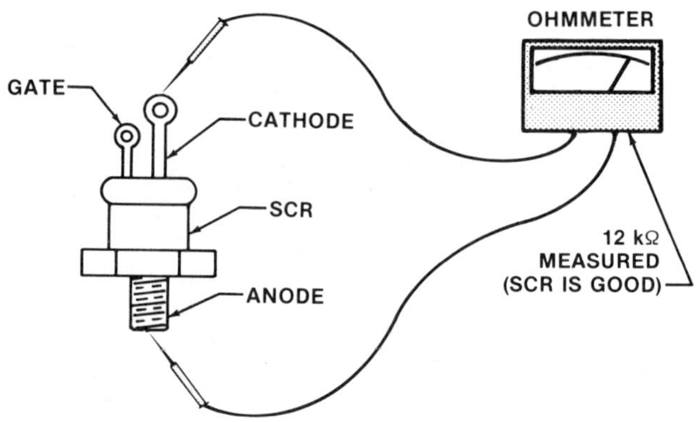

Figure 3-55. SCR's are tested by using an ohmmeter. The ohmmeter will measure 12,000 ohms or the needle will peg if the SCR is good.

REVIEW—CHAPTER 3

Name ____________________ Date ____________

True-False

T F **1.** Five types of motors are listed in Table 430-152.

T F **2.** Split-phase motors listed in Table 430-152 are equipped with a running winding and a starting winding.

T F **3.** The resistor banks used to start wound-rotor motors are never to be used for control.

T F **4.** Synchronous motors can only be started by the aid of an additional motor.

T F **5.** DC motors can only be connected to operate in series.

T F **6.** DC motors connected in series will provide a very high starting torque.

T F **7.** In full-voltage starting, the motor is started directly from the voltage supplied by the utility company.

T F **8.** Resistor starting utilizes coils to reduce the starting current to a motor.

T F **9.** When additional components are used to reduce the starting current to a motor, the starting torque is reduced.

T F **10.** The full-load starting current is usually less than the values listed in Table 430-152 when using the code letters of Table 430-7(b) for A through G.

T F **11.** Full-voltage starting is the most expensive starting method used to start and run motors.

T F **12.** Resistor starting does not provide a starting torque as high as an autotransformer.

T F **13.** Reactor starting is usually installed to start and run motors rated over 600 volts.

T F **14.** Torque is the turning or twisting force of the motor when the motor windings are energized with electrical power.

T F **15.** The torque efficiency of reactor starting is the same as that of full-voltage starting.

T F **16.** The principle of the eddy-current drive is to transmit power from an AC induction motor to the driven load.

T F **17.** Clutch input and output members have physical contact with each other and wear out easily due to friction.

T F **18.** The AC motor rotates the drum at a constant speed.

T F **19.** The most durable prime mover is the AC induction motor.

T F **20.** Adjustable frequency drives have become very popular as a method of speed control.

T F **21.** Existing AC motors cannot be converted to adjustable speed systems.

T F 22. Adjustable frequency drives provide motors with a soft start.

T F 23. Inverters used for speed control must be sized to provide the current required by the motor at maximum operating torque.

Completion

__________ 1. An autotransformer provides __________ of 50, 65, and 80% to reduce the inrush current to motors.

__________ 2. The starting current and torque using an autotransformer is determined by __________ the percentage of the tap.

__________ 3. Wye-delta starting reduces the starting torque to __________ % of the normal.

__________ 4. Code letter F has a maximum kVA of __________ per horsepower.

__________ 5. The locked-rotor values found in Table 430-151 are based on __________ times the full-load current rating of Table 430-150.

__________ 6. A nontime-delay fuse holds __________ times its rating for about one-quarter second.

__________ 7. A 100-amp frame circuit breaker holds __________ times its rating for settings of 15 amps to 100 amps.

__________ 8. Instantaneous trip circuit breakers can be adjusted up to __________ percent per Table 430-152.

__________ 9. The maximum setting of time-delay fuses used to allow motors to start and run due to large inrush current is __________%.

__________ 10. Time-delay fuses can be used to provide backup overload protection if they are sized at 115% to __________%.

__________ 11. Two or more motors of different ratings can be connected to a branch circuit as long as the maximum rating of the overcurrent protection device will __________ the smallest motor and allow the largest to start.

__________ 12. Fractional-horsepower motors are rated at less than __________ horsepower.

__________ 13. The circuit components of motors installed as a group must be approved and installed as __________ listed assembly.

__________ 14. Compound DC motors provide high starting torque of 180% to __________% of the full-load torque.

__________ 15. Care must be taken when selecting a reduced starting method to ensure enough __________ are provided to accelerate the driven load.

__________ 16. The __________ is used to vary the field coil current to increase or decrease speed.

__________ 17. __________ generated in the drum have magnetic poles that interact with the electromagnetic poles of the field assembly to produce torque.

__________ 18. Energizing the field coil with __________ voltage produces magnetic flux between the field assembly and drum.

__________ **19.** Adjustable-frequency drives vary the __________ to an AC motor for speed control.

__________ **20.** AC induction motors have maximum inrush current when __________ frequency is applied to the windings.

Multiple Choice

__________ **1.** Resistor starting provides a starting current of about __________% of normal.
A. 45
B. 55
C. 65
D. 70

__________ **2.** Reactor starting reduces the starting current to 65% of normal starting current and reduces the starting torque to __________% of normal.
A. 32
B. 42
C. 50
D. 65

__________ **3.** The motor most widely used in industry is a class __________ motor.
A. A
B. B
C. C
D. D

__________ **4.** A class C motor increases the starting of an induction motor to about _______% of the full-load torque.
A. 125
B. 150
C. 175
D. 225

__________ **5.** An autotransformer with a 50% tap provides a starting current of about ______% of normal.
A. 15
B. 25
C. 30
D. 45

__________ **6.** An autotransformer with a 65% tap reduces the voltage of a 240-volt supply to __________ volts.
A. 156
B. 180
C. 192
D. 208

__________ **7.** Part-winding starting has the disadvantage of using only __________ the horsepower when starting.
A. one-fourth
B. one-third
C. one-half
D. two-thirds

_____ 8. The line voltage of the wye winding in wye-delta starting is _____ times the phase voltage.
A. 1.15
B. 1.25
C. 1.50
D. 1.732

_____ 9. The starting current where wye-delta starting is used is reduced to _____ the normal.
A. one-fourth
B. one-third
C. one-half
D. two-thirds

_____ 10. The torque efficiency of an autotransformer is the same value as the _____.
A. part-winding
B. resistor
C. reactor
D. full voltage

_____ 11. The difference between the speed of the clutch drum and field assembly is the _____.
A. slip
B. torque
C. starting speed
D. rotating speed

_____ 12. The output speed of an eddy-current drive is always _____ the speed of the AC motor.
A. more than
B. less than
C. the same as
D. 15% more than

_____ 13. During operation, the only parts subject to wear are the slip rings, brushes, and _____.
A. clutch
B. output member
C. drum
D. bearings

_____ 14. Where starting a motor using adjustable frequency drive units, the amount of frequency applied is _____ Hz or less.
A. 2
B. 10
C. 15
D. 60

_____ 15. As frequency to the windings of a motor is increased, the starting torque will _____.
A. increase
B. decrease
C. remain the same
D. none of the above

__________ **16.** Motors are sometimes __________ to provide a wider range of speed control.
A. mounted
B. undersized
C. oversized
D. none of the above

__________ **17.** A basic adjustable frequency drive system consists of an AC motor, an inverter, and a __________.
A. start button
B. stop button
C. disconnecting means
D. all of the above

Problems

1. What is the locked-rotor current of a three-phase, 440-volt, 75-horsepower induction motor?

2. What is the locked-rotor current of a three-phase, 230-volt, 40-horsepower motor with code letter D, using the maximum kVA per horsepower?

3. What is full-load torque of a three-phase, 230-volt, 125-horsepower motor turning at 3450 rpm?

4. Based on the full-load torque rating in problem 3, what is the starting torque using a class B motor?

5. What is the maximum starting torque of a DC shunt motor that has a full-load torque rating of 180 foot-pounds?

6. A 100-horsepower part-winding motor has a locked-rotor current of 744 amps. What is the reduced starting current?

7. What is the reduced starting current for a motor having a locked-rotor current of 1080 amps using wye-delta starting?

8. The utility company has requested that a motor having a locked-rotor current of 858 amps be reduced to 560 amps by applying reduced starting. Would resistor starting provide the reduced starting current needed?

9. What size and class motor is required to start and drive a load of 70 foot-pounds with the rotor turning at 1760 rpm and a starting torque of 85 foot-pounds?

10. An induction motor has a locked-rotor current of 1488 amps, and an autotransformer with 65% tap is used to start the load. Find the following: (a) winding current, (b) line current, (c) transformation current.

11. An induction motor has a full-load current rating of 275 amps with a starting torque of 190 ft lb and is started using solid-state starting. Find the following: (a) starting current using maximum percentage, (b) starting torque using maximum percentage.

12. An induction motor has a full-load current rating of 156 amps, with a starting torque of 365 ft lb and is started with an adjustable frequency drive unit. Find the following: (a) starting current by varying the frequency 10 Hz, (b) starting torque by varying the frequency 10 Hz.

MOTOR FEEDER AND BRANCH-CIRCUIT CONDUCTORS

Chapter 4

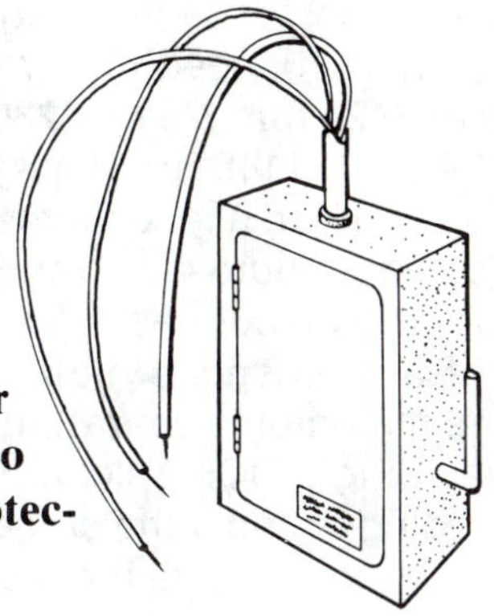

Branch-circuit conductors are selected by the loads they supply. The percentages used to size the conductors are based on the use of the loads. The loads are either continuous or noncontinuous. Continuous loads are calculated at 125% while noncontinuous loads are calculated at 100%. Where the power factor is low and currents are high, capacitors are installed to correct the power factor and reduce currents. Overcurrent protection devices are sized according to these loads.

SIZING BRANCH-CIRCUIT CONDUCTORS—430-22(a)

Conductors supplying a single motor used for a continuous-duty load must have a current-carrying capacity of not less than 125% of the motor's full-load current (flc) rating according to Tables 430-147, 430-148, 430-149, and 430-150.

When a motor is started, there is a momentary inrush current of four to six times the full-load current of the motor. This creates heat and strain on the conductors feeding the motor circuit. A current-carrying capacity of 125% of the motor's full-load current serves to limit the full-load conditions of the motor to 80% ampacity of the conductors and helps eliminate the problem of heat and strain on the conductors. See Figure 4-1.

The full-load current rating of a motor must be

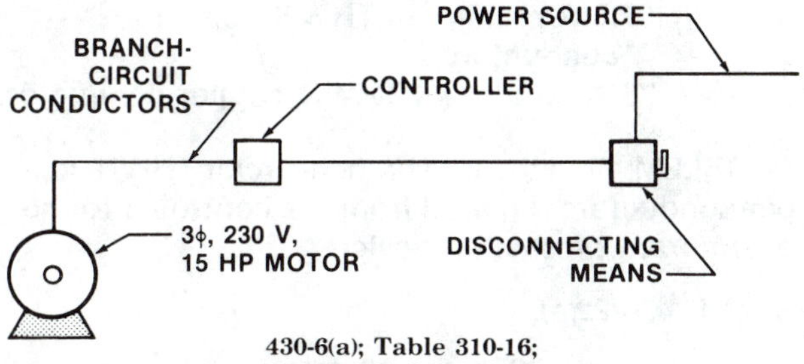

430-6(a); Table 310-16; 430-22(a); Table 430-150

PROBLEM: Find the size THWN copper conductors required to supply this motor.

Step 1: *Table 430-150, Col. 2.*
A 3ϕ, 230-V, 15-hp motor has a 42-A flc rating.

Step 2: *430-22(a).*

42 A × 125% = 52.5 A

Step 3: *Table 310-16, Col. 2.*

Answer: **#6, 75°C rated THWN copper conductors**

Figure 4-1. Sizing branch-circuit conductors for a single motor.

increased by 125% to select conductors because the running overload protection can be sized at 125% of the running current (amperes) listed on the motor's nameplate per 430-32(a)(1) or 430-34.

Circuit conductors for multispeed motors must be sized to the controller large enough to supply the highest nameplate full-load current rating of the motor windings involved. The conductors for each speed are sized and selected according to the rpm's and full-load current rating (amperes) for each speed of the motor. For example, a multispeed motor with four speeds turning at 3420 rpm (28 A), 2280 rpm (37 A), 1710 rpm (46 A), and 1140 rpm (61 A) must have each amperage of each speed increased by 125% per 430-22(a). See Figure 4-2. Branch-circuit conductors from the disconnecting means to the controller of a multispeed motor are sized based upon the maximum rpm rating of the motor per 430-22(a). Single OCPD's can be used to protect each winding speed if the branch circuit conductors and the conductors for each speed are sized from the winding with the greater current rating. See 430-52(a), Ex.

Varying Duty

When a motor is not continuous duty but operates on a short-time duty cycle, the motor windings and circuit conductors have a chance to cool during periods of rest. These motors do not require conductors with a current-carrying capacity of 125% of the motor's full-load current rating. Conductors are sized from the percentage of nameplate current rating listed in Table 430-22(a), Exception according to the classification of service of the motor. Table 430-22(a), Exception includes requirements for sizing conductors to supply individual motors used for short-time, intermittent, periodic, or varying duty. Conductor sizing changes with the application, as the starting and stopping duration of operation cycles imposes varying heat loads on the conductors.

A motor used to drive a load for 15 minutes must have 45 minutes rest unless the nameplate

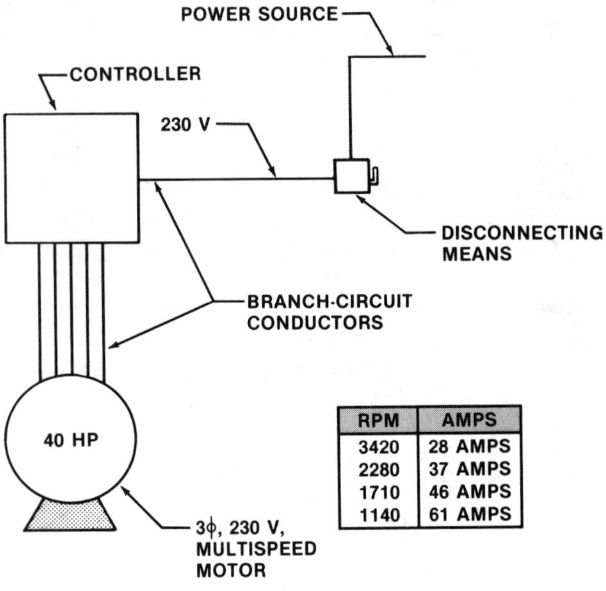

RPM	AMPS
3420	28 AMPS
2280	37 AMPS
1710	46 AMPS
1140	61 AMPS

430-6(a);
Table 310-16;
430-22(a)(b)

PROBLEM: Find the size branch-circuit THWN copper conductors required for various speeds of the multispeed motor.

Step 1: *430-22(a).*

28 A × 125% = 35 A
37 A × 125% = 46 A
46 A × 125% = 57.5 A
61 A × 125% = 76 A

Step 2: *Table 310-16, Col. 2.*

Answer: **35 A require #10 THWN copper conductors**
46 A require #8 THWN copper conductors
57.5 A require #6 THWN copper conductors
76 A require #4 THWN copper conductors

PROBLEM: Find the size branch-circuit THWN copper conductors required from the controller to the overcurrent protection device.

Step 1: *430-22(a).*

61 A × 125% = 76 A

Step 2: *Table 310-16, Col. 2.*

Answer: **#4 THWN copper conductors**

Figure 4-2. Sizing branch-circuit conductors for a multispeed motor based on the maximum rpm rating.

of the motor indicates otherwise. When a motor is used for a duty cycle, the work period is subtracted from 60 minutes. For example, a motor used for a 30-minute work period must rest 30 minutes (60 − 30 = 30). A Class B motor can be used to rotate a turntable for 15 minutes with a 45-minute rest period. See Figure 4-3.

CONDUCTORS SUPPLYING SEVERAL MOTORS—430-24

To find the size of conductors for a feeder supplying two or more motors, multiply the full-load current rating of the largest motor of the group by 125%. Add the full-load current rating of the remaining motors to this total.

The feeder conductors can be routed to a subfeed panelboard and terminated. Individual circuit breakers are installed in the panel to supply each motor per Table 430-152. The conductors to supply each motor are sized per 430-22(a).

The feeder conductors can be terminated in an auxiliary gutter and taps made to a feeder or transformer secondary per 430-28(1)(2). A fusible disconnect or panelboard with circuit breakers can be installed under the gutter for each motor. Overcurrent protection devices are selected per 430-52 and Table 430-152. See Figure 4-4. See 430-28 for taps to motors.

If one of the motors is classified as intermittent duty or another classification of service, find the percentage that the motor's full-load current rating can be reduced per Table 430-22(a), Exception. The feeder conductor can be determined by applying percentages to each motor based on the type of duty and selecting the conductors from Table 310-16. The conductors are selected on the total amperage of the motors after demand. See Figure 4-5.

WOUND-ROTOR SECONDARY

Wound-rotor motors are three-phase motors that operate in the same manner as squirrel-cage induction motors. The only difference between the two is that the wound-rotor motor has two sets of leads extending from the controller and a bank of resistors to slip rings that are connected to the rotor. As the amount of resistance in the motor circuit varies, the speed of the motor also varies. The greater the resistance in the rotor, the slower the motor will run, and vice versa. The resistor banks may be separate from the motor or the resistances may be incorporated in the controller.

Continuous Duty—430-23(a)

The secondary leads between the controller and the motor must be sized at not less than 125% of the secondary full-load current of the motor. This rating is obtained from the manufacturer or found on the nameplate of the motor.

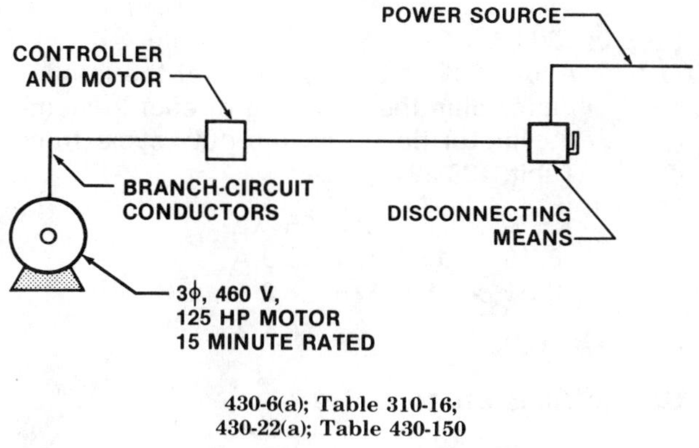

PROBLEM: Find the size THWN copper conductors required to supply this intermittent-duty motor.

Step 1: *Table 430-150, Col. 3.*
A 3ϕ, 460-V, 125-hp motor has a 156-A flc rating.

Step 2: *Table 430-22(a), Ex., Col. 2.*
A 15-minute rated motor allows 85% of the nameplate current rating.

156 A × 85% = 132.6 A

Step 3: *Table 310-16, Col. 2.*

Answer: **#1/0 THWN copper conductors**

NOTE: Motor must rest for 45 minutes when used at continuous duty for 15 minutes.

Figure 4-3. Sizing branch-circuit conductors for a duty-cycle motor.

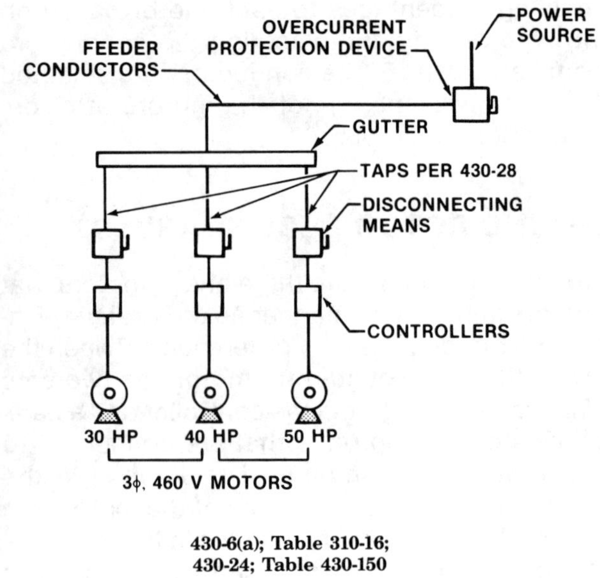

PROBLEM: Find the size THWN copper feeder conductors required to supply these motors.

Step 1: *Table 430-150, Col. 3.*
The flc ratings of the motors are:

Horsepower	Amps
30	40
40	52
50	65

Step 2: *430-24.*
Add 125% of the flc rating of the largest motor plus the flc rating of the remaining motors.

$$125\% \times 65\text{ A} = 81.25\text{ A}$$
$$52\text{ A}$$
$$40\text{ A}$$
$$173.25\text{ A}$$

Step 3: *Table 310-16, Col. 2.*

Answer: **#2/0 THWN copper conductors**

Figure 4-4. Sizing feeder conductors for two or more motors.

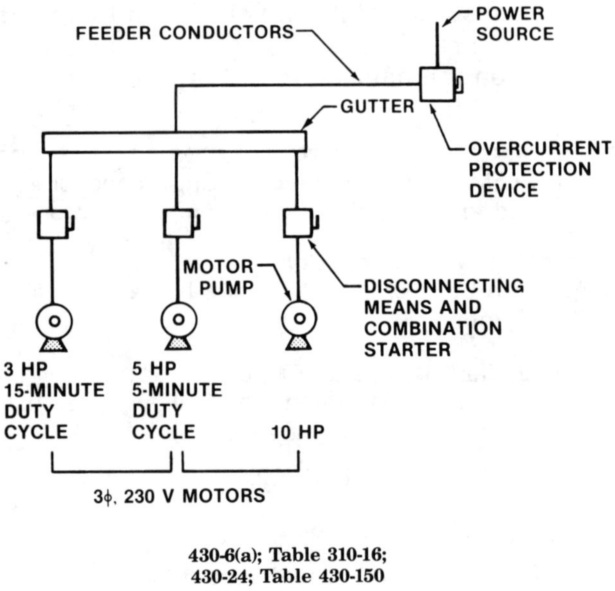

PROBLEM: Find the size THWN copper conductors required to supply these motors.

Step 1: *Table 430-150, Col. 3.*
The flc ratings of the motors are:

Horsepower	Amps
3	9.6
5	15.2
10	28

Step 2: *430-24.*
Add 125% of the flc rating of the largest motor plus the flc rating of each remaining motor based on the duty cycle from Table 430-22(a), Ex.

$$125\% \times 28\text{ A} = 35\text{ A}$$
$$85\% \times 15.2\text{ A} = 12.9\text{ A}$$
$$85\% \times 9.6\text{ A} = 8.2\text{ A}$$
$$56.1\text{ A}$$

Step 3: *Table 310-16, Col. 2.*

Answer: **#6 THWN copper conductors**

Figure 4-5. Sizing feeder conductors for two or more motors with duty cycles.

Other Than Continuous Duty—430-23(b)

The secondary conductors may be sized at less than 125% of the secondary current when the motor is used for periodic duty. Table 430-22(a), Exception must be followed, with the conductors sized by the classification of service and the correct percentages applied.

Resistor Separate From Controller—430-23(c)

When the resistor bank is separately installed and is not part of the controller, the leads between the controller and resistor bank must be sized according to the resistor duty classification listed in Table 430-23(c). See Figure 4-6.

SIZING CONDUCTORS TO A CAPACITOR—460-8

The ampacity of conductors feeding a capacitor must be at least 135% of the full-load current of the capacitor. If the capacitor supplies a motor, the leads for the capacitor must be at least one-third the size of the power leads to the motor. Capacitors are mostly used to raise the power factor of a motor circuit. Capacitors must be selected no larger than the size needed to correct the circuit power factor to unity. Capacitors raise the power factor to the motor and lower the amount of current the motor pulls. Running overload protection must be sized for this new value of current. See Figure 4-7.

Capacitors can be checked by using an ohmmeter and placing the maximum resistance of the meter between the terminals of the capacitor. If the ohmmeter needle pegs to the right and

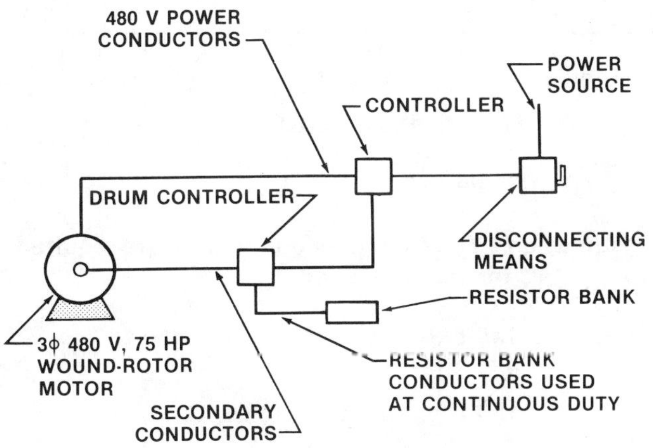

430-6(a); Table 310-16; 430-22(a); Table 430-150

PROBLEM: This motor has a rating of 60 amps in the secondary conductors. The resistor bank, mounted separately from the drum controller, controls motor speed. Find the sizes of the THWN copper conductors required for:

A. The motor's power conductors
B. The motor's secondary conductors
C. The resistor bank conductors

A. MOTOR'S POWER CONDUCTORS

Step 1: *Table 430-150, Col. 3.*
The flc rating of the motor is 96 amps.

Step 2: *430-22(a).*

125% × 96 A = 120 A

Step 3: *Table 310-16, Col. 2.*

Answer: **#1 THWN copper conductors**

B. MOTOR'S SECONDARY CONDUCTORS

Step 1: *430-23(a).*

125% × 60 A = 75 A

Step 2: *Table 310-16, Col. 2.*

Answer: **#4 THWN copper conductors**

C. RESISTOR BANK CONDUCTORS

Step 1: *Table 430-23(c).*

110% × 60 A = 66 A

Step 2: *Table 310-16, Col. 2*

Answer: **#4 THWN copper conductors**

Figure 4-6. Sizing power, secondary, and resistor bank conductors for a wound-rotor motor.

slowly returns to the left, the capacitor is good. A ¼-amp fuse can be used in the line. If the fuse blows, the capacitor is shorted. A screwdriver can be used to short across the terminals of the capacitor. If the capacitor discharges with a hot spark, the capacitor is good. Note that the power must be disconnected to apply this method. A very simple method is to charge the capacitor and then discharge it. A hot spark indicates the capacitor is good.

CORRECTING POWER FACTOR FOR MOTOR LOADS

Figure 4-8 illustrates a step-by-step procedure for sizing and selecting a capacitor to raise the power factor for a circuit supplying a motor with poor power factor. An overcurrent protection device must be placed in each ungrounded conductor to protect the capacitor circuit. The overcurrent protection device can be the device protecting the power circuit to the motor if the conductors supplying the capacitor bank are on the load side of the magnetic starter controlling the motor.

A disconnecting means must be provided to disconnect each ungrounded conductor to a bank of capacitors. If the bank of capacitors is connected to the load side of the magnetic starter, the overcurrent protection device supplying the motor circuit can serve as the disconnect. See Figure 4-9.

DESIGNING MOTOR CURRENTS NOT LISTED IN TABLES 430-147–430-150

For a motor that has a horsepower rating not listed in Tables 430-147 through 430-150, the full-load current rating may be found as follows:

(1) Determine the horsepower rating immediately below the unlisted motor.

(2) Divide the listed motor's full-load current rating by its horsepower rating.

(3) Multiply these values by the horsepower rating of the unlisted motor to obtain its full-load current.

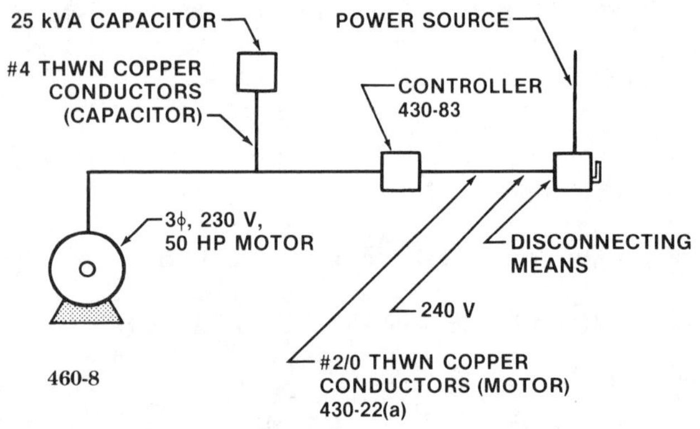

PROBLEM: What size THWN copper conductors are required to supply the capacitor?

Step 1: $A = \dfrac{kVA}{V \times \sqrt{3}}$

$I = \dfrac{kVA \times 1000}{240\ V \times 1.732}$

$I = \dfrac{25{,}000\ VA}{416\ V}$

$I =$ **60 A**

Step 2: *460-8.*

60 A × 135% = 81 A

Step 3: *Table 310-16, Col. 1.*

Answer: **#4 THWN copper conductors are required for the capacitor.**

Step 4: *Table 310-16, Col. 1; Table 430-150; 430-22(a).*

130 A × 125% = 162.5 A

Answer: **#2/0 THWN copper conductors are required for the motor circuit (175 A).**

Step 5: *460-8.*
one-third of 175 A = 58 A

Step 6: *Table 310-16, Col. 1.*

Answer: **#6 THWN copper conductors are required for the capacitor.**

NOTE: The larger of the two conductors found is required.

Figure 4-7. Sizing conductors for a capacitor.

Figure 4-10 shows the step-by-step procedure to follow when determining the full-load current of a motor not listed in the tables.

A simple method of finding the full-load current of a motor can be applied by using the following rules of thumb:

Rules of Thumb

At 575 V, a 3ϕ motor draws 1 A per hp
At 460 V, a 3ϕ motor draws 1.27 A per hp
At 230 V, a 3ϕ motor draws 2.5 A per hp
At 230 V, a 1ϕ motor draws 5 A per hp
At 115 V, a 1ϕ motor draws 10 A per hp

The values listed are not as accurate as the method shown in Figure 4-10. The method for finding the full-load current rating using a multiplier times the horsepower is based on the motor's phase and voltage. See Figure 4-11. Multiply the horsepower of the motor by 1 when using three-phase motors with 550, 575, or 600 volts. Multiply the horsepower of the motor by 127% when using three-phase motors with 440, 460, or 480 volts. The full-load current rating for three-phase motors with 220, 230, or 240 volts can be determined by multiplying the horsepower by 250%. The full-load current rating for single-phase motors with 220, 230, or 240 volts can be determined by multiplying the horsepower

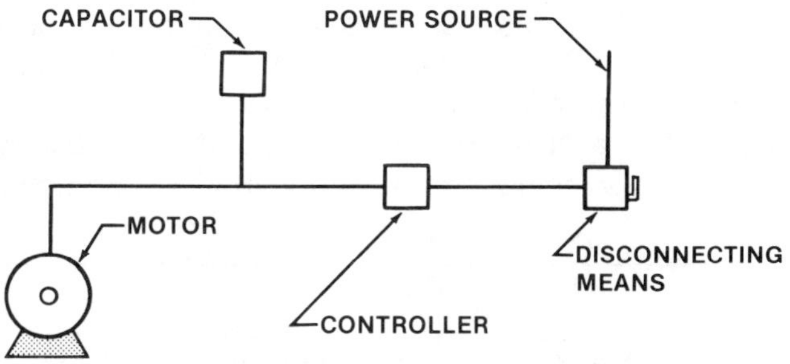

PROBLEM: What size capacitors are required to raise the power factor for a three-phase, 240-volt, 30-horsepower motor pulling 110 amps? The motor has a load rating of 33,280 volt-amps (80 amps) and the power factor is to be raised to 85%.

Step 1: $kVA = \dfrac{V \times \sqrt{3} \times A}{1000}$

$kVA = \dfrac{240\text{ V} \times 1.732 \times 110\text{ A}}{1000}$

$kVA = 45.76$ (apparent power)

Step 2: $PF = \dfrac{W}{VA}$

$PF = \dfrac{33{,}280\text{ W (true power)}}{45{,}760\text{ VA (apparent power)}}$

$PF = 73\%$

Step 3: Raise PF to 85%.

$VA = \dfrac{W}{85\%}$

$VA = \dfrac{33{,}280\text{ W}}{85\%}$

$VA = 39{,}153$ or 39.2 kVA

Step 4: $\textit{Reduced amps} = \dfrac{VA}{V}$

$\textit{Reduced amps} = \dfrac{39{,}153\text{ VA}}{240\text{ V} \times 1.732}$

$\textit{Reduced amps} = 94$ A

Step 5: Angle of cosine = 85% or 32°
Sine of 32° = 53%

Step 6: $kVAR = \dfrac{\text{Sine} \times kVA}{1000}$

$kVAR = \dfrac{.53 \times 39{,}153\text{ VA}}{1000}$

Corrected kVAR = 20.75

Step 7: Angle of cosine = 73% or 43°
Sine of 43° = 68%

Step 8: $kVAR = \dfrac{.68 \times 45{,}760\text{ VA}}{1000}$

Existing kVAR = 31

Step 9: *Existing kVAR* − *Corrected kVAR*

31 − 20.75 = 10.25 kVAR

Answer: **10.25 kVAR**

Figure 4-8. Sizing capacitors to raise power factor for a motor. *NOTE:* See Trigonometric Chart on inside back cover and Capacitor Calculating Chart on page 19.

by 500%. The full-load current rating for single-phase motors with 110, 115, or 120 volts can be determined by multiplying the horsepower by 1000%.

CONDUCTORS SUPPLYING MOTORS AND LOADS—430-25

When combination loads are involved and one or more motor loads are on the same circuit with lights or appliances, calculate the motor load according to either 430-22(a) or 430-24. Figure other loads according to Article 220 and other applicable articles. The total of all the loads equals the ampacity required for the circuit. See Figure 4-12.

SIZING OVERCURRENT PROTECTION DEVICES FOR FEEDER CIRCUITS 430-52 AND TABLE 430-152

Overcurrent protection devices can be sized by the total current rating of the loads or by the ampacity of the largest overcurrent protection device plus the ampacities of the remaining motors.

When the overcurrent protection device is selected by the total current rating of the loads per 430-62(b), 240-3(b) permits the next higher size overcurrent device to be selected when the device does not correspond to the ampacity of the conductors used. For example, the total load of a feeder is 216 amps. From Table 310-16, a #4/0 THW copper conductor has an ampacity of 230 amps. The next higher standard size device is 250 A per 240-6.

For the second method of sizing an overcurrent protection device, first find the full-load current of the motor per Table 430-150. Multiply this value by the percentage selected from Table 430-152 based on the characteristics of the motor. Add this figure to all other motor loads plus any other loads, and select the next lower rating per 240-3(f) and 240-6. For example, a feeder circuit has the following loads: two single-phase, 240-volt, 7½-horsepower

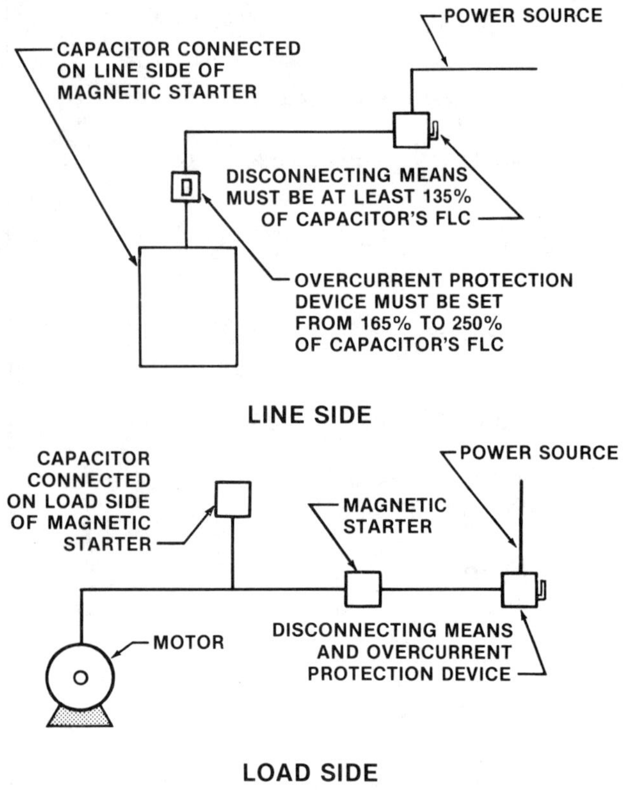

Figure 4-9. When a capacitor is connected to the line side of a magnetic starter, an overcurrent protection device is required to protect the capacitor. When the capacitor is connected on the load side, the motor's overcurrent protection device protects the capacitor.

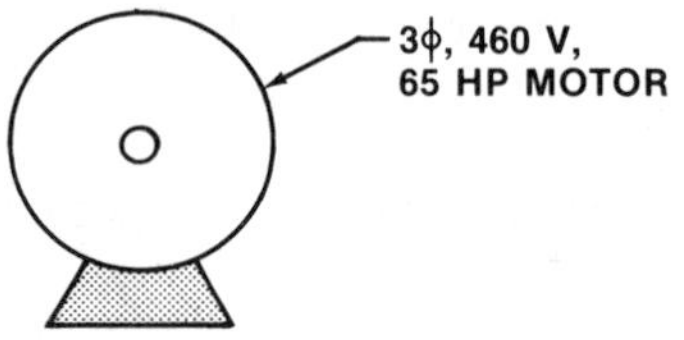

Table 430-150

PROBLEM: Find the full-load current rating of this motor.

Step 1: *Table 430-150, Col. 3.*
The flc rating of a listed 60-hp motor is 77 A.

Step 2: $\textit{Multiplier} = \dfrac{\textit{amps}}{\textit{horsepower}}$

$\textit{Multiplier} = \dfrac{77\text{ A}}{60\text{ hp}}$

$\textit{Multiplier} = 1.283$

Step 3: *Multiplier × horsepower of unlisted motor*

$1.283 \times 65 = 83\text{ A}$

Answer: **83 A of flc rating for a 65-hp motor.**

Figure 4-10. Determining the full-load current rating of a motor not listed in the tables.

motors, one single-phase, 120-volt, 3-horsepower motor on each phase, one single-phase, 240-volt, 37.5-amp lighting load and 31-amp appliance load. Find the size overcurrent protection device required.

Step 1: *Table 430-148.*
$7\frac{1}{2}$ hp = 40 A
$7\frac{1}{2}$ hp = 40 A
3 hp = 34 A
2 hp = 24 A

Step 2: *Table 430-152.*
40 A × 250% = 100 A (largest overcurrent protection device)

Step 3: Add largest overcurrent protection device to amperages of remaining motors
100 A + 40 A + 34 A + 24 A + 31 A + 37.5 A = 266.5 A

Answer: *240-3(f); 240-6; 430-63.*
250 A cb

A 250-amp circuit breaker is required when using either method. *NOTE:* The procedure giving the largest overcurrent protection device should be used when sizing overcurrent protection devices for the service-entrance conductors or feeder circuit conductors.

SELECTING CONDUCTORS AND OVERCURRENT PROTECTION DEVICES

The conductors for a feeder or service entrance must be selected by adding all loads of the premise plus 125% of the largest motor. The first step is to calculate the lighting load. All the lighting of the premise must be figured at continuous loads or noncontinuous loads. This includes inside lighting and outside lighting. The second step is to compute the receptacle load. The third step is to add all the special loads. The fourth step is to take 125% of the largest motor and add this total to all the remaining motors of the group. The final step is to total all the computed loads and divide by the voltage supplying the premise. For example, a commercial building has calculated lighting loads of 62 amps per phase. Receptacle loads are computed at 58

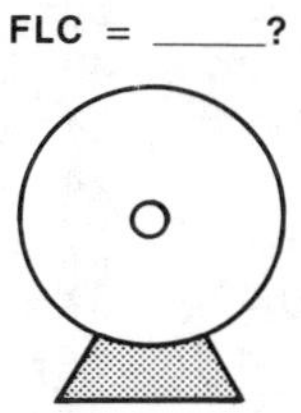

PROBLEM: What is the flc rating for a 3ϕ, 240-V, 30-hp motor?

Step 1: 30 hp × 250% = 75 A

Answer: **The flc rating is 75 A.**

PROBLEM: What is the flc rating for a 3ϕ, 575-V, 30-hp motor?

Step 1: 30 hp × 1 = 30 A

Answer: **The flc rating is 30 A.**

PROBLEM: What is the flc rating for a 1ϕ, 240-V, 5-hp motor?

Step 1: 5 hp × 500% = 25 A

Answer: **The flc rating is 25 A.**

PROBLEM: What is the flc rating for a 3ϕ, 480-V, 30-hp motor?

Step 1: 30 hp × 127% = 38 A

Answer: **The flc rating is 38 A.**

PROBLEM: What is the flc rating for a 1ϕ, 120-V, 5-hp motor?

Step 1: 5 hp × 1000% = 50 A

Answer: **The flc rating is 50 A.**

Figure 4-11. To find the full-load current of a motor, a multiplier based on the motor's phase and voltage is multiplied by the horsepower.

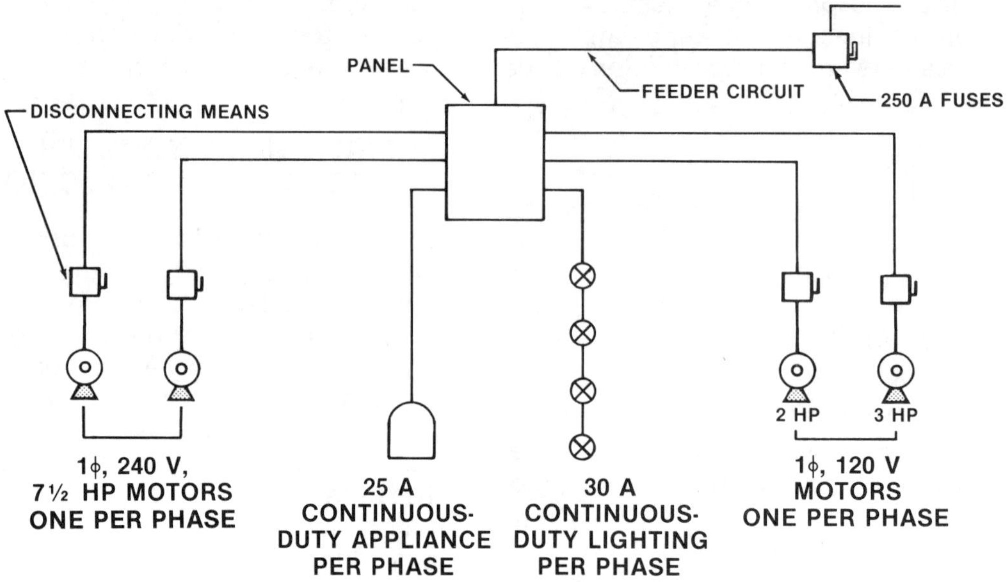

220-10(a)(b); Table 310-16
430-24; 430-25
Table 430-148

PROBLEM: What size THWN copper conductors are required for the feeder circuit?

Step 1: *Table 430-148, Col. 2.*
240-V motors
7½-hp motors pull 40 A each.

Table 430-148, Col. 1.
120-V motors
3-hp motor pulls 34 A
2-hp motor pulls 24 A

Step 2: *220-10(a)(b).*
appliance loads

25 A × 125% = 31 A

Step 3: *220-10(a)(b).*
lighting loads

30 A × 125% = 37.5 A

Step 4: *430-24; 430-25.*
all loads

	OCPD	CONDUCTORS
40 A × 125% =	50 A	50 A
remaining motors	40 A	40 A
	34 A	34 A
	24 A	24 A
appliance load	31 A	25 A
lighting load	37.5 A	30 A
	216.5 A	203 A

Step 5: *Table 310-16, Col. 2.*

Answer: **#4/0 THWN copper conductors for Phases A and B (Based on OCPD)**

Step 6: *220-22* and *430-24.*

34 A × 125% =	34 A
remaining motors	24 A
appliance loads	25 A
lighting loads	30 A
	113 A

Step 7: *Table 310-16, Col. 1.*

Answer: **#2 THWN copper conductors for neutral**

Figure 4-12. Conductors and their loads are added at 100% and may have to be increased to match the OCPD, which is calculated at 125% for continuous duty. This total is added to the motor load per 430 and 430-25.

amps per phase. Special loads are figured at 88 amps per phase, and the building is equipped with three 208-volt motors rated at 42 amps, 28 amps, and 17 amps. Figure 4-13 illustrates the step-by-step procedure for determining the size conductors for a building with motors and other loads.

The overcurrent protection device is selected by applying one of two methods. The first method is to select the overcurrent protection device based on the ampacity of the service-entrance conductors. The second method is to select the largest overcurrent protection device for the group of motors and add the other motor loads plus the other loads of the premises. See Figure 4-14.

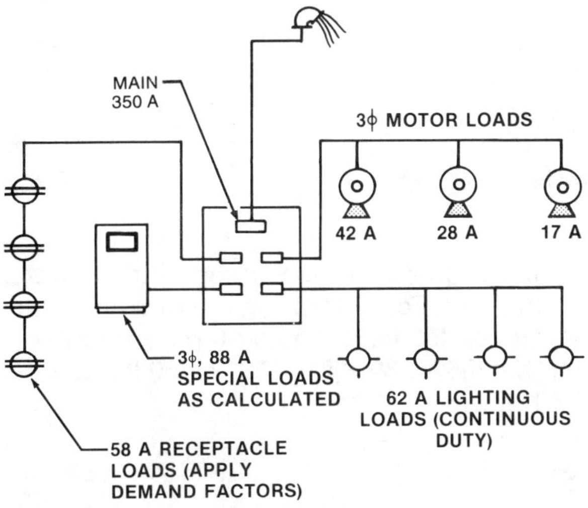

220-10(a)(b); 430-25
220-22; Table 310-16

PROBLEM: What size THWN copper conductors are required to supply a commercial building with motors and other loads?

Step 1: *220-10(a)(b).*
Lighting load

62 A × 125% = 77.5 A

Step 2: *220-3(c)(6).*
Receptacle load

58 A × 208 V × 1.732 = 20,880 VA

Table 220-13.
Applying demand factors

First 10,000 VA × 100% = 10,000 VA
Next 10,880 VA × 50% = 5,440 VA
15,440 VA

$$I = \frac{VA}{V}$$

$$I = \frac{15{,}440}{208 \times 1.732}$$

$$I = 42.9 \text{ A}$$

Step 3: *220-10(a)(b).*
Special loads = 88 A (Already calculated at continuous or noncontinuous loads)

Step 4: *430-24* and *430-25.*
Motor loads

42 A × 125% = 52.5 A
52.5 A + 28 A + 17 A = 97.5 A

Step 5: Total the loads

Total the loads	OCPD	CONDUCTORS
Lighting loads	77.5 A	62 A
Receptacle loads	42.9 A	42.9 A
Special loads	88 A	88 A
Motor loads	97.5 A	97.5 A
Total load	305.9 A	290.4 A

Answer: *Table 310-16.*

Phases A, B, and C = three #350 kcmil THWN copper conductors (Based on OCPD)

Step 1: *220-22.*
Neutral

Lighting load	62 A
Receptacle load	42.9 A
	104.9 A

Answer: **Neutral = one #2 THWN copper conductor**

Figure 4-13. Determining the size conductors for a building with motors and other loads.

Method 1: *240-3(b); 430-62(b); 430-63.*
Selecting overcurrent protection device per service conductors.
Table 310-16.
#350 kcmil THWN = 310 A

Answer: **Next size device per 240-6 is 350 A.**

Method 2: *430-62(a); 430-52(a), Ex. 1; Table 430-152.*
Selecting largest device.

42 A × 250% =	100 A CB
Adding loads	28 A
Motor loads	17 A
Lighting loads	77.5 A
Receptacle load	42.9 A
Special loads	88 A
	353.4 A

Answer: **350-A cb or fuses (next lower rating must be used)**

NOTE: The overcurrent protection device must be selected by the method that allows the largest motor to start.

Figure 4-14. Selecting overcurrent protection devices for combination loads.

OPERATING 230 V MOTORS ON 208 V SUPPLY CIRCUITS

Slight changes in voltage, torque, slip, and current occur where 230 V, 3ϕ, 60 Hz motors operate on 208 V, 3ϕ, 60 Hz supply circuits. For example, the full-load current for a motor operating 10%+ above rated voltage is 7% below normal. See Figure 4-15.

NEMA standards allow motor terminal voltage to vary 10% below or above rated voltage. For example, a 230-volt motor can operate between 207 volts (230 V × 10% = 230 V − 23 V = 207 V) and 253 volts (230 V × 110% = 253 V).

Characteristic	+10%	−10%
Torque	Up 21%	Down 19%
Full-load speed	Up 1%	Down 2%
Power factor	Down 4%	Up 3%
Full-load current	Down 7%	Up 11%
Temperature	Down 10%	Up 17%
Maximum overload	Up 21%	Down 19%
Efficiency	Up 1%	Down 2%

Figure 4-15. Characteristics of induction motors operating 10% above or 10% below supply voltage.

Torque

The torque of a motor varies with the square of the voltage. The starting torque and maximum running torque of motors running on 208 volts are determined by squaring the supply voltage (208 V) and dividing by the voltage of the motor (230 V) squared. For example, a motor with a starting torque of 150 lb. ft. has a starting torque of 119 lb. ft. [(208 V/230 V)^2 = .79 × 150 lb. ft. = 118.5 lb. ft.)].

Slip

The slip of induction motors varies inversely with the square of the voltage. The slip of a 230-volt motor operating on a 208 volt supply is 1.12 times the slip of the 230-volt rating on the motor's nameplate [(230 V/208 V)^2 = 1.12]. For example, if the synchronous speed is 1800 RPM and the actual speed is 1725 RPM, the slip is 75 RPM (1800 RPM − 1725 RPM = 75 RPM). The new slip of RPM for a 230-volt motor operating on 230 volts is 84 RPM [(230 V/208 V)^2 = 1.12 × 75 RPM = 84 RPM].

Current

Motors are more efficient at higher voltage because they draw less current and run slightly cooler. Motors operating at lower voltage pull more current and are less efficient. Motors rated at 230 volts and operating on a 208-volt supply draw approximately 11% more current than when operating at 230 volts. For example, a 230-volt motor operating at 30 amps on 208 volts draws approximately 33.3 amps (30 A × 111% = 33.3 A).

REVIEW—CHAPTER 4

Name ____________________ Date ____________________

True-False

T F **1.** Conductors supplying a single motor must be computed at 135% of the motor's full-load current rating.

T F **2.** Conductors supplying multispeed motors must be sized and selected from the largest amperage of the motor for the circuit from the panel to the starter.

T F **3.** Where motors are used on a duty cycle, the windings must have a chance to rest during the OFF cycle.

T F **4.** Conductors supplying individual motors used for a duty cycle must be sized and based on the running time.

T F **5.** If a motor is used for 15 minutes to drive a load, the motor must rest for 60 minutes.

T F **6.** Conductors supplying two or more motors must be sized by multiplying all the motors in the group by 125%.

T F **7.** Feeder conductors supplying two or more motors can be routed to a gutter and spliced to the conductors feeding each motor.

T F **8.** Full-load current ratings selected to size conductors for motor circuits are taken from the motor's nameplate.

T F **9.** The ampacity of conductors used to supply motor circuits is selected from Table 310-16 for conductors routed in a building.

T F **10.** Conductors supplying power to a wound-rotor motor are sized at 110% of the full-load current rating.

T F **11.** Resistor leads are the conductors between the resistor bank and drum controller.

T F **12.** Conductors supplying capacitors used for power factor correction are sized at 125% of the full-load current rating of the capacitor.

T F **13.** Additional overload protection is not required for a capacitor connected on the load side of the magnetic starter.

T F **14.** An additional disconnecting means is not required for capacitors installed on the load side of the magnetic starter.

T F **15.** The rating of a nonfusible disconnecting means must be sized at least 125% of the motor's full-load current rating.

Completion

__________ **1.** Conductors supplying an individual motor driving a pump must be sized at least _____% of the motor's full-load current rating.

______ 2. Feeder circuit conductors must be sized at 125% of the ______ motor of the group plus 100% of the remaining motors.

______ 3. Circuit conductors to a multispeed motor must be sized from the highest ______ selected from the speeds.

______ 4. The windings of a duty-cycle motor have a period of rest, allowing the windings to ______.

______ 5. Feeder conductors supplying a group of motors can be terminated to the lugs of a(n) ______, and individual circuit breakers can be sized for each motor.

______ 6. The full-load current ratings of motors are selected from Table 430-150 for sizing ______ for motor circuits.

______ 7. Conductors between the controller and a wound-rotor motor must be sized at 125% of the full-load current rating for ______ duty.

______ 8. The full-load current rating for sizing the secondary conductors is selected from the ______ or is found on the nameplate of the wound-rotor motor.

______ 9. Where capacitors are used for power-factor correction, the new value of ______ must be used to select overload protection.

______ 10. Continuous loads must be computed at ______%.

______ 11. The full-load current rating of a three-phase, 460-volt, 20-horsepower motor can be determined by applying the rule-of-thumb method, which is multiplying the horsepower by ______.

______ 12. The full-load curent rating of a three-phase, 230-volt, 10-horsepower motor is ______ amps.

______ 13. Capacitors can be checked by using a(n) ______.

______ 14. A(n) ______ is used to check the discharge of older capacitors without damaging the capacitor being tested.

______ 15. A motor that is used for 15 minutes to raise a drawbridge must have a rest period of ______ minutes.

Multiple Choice

______ 1. A motor used for a pump with a 15-minute intermittent-duty cycle will have conductors sized at ______% of the full-load current.

A. 85
B. 110
C. 115
D. 120

______ 2. Resistors separated from the controller of a wound-rotor motor and used for heavy-duty starting can be reduced to ______%.

A. 35
B. 45
C. 55
D. 60

_______ **3.** The load rating to size the conductors for a three-phase, 230-volt, 25-horsepower motor is _____ amps.
A. 68
B. 85
C. 95
D. 100

_______ **4.** The full-load current rating of a three-phase, 575-volt motor can be determined by multiplying the horsepower by _____ amp(s).
A. 1
B. 2
C. 2.5
D. 3

_______ **5.** The full-load current rating of a single-phase, 230-volt motor can be determined by multiplying the horsepower by _____ amps.
A. 2
B. 3
C. 4
D. 5

_______ **6.** The size THWN copper conductors required to supply a motor load calculated at 240 amps is _________.
A. #4/0
B. 250 kcmil
C. 300 kcmil
D. 350 kcmil

_______ **7.** A capacitor with a full-load current rating of 84 amps requires a disconnecting means of _____ amps.
A. 100
B. 114
C. 125
D. 150

_______ **8.** THHN insulation can be used only in _________ locations.
A. damp
B. wet
C. dry
D. damp and dry

_______ **9.** An in-line fuse with a rating of _____ amps may be used to test a capacitor.
A. ¼
B. ⅓
C. ½
D. 1

_______ **10.** Overcurrent protection for a capacitor can be rated from _____% to _____% of the capacitor's full-load current.
A. 125; 150
B. 150; 175
C. 165; 250
D. 250; 300

Problems

1. What size THWN copper conductors are required to supply a single-phase, 230-volt, 7½ horsepower motor?

2. What size THWN copper conductors are required to supply a multispeed motor that has speeds rated at 32, 45, 53, and 65 amps, respectively?

3. What size THWN copper conductors are required to supply a three-phase, 440-volt, 60-horsepower motor that drives a coal-handling machine? The motor operates every 30 minutes with a periodic duty rating.

4. What size THWN aluminum conductors are required to supply four motors at three-phase, 460 volts, and rated at 10, 15, 20, and 30 horsepower?

5. A three-phase, 230-volt, 40-horsepower wound-rotor motor has a secondary current rating of 85 amps. Find the size of the following using THW copper conductors:
   A. main conductors
   B. secondary conductors
   C. resistor conductors (the resistor bank is used for starting and controlling the motor).

6. What size 75°C XHHW aluminum conductors are required to supply a 30-kVA capacitor that is installed to correct the power factor for a three-phase, 440-volt, 60-horsepower motor supplied by #1 75°C XHHW aluminum conductors?

7. What is the minimum rating of the disconnecting means for a 35-kVA capacitor used for the power factor correction on a three-phase, 440-volt system?

8. What is the maximum size overcurrent protection device permitted for the capacitor in problem 7?

9. Use the rule-of-thumb method to determine the full-load current rating of a three-phase, 230-volt, 25-horsepower motor.

10. Use the rule-of-thumb method to determine the full-load current rating of a three-phase, 460-volt, 200-horsepower motor.

MOTOR PROTECTION

Chapter 5

A disconnecting means and controller must be provided for each motor. Overload protection must be provided to protect the windings of a motor from overload conditions. Many types of devices are commercially available that can be used to control, protect, and disconnect motors. Article 430 gives the requirements for connecting these devices.

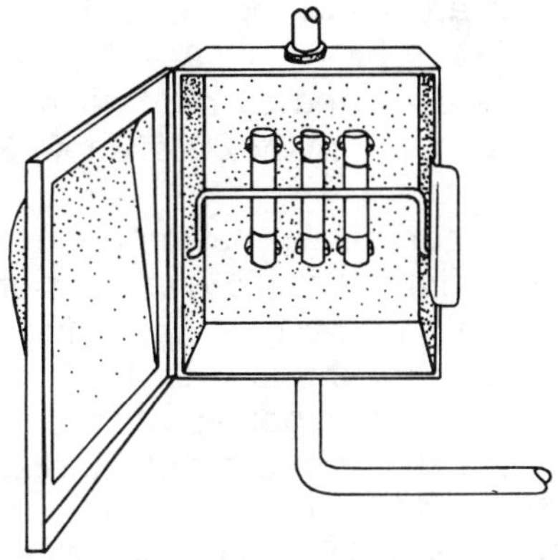

SIZING RUNNING OVERLOAD PROTECTION FOR MOTORS—430-32

Running overload protection for motors of more than 1 horsepower may be provided by separate current-interrupting devices such as terminal cutouts, thermal protectors, thermal relays, or fusetrons. If the motor inlets and outlets are covered by a blanket of lint or if a bearing should begin to lock, excessive heating of the motor windings will cause operating overcurrent that could damage the motor. The devices are set to open at 115% or 125% of the motor's full-load current rating, depending upon the service factor or temperature rise of the motor. If the motor is not marked 115% is used.

The temperature rise of a motor is the rise of temperature after the motor has been running as a result of the current passing through motor windings. The *ambient temperature* is the temperature of the outside air around the motor after the motor has been installed and operating. The difference between the temperature rise of the motor windings and the ambient temperature is the *temperature difference*. A motor operating with a 25°C temperature rise and installed in a location with 30°C ambient temperature will have a 5°C (30°C − 25°C = 5°C) temperature difference. Motors without special design are built to operate at 40°C (104°F) ambient temperature. A motor with 40°C can heat up to, and above, 30°C and not be damaged by excess heat. Higher temperature due to increased current in the motor windings or high temperature surrounding the motor enclosure can cause insulation failure. Heat from the motor windings can only be dissipated when the heat rise of the motor is above the ambient temperature. The factor of heat rising determines the dissipation rate of heat from the motor. Heat will always move to a cooler location.

Temperature Rise

Altitude affects the temperature rise in a motor. In higher altitudes, thinner air allows less heat to be carried from the windings of the motor through the inlets and outlets. Air thick enough to carry heat effectively can be obtained at elevations of 3300′ or less. A motor's horsepower rating must be derated at elevations above 3300′.

For every 330′ above 3300′ a motor must be derated 1%. For example, a 25-horsepower motor pulling 68 amps and installed at an altitude of 3960′ must be derated to a running current of 66.6 amps (68 A × 2% = 1.36 A; 68 A − 1.36 A = 66.6 A). For every 1000′ above 3300′ a motor must be derated 10%. For example, a 25-horsepower motor pulling 68 amps and installed at an altitude of 11,300′ must be derated to an operating current of 13.6 amps (68 A × 80% = 54.4 A; 68 A − 54.4 A = 13.6 A). The 80% is obtained by multiplying 8 by 10%, which is derived from the 8000′ above 3300′ (8000′ + 3300′ = 11,300′).

A motor marked with a temperature rise not over 40°C is figured at 125% of the motor's nameplate full-load current rating. A motor marked with a temperature rise over 40°C is figured at 115% of the motor's nameplate full-load current rating. Note that motors not marked must be calculated at 115% of the motor's full-load current rating per 430-6(a) and 430-32(a)(1). See Figure 5-1. The overloads should trip open due to the starting of the motor, and 430-34 can be applied using a higher percentage. Apply 430-25 if the overloads trip open when using 430-34.

Service Factor

For motors marked with a service factor of not less than 1.15 (115%), 125% of the full-load current rating of the motor is used. For a motor with a service factor less than 1.15 (115%), 115% of the full-load current must be used. A service factor of 1.15 (115%) indicates that the motor will operate at 115% of the full-load current rating without damage to the motor's insulation. A service factor of 1.10 (110%) indicates that the motor will operate at 110% of the full-load current rating without damage to the motor's insulation. For example, a motor with a full-load current rating of 28 amps (nameplate rating) requires an overload of 30.8 amps (28 A × 1.10 = 30.8 A). See Figure 5-2.

SELECTING OVERLOADS FROM CONTROLLER COVER

When the sizes of overloads are selected from the cover of a magnetic starter or controller, the nameplate full-load running current of the motor is used. The full-load running current is not increased by 125% when the overloads are selected in this manner.

SELECTING OVERLOADS FROM MANUFACTURER'S CHART

The size overloads required to protect the windings of a motor can be determined by taking the motor's full-load current rating and selecting the

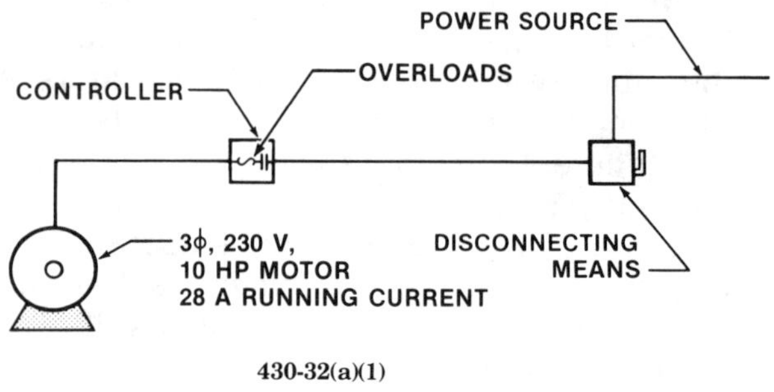

PROBLEM: Find the minimum size overloads to protect this motor with a 40°C temperature rise.

Step 1: *430-32(a)(1).*
The maximum allowed is 125%.
28 A × 125% = 35 A

Answer: **Overloads may be sized at 35 A.**

PROBLEM: Find the minimum size overloads to protect this motor with a 50°C temperature rise.

Step 1: *430-32(a)(1).*
The maximum allowed is 115%.
28 A × 115% = 32.2 A

Answer: **Overloads may be sized at 32.2 A.**

Figure 5-1. To find minimum size overloads for a motor not exceeding 40°C temperature rise, multiply the full-load current rating by 125%. For minimum size overloads for motors exceeding 40°C temperature rise, multiply the full-load current rating by 115%. For overloads selected from the controller cover, size from nameplate rating of motor's FLC per 430-6(a). *NOTE:* See Figure 5-3.

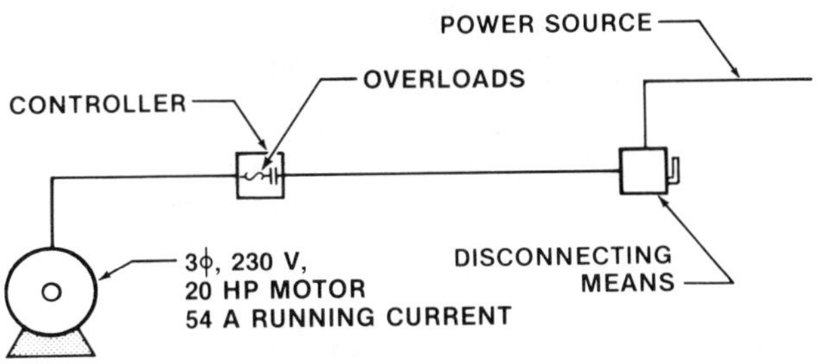

430-32; 430-34

PROBLEM: Find the minimum size overloads required to protect this motor with a service factor of 1.15.

Step 1: *430-32.*
The maximum allowed is 125%.

54 A × 125% = 67.5 A

Step 2: *430-32.*

Answer: **Overloads may be sized at 67.5 A.**

PROBLEM: Find the maximum size overloads required to protect this motor with a service factor of 1.15.

Step 1: *430-34.*
The maximum allowed is 140%.

54 A × 140% = 75.6 A

Step 2: *430-34.*

Answer: **Overloads may be sized at 75.6 A.**

NOTE: A 1.15 service factor will permit the motor to deliver 115% (23 hp) of the 20-hp rating without damage.

Figure 5-2. To find the minimum size overloads when using a service factor of not less than 1.15, multiply the full-load current rating by 125%. To find maximum size overloads multiply by 140%.

size overloads from the cover of a magnetic starter, a motor control center, or the manufacturer's catalog. For example, per the chart in Figure 5-3, a three-phase motor with a full-load current rating of 39 amps requires three overload units with catalog number H1047. Overload units number H1047 are selected because the 39-amp full-load current rating of the motor is between 38.0 and 42.9 amps. See Figure 5-3 for a step-by-step procedure for selecting the size overloads from the motor's full-load current rating, and the cover of a starter, a control center, or catalog.

SINGLE-PHASING

When overloads are used in a magnetic starter to protect the motor windings from overheating, they must be selected at 54 amps or less per 430-6(a). This percentage protects the motor windings from single-phasing or overheating due to the driven load. However, fuses or a circuit breaker can be used where they do not exceed the percentages per 430-32(a)(1). For example, a 60-amp fuse or circuit breaker can be used to provide overload protection for a motor. Usually a time-delay fuse is selected because it holds five times its rating for 10 seconds (100 A × 5 = 500 A).

Single-phasing occurs when one phase of a three-wire system is lost. A motor pulling 54 amps pulls 93.5 amps (54 A × 1.732 = 93.5 A) if single-phasing occurs. *NOTE:* 93.5 amps remain on each remaining phase.

A 60-amp overcurrent protection device does not exceed 125% of the motor's full-load current rating obtained from the motor's nameplate. The 93.5 amps remaining on phases A and B are higher in rating than a 60-amp overcurrent protection device and trips open due to an overload exceeding 60 amps. See Figure 5-4.

The percentages listed in 430-32(a)(1) can be applied to

(1) circuit breakers
(2) time-delay fuses
(3) thermal relays
(4) thermal cutouts
(5) thermal devices designed into motors
(6) motor switches with thermal devices

Time-delay fuses are used to provide backup protection to the overloads in a magnetic starter. However, the rating and setting must not exceed 125% of the motor's full-load current listed on the nameplate.

MOTOR NAMEPLATE

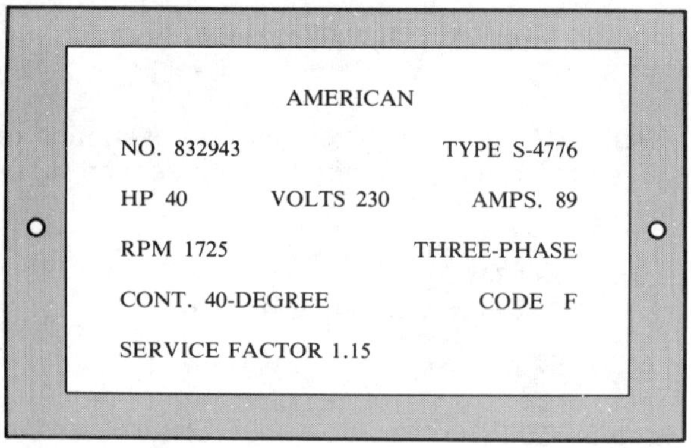

OVERLOAD CHART

AMPERAGE	OVERLOAD UNITS
20.6-23.3	H1042
23.4-26.0	H1043
26.1-30.5	H1044
30.6-33.6	H1045
33.7-37.9	H1046
38.0-42.9	H1047
43.0-48.2	H1048
48.3-54.6	H1049
54.7-61.2	H1050
61.3-67.6	H1051
67.7-75.9	H1052
76.0-87.1	H1054
87.2-97.5	H1055
97.6-109.0	H1056
110.0-122.0	H1057
123.0-135.0	H1058

PROBLEM: What size overload units from the overload chart are required to protect an induction motor with the above nameplate?

Step 1: *430-6(a); 430-7(a)(4); 430-32(a)(1).*
Motor nameplate
FLC = 89 A

Step 2: Overload chart
87.2 A to 97.5 A = H1055

Answer: **H1055**

Figure 5-3. Selecting overloads for a motor from an overload chart based on the full-load current rating listed on the motor's nameplate.

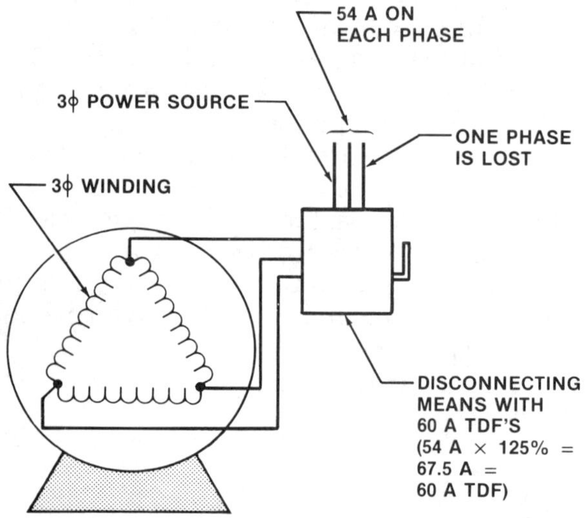

PROBLEM: What is the current in phases A and B due to a single-phase condition?

Step 1: Phases A, B, and C pull 54 A each.

Step 2: 54 A × 1.732 = 93.5 A

Answer: **Phases A and B pull 93.5 A each.**

Figure 5-4. When one phase of a three-phase system is lost due to single-phasing, the remaining phases will carry 1.732 times the original current.

MAXIMUM SIZING OF OVERLOAD PROTECTION—430-34

If the percentage that 430-32 allows is not sufficient, 430-34 permits higher percentages for a service factor of not less than 1.15 (115%) or for a temperature rise of not more than 40°C. Also, 140% of the full-load current rating of the motor must not be exceeded. For all other motors, a maximum of 130% is allowed. Note that the maximum rating of 430-34 can be applied only for overloads and never for fuses or circuit breakers. See Figure 5-5. Overload relays are selected per 430-34 to provide running overcurrent protection for motors. See 430-35 for shunting out overload protection. Section 430-34, FPN allows overloads with longer time delays to be used instead of applying oversize overloads to compensate for high inrush currents.

SIZING CONTROLLERS TO START AND STOP MOTORS—430-81 AND 430-83

Basic requirements for sizes and types of motor controllers are given in 430-81 and 430-83. All controllers must have at least the horsepower rating of the motor to be controlled.

The branch-circuit protection device serves the controller for motors of ⅛-horsepower or less. These motors have high-impedance windings and deliver very little fault current. The branch-circuit overcurrent protection device can be a fuse or circuit breaker located in the service panelboard or in a subfeed panel. A 15- or 20-amp device can be used with number 14 or 12 copper conductors to supply a plug outlet, and a cord can be used to plug in a wall-mounted clock of ⅛-horsepower or less. See Figure 5-6.

A plug and receptacle serve as the controller for portable motors of up to ⅓-horsepower. The cord and plug can be utilized to connect and disconnect the motor without the aid of a controller and switch. A ⅓-horsepower motor does not pull enough current when started to create an arc that would be hazardous to a person's hand when disconnecting a motor under a load.

CONTROLLER OTHER THAN HORSEPOWER-RATED—430-83

All controllers must be horsepower-rated. However, there are exceptions to this basic rule that permit other than horsepower-rated controllers to be used where they comply with certain requirements as provided by the exceptions to 430-83. See Figure 5-7.

Controllers with a straight marking (480 V) can be used in grounded or ungrounded systems. Controllers with a slash voltage marking (480/277 V) can be used only in grounded systems per 430-83(b), FPN.

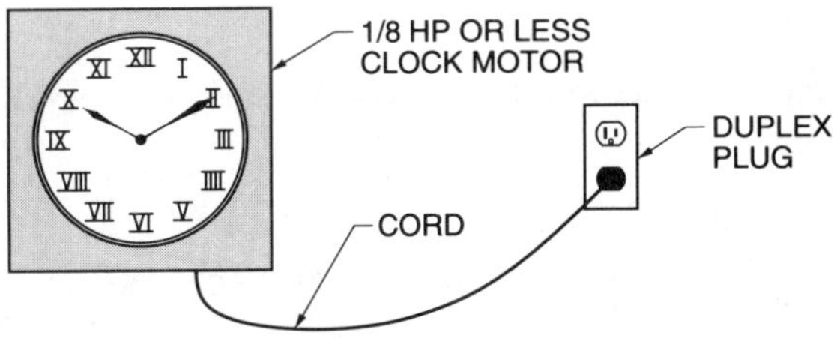

Figure 5-6. The branch-circuit overcurrent protection device may serve as a controller for motors of ⅛ horsepower or less.

430-83, Ex. 1

A general-use switch rated for at least twice the motor's full-load current may serve as the controller if the motor's rating is 2 horsepower or less. For example, a single-phase, 115-volt,

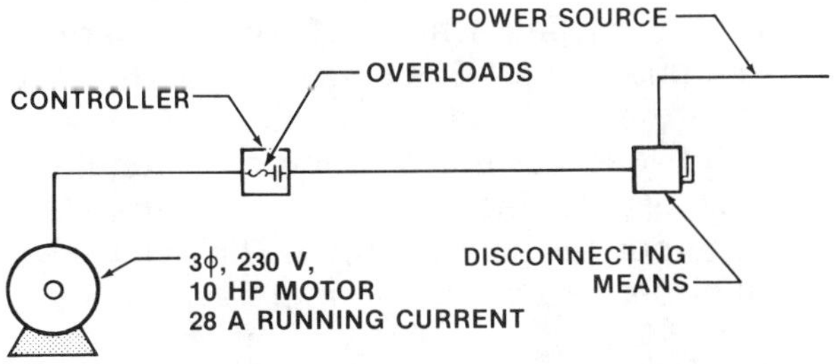

430-34

PROBLEM: Find the maximum size overloads to protect this motor with a 40°C temperature rise.

Step 1: *430-34.*
The maximum allowed is 140%.

28 A × 140% = 39.2 A

Answer: **Overloads may be sized at 39.2 A.**

PROBLEM: Find the maximum size overloads to protect this motor with a 50°C temperature rise.

Step 1: *430-34.*
The maximum allowed is 130%.

28 A × 130% = 36.4 A

Answer: **Overloads may be sized at 36.4 A.**

Figure 5-5. To select overloads for a motor with a temperature rise not exceeding 40 °C, multiply the full-load current rating times 140%. For all other motors, multiply by 130%.

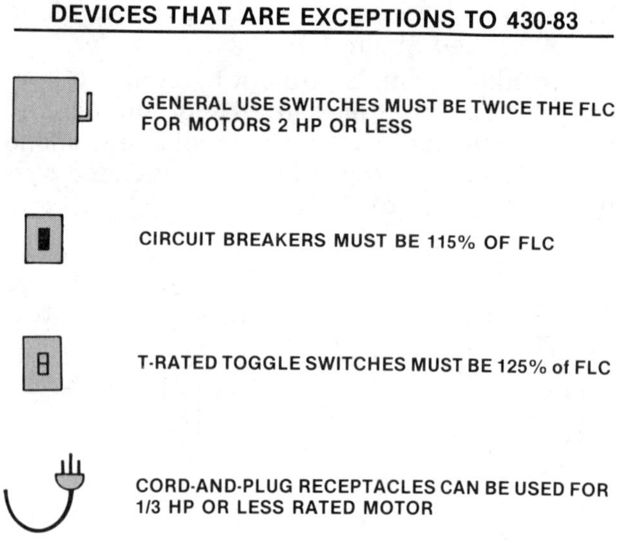

Figure 5-7. All controllers must be horsepower-rated unless they comply with one of the exceptions to 430-83.

2-horsepower motor has a full-load current rating of 24 amps. The size general-use switch can be determined as follows:

Step 1: *Table 430-148.*
1ϕ, 115 V, 2 hp = 24 A

Step 2: *430-83, Ex. 1.*
24 A × 2 = 48 A

Answer: **60-A general-use switch**

For a single-phase, 230-volt, 2-horsepower motor, the full-load current rating is 12-amp per Table 430-148. Applying the same procedure:

Step 1: *Table 430-148.*
1ϕ, 230 V, 2 hp = 12 A

Step 2: *430-83, Ex. 1.*
12 A × 2 = 24 A

Answer: **30-A general-use switch**

A general-use switch is not rated in horsepower. Note that 430-83, Exception 1 can be applied to both single-phase and three-phase motors per Table 430-148 or 430-150. The requirement that the switch must be rated at least twice the full-load current rating of the motor can be applied to single-phase or three-phase motors rated at 2 horsepower or less. A 20-amp, general-purpose, single-pole AC snap switch can be used. A snap switch must be at least 125% of the full-load current rating of the motor it controls.

If a motor is rated at $\frac{1}{2}$ horsepower and pulls 9.8 amps on a single-phase, 120-volt, 15-amp circuit, the snap switch can be sized and selected by the following procedure:

Step 1: *Table 430-148.*
1ϕ, 115 V, $\frac{1}{2}$ hp = 9.8 A

Step 2: *430-83, Ex. 1.*
9.8 A × 125% = 12.25 A

Answer: **15-A, single-pole snap switch**

430-83, Ex. 2

A circuit breaker rated in amps only may serve as the controller. The circuit breaker must be equivalent to the motor's full-load current rating. For example, a circuit breaker starting and stopping a motor drawing 16 amps must be rated at 16 amps. Therefore, a 20-amp circuit breaker is required. If the circuit breaker is used to provide running overload protection for the motor, the requirements of 430-32 must be applied.

If a motor has a full-load running current of 13.5 amps and a 1.32 service factor, the circuit breaker can be rated no higher than 15 amps. The full-load current of a motor is required by 430-32(a)(1) to increase no more than 125% based upon the service factor. The following procedure demonstrates the method:

Step 1: *430-6.*
flc = 13.5 A

Step 2: *430-109, Ex. 2; 430-32(a)(1).*
13.5 A × 125% = 16.88 A

Answer: **A 15-A cb is the next lower standard size.**

NOTE: 125% of the motor's flc rating must not be exceeded.

430-83, Ex. 3

The controller for a torque motor must have a continuous-duty, full-load current rating not less than the nameplate current rating of the motor. Torque motors are intended to operate in the stalled position and are special case types of motors. If the nameplate current rating of a torque motor is 38 amps, the controller must be rated at least 38 amps.

NUMBER OF MOTORS SERVED BY EACH CONTROLLER—430-87

Each controller must have an adjacent disconnecting means. This prevents unauthorized energizing of a power system during maintenance operations. See Figure 5-8. The exception permits any number of motors of 600 volts or less to be served by a single controller. In this case, the controller could serve motors used to drive different parts of the same machine.

A processing machine with a number of motors can be controlled by one controller as long as the motors drive the same machine or apparatus. The controller must be rated for the total horsepower of all the motors.

A single controller can serve a group of motors located in a room. Each motor in the group must be within sight and visible to the user. For example, a number of exhaust fans can be installed in the ceiling of a large room. If the fans are within sight of the controller, one controller can be used to start and stop the exhaust fan's motors.

A number of motors driving an apparatus can be located within sight in a room and operated by one controller. The controller must be within sight (50′) of all motors and comply with the requirements of 430-87, Exception.

DISCONNECTING CONTROLLERS AND MOTORS FROM UNGROUNDED SUPPLY CONDUCTORS—430-109 AND 430-110

The disconnecting means must be horsepower-rated and the rating must be at least 115% of the motor's current to be disconnected. See 430-110(a). Section 430-57 requires the fuseholder and disconnect to be selected to hold the size fuses needed to start and run a motor. Section 430-52 requires the fuses to be sized by the percentages listed in Table 430-152 for time- and nontime-delay fuses. The maximum percentage for a nontime-delay fuse is 300% and for a time-delay fuse 175%. These percentages times the motor's full-load current rating per Table 430-150 require a disconnect well above 115%. The

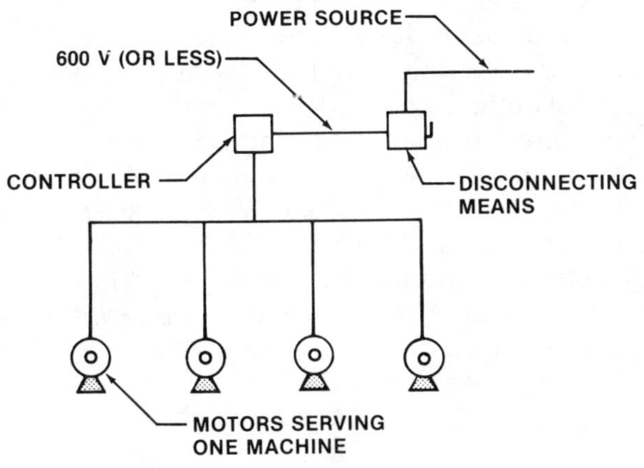

CONTROLLERS MAY SERVE MORE THAN ONE MOTOR PROVIDED THEY ARE WITHIN SIGHT

430-87; 100

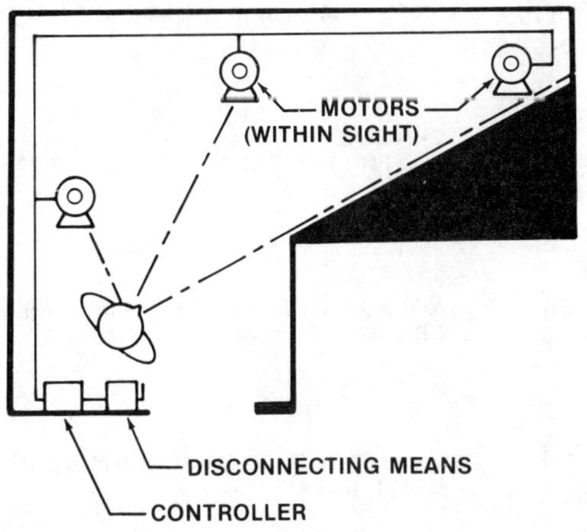

MOTORS MUST BE WITHIN SIGHT OF THEIR CONTROLLER

430-87, Ex.; 100

Figure 5-8. Controllers must be located to prevent unauthorized energizing.

115% is used for selecting a nonfusible disconnect or a nonautomatic circuit breaker to disconnect an AC unit or motor. For example, what size nonfusible disconnect or nonautomatic circuit breaker is required to disconnect an AC unit or motor rated at 42 amps?

Step 1: *430-110(a).*

42 A × 115% = 48 A

Answer: **60-A disconnect**

NOTE: A 60-A disconnect holds fuse ratings from 35 A to 60 A.

DISCONNECTS OTHER THAN HORSEPOWER-RATED—430-109

Switches that are horsepower-rated are made by the manufacturer with a horsepower rating to match the maximum size motor that the switch can disconnect. See Figure 5-9. Switches are the general-use type when they are made in amperes only. Exceptions to the basic rule exist that permit other than horsepower-rated disconnecting means to be used.

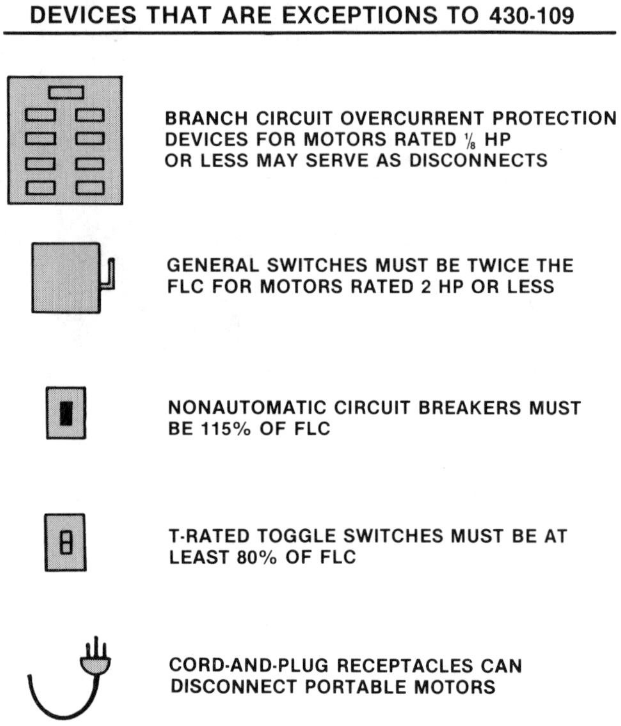

Figure 5-9. All disconnects must be horsepower-rated unless they comply with one of the exceptions to 430-109.

430-109, Ex. 1

The branch-circuit overcurrent protection device serves as the disconnecting means for stationary motors of 1/8-horsepower or less. The branch-circuit overcurrent protection device is usually located in the service equipment. This overcurrent protection device is not required to be within sight of the 1/8-horsepower motor.

430-109, Ex. 2

A general-use switch rated at not less than twice the motor's full-load current serves as the disconnecting means for motors of 2 horsepower or less. A general-use switch is rated in amperes and not horsepower. For example, what size disconnect is required to disconnect a single-phase, 115-volt, 1½-horsepower motor?

Step 1: *Table 430-148.*
1½ hp = 20 A

Step 2: *430-109, Ex. 2.*

20 A × 2 = 40 A

Answer: **60-A general-use disconnect**

A general-use, AC-rated snap switch serves as the disconnecting means for motors of 2 horsepower or less if the full-load current of the motor does not exceed 80% of the amperage of the switch. Section 380-14 lists the requirements and percentages for sizing and selecting snap switches to disconnect and control the various loads. General-use snap switches are rated and marked either AC, DC, or AC and DC. The load of the switch is limited by the marking. For example, what size general-use snap switch is required to disconnect a single-phase, 115-volt, ¾-horsepower motor?

Using a 15-A AC snap switch:

Step 1: *Table 430-148.*
¾ hp = 13.8 A

Step 2: 430-109, Ex. 2.
15 A × 80 % = 12 A

Answer: **A 15-A AC snap switch shall not be used.**

Using a 20-A AC snap switch:

Step 1: *Table 430-148.*
¾ hp = 13.8 A

Step 2: *430-109, Ex. 2.*
20 A × 80% = 16 A

Answer: **A 20-A AC snap switch shall be used.**

430-109, Ex. 3

For AC or DC motors over 2-horsepower to 100-horsepower, the disconnecting means may be a motor-circuit, horsepower-rated switch. If a motor is rated at 60 horsepower on a three-phase, 480-volt system, the disconnect must be rated at least 60 horsepower to serve as a disconnecting means.

The disconnecting means for motors above 100-horsepower may be rated in horsepower or amps. A three-phase, 440-volt, 125-horsepower motor per Table 430-150 has a full-load rating of 156 amps. The disconnecting means must have a rating of at least 156 amps. A 200-amp disconnect is required to handle the 156 amps.

430-109, Ex. 5

A plug and receptacle serves as the disconnecting means for portable motors. A portable circular saw or a portable sander is cord-and-plug connected. Any size portable motor can be connected with a cord. Care must be exercised when disconnecting the motor under load.

430-109, Ex. 7

An ITCB may be used as a disconnecting means in a listed combination motor controller per 430-102.

DISCONNECTING MEANS FOR CONTROLLERS AND MOTORS—430-102 AND 430-107

A motor and its driven machinery or load must be within sight of the controller for the motor. If the motor does not have a disconnecting means within sight, other provisions must be made. If the the disconnecting means is within sight of the motor and not more than 50′ from the motor, it is considered within sight per Article 100.

Disconnecting Means within Sight—430-102 and 430-106

The motor disconnecting means must be within sight of the motor controller. It must disconnect both motor and controller from all of the ungrounded conductors supplying the motor circuit. This disconnect by the controller provides safety for a person serving the controller. With all ungrounded conductors disconnected from the controller and the disconnect within sight, no one can accidentally connect power to the controller. See Figure 5-10.

If the disconnect by the controller is out of sight from the motor, an additional disconnect must be placed within sight of the motor. This additional disconnect in the OFF position prevents the power from being energized while a worker is pulling maintenance on the motor. See Figure 5-11. *NOTE:* See 440-14 for the disconnecting means by an AC unit.

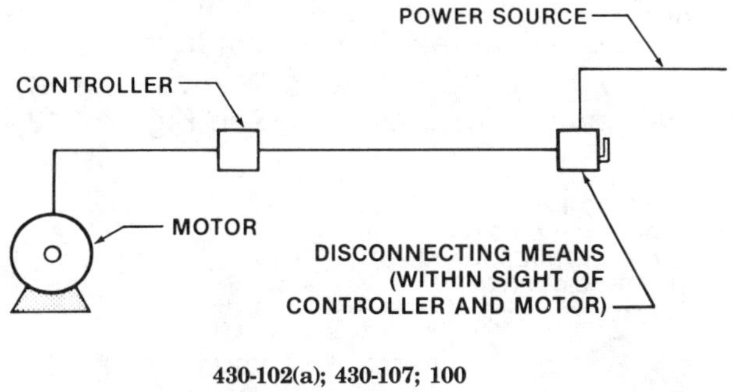

Figure 5-10. Disconnecting means must be within sight of the motor and controller.

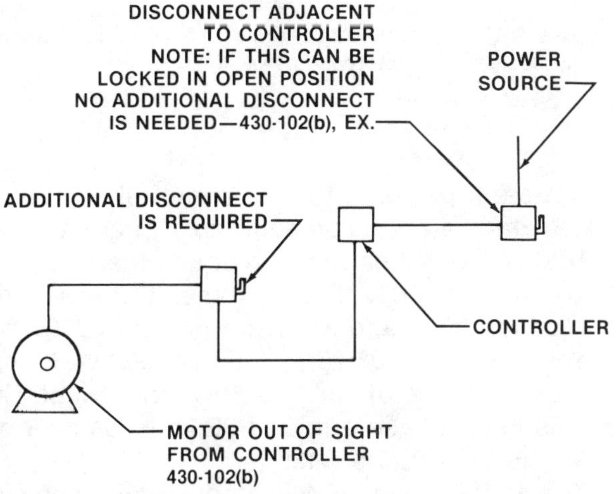

Figure 5-11. An additional disconnect is required when the motor is out of sight from the disconnect and controller and the controller cannot be locked in the open position.

POWER FACTOR

Power factor (PF) is the ratio of actual power used in a circuit to the apparent power drawn from the line. *Actual power* is the true power used to produce heat or work. Actual power is also known as real, true, or useful power. It is measured in watts and kW with a wattmeter. Apparent power is measured in VA and kVA with a voltmeter and an ammeter. The voltage and current waves may be in-phase or out-of-phase when measuring. In a pure resistance circuit, actual and apparent power are the same and the power factor is 100%.

Circuits with motors and transformers having coils produce iron effects which consume less actual power. The power factor of these circuits is less than 100%. Inductance of the coils causes the inequality between actual power and apparent power. The actual power never exceeds the apparent power.

Reactive power exists when kVA exceeds kW. Operating current consists of useful work and reactive current. *Reactive current* is the current drawn by an inductive load to develop magnetic fields required for operation. See Figure 5-12.

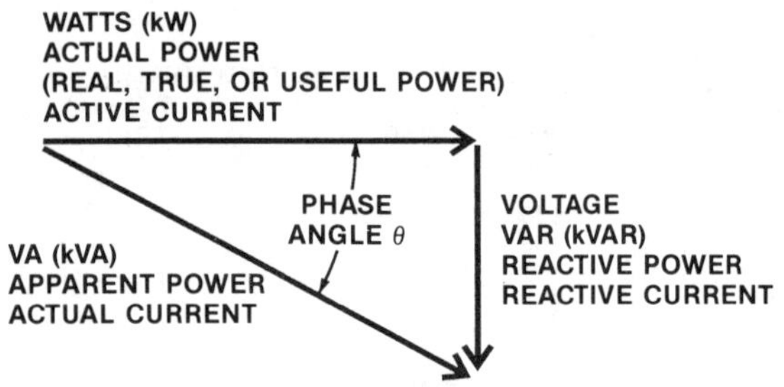

Figure 5-12. A right angle shows the relationship of terms used to demonstrate power factor.

Apparent power is found by multiplying volts times amps and can be equal to or greater than actual power. When they are equal, their ratio is 1 to 1, 1.0, or 100%. For example, if power consumed is 800 W and apparent power is 2000 W, the ratio is .4 or 40% (PF = 80 W/2000 W = .4). A power factor of 40% is low. Trigonometric charts list a phase angle of 66° for cosine θ of 40% PF. See Figure 5-13.

The actual power in W in a pure resistance circuit is found by multiplying V × A. For example, a 240 V, 1ϕ water heater with a 20 A heating element has a load of 4800 W.

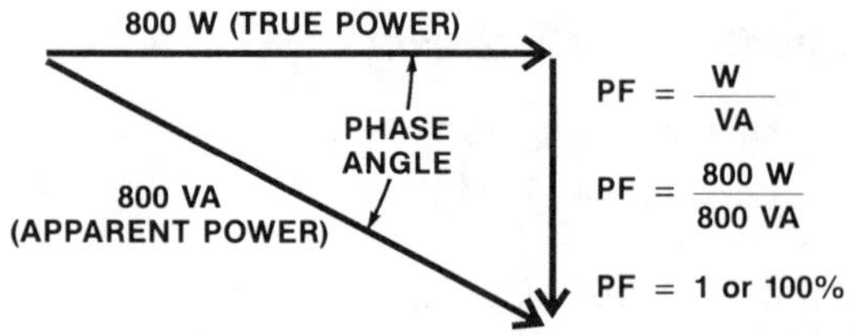

CIRCUIT WITH HIGH POWER FACTOR

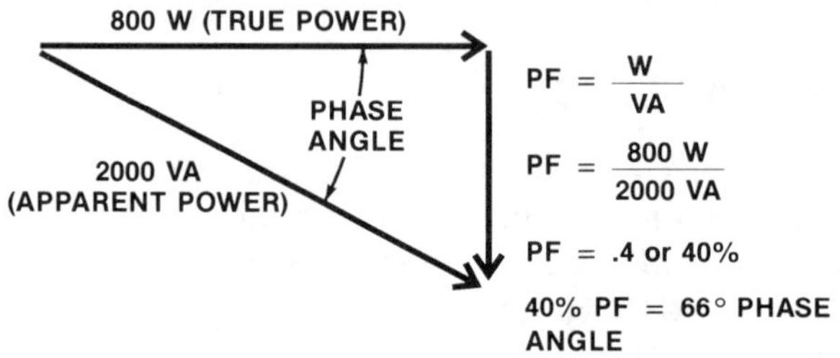

CIRCUIT WITH LOWER POWER FACTOR

Figure 5-13. Circuits with watts (or kW) closer to the VA (or kVA) rating of the apparent power have a higher power factor. Circuits with watts (or kW) less than the VA (or kVA) rating of the apparent power have a lower power factor. See trigonometric charts to match power factor to the phase angle.

Step 1: $W = V \times A$
$W = 240\text{ V} \times 20\text{ A} = 4800\text{ W}$
Answer: **Actual power = 4800 W**

The actual power in W in a 1ϕ, A-C circuit with iron effects is found by multiplying V × A × cosine θ (PF). The cosine of the phase angle accounts for reactive power. For example, a 240 V, 1ϕ motor with an FLC of 20 A and 80% power factor has a load of 3840 W.

Step 1: $W = V \times A \times PF$
$W = 240\text{ V} \times 20\text{ A} \times 80\% = 3840\text{ W}$
Answer: **Actual power = 3840 W**

The actual power in W in a 3ϕ, A-C circuit with iron effects is found by multiplying V × $\sqrt{3}$ × A × cosine θ (PF). For example, the actual power in W for a motor with an FLC of 30 A supplied by a 480 V, 3ϕ circuit with a power factor of 75% is 18,697.5 W.

Step 1: $W = V \times \sqrt{3} \times A \times PF$
$W = 480\text{ V} \times 1.732 \times 30\text{ A} \times 75\% = 18{,}697.5\text{ W}$
Answer: **Actual power = 18,697.5 W**

NOTE: Trigonometric charts list a phase angle of 41° for a cosine θ of 75% PF. Also note that 831 is used for 480 V × 1.732.

REVIEW—CHAPTER 5

Name ______________________________ Date ______________

True-False

T F 1. The temperature rise of a motor occurs after the motor is running due to the current passing through the motor windings.

T F 2. Motors with a 40°C temperature rise should have overload protection computed at 150%.

T F 3. A motor with a service factor of 1.10 requires the running overload protection to be computed at a value not exceeding 115% of the motor's nameplate current rating.

T F 4. Temperature difference is the difference between the windings of the motor and the outside air around the motor.

T F 5. Motors are rated to operate at any temperature without serious damage.

T F 6. The windings of induction motors will heat up due to an increased load.

T F 7. The windings of a motor will dissipate heat regardless of the surrounding temperature.

T F 8. Altitude affects the operation and temperature rise of a motor.

T F 9. Air is thinner in higher altitudes and carries less heat from a motor's windings.

T F 10. The horsepower rating of a motor must be derated where the elevation exceeds 3300′.

T F 11. It is not necessary for controllers to have a horsepower rating equal to the motor that it controls where the rating is above 2 horsepower.

T F 12. Motors with high-impedance windings have high fault current.

T F 13. A cord and plug can be used as a controller for certain size motors under special conditions.

T F 14. A circuit breaker rated in amps can be used as a controller.

T F 15. An AC inductive snap switch cannot be used to control a motor.

T F 16. Start-and-stop stations must not be used to serve as a disconnecting means for a motor.

Completion

__________ 1. The general rule requires each motor to have an individual __________ means.

__________ 2. Disconnects used to disconnect motors must be rated at least ______% of the motor's full-load current.

__________ 3. Motors located in the same room and within ________ can be served by the same controller.

___________ **4.** Where the disconnecting means is located by the controller and can be locked in the open position, an additional ___________ is not required near the motor.

___________ **5.** Where motors are out of sight from the controller, a disconnecting means must be provided at the _________.

___________ **6.** For motors rated over 600 volts, the disconnecting means can be located out of sight from the ___________.

Multiple Choice

___________ **1.** If a three-phase motor pulls 15 amps on each phase conductor, and a fuse blows, the remaining two phases will pull _______ amps.
- A. 20
- B. 22
- C. 25.98
- D. 30

___________ **2.** A motor with a full-load current of 12.5 amps and 1.15 SF can be protected from running overload protection by a _______-amp fuse. (Use minimum rating.)
- A. 10
- B. 12
- C. 15
- D. 20

___________ **3.** A general-use switch can be used to control motors rated at _______ horsepower or less.
- A. 1/2
- B. 3/4
- C. 1
- D. 2

___________ **4.** A cord and plug can be used as a controller for portable motors rated at _______ horsepower or less.
- A. 1/8
- B. 1/3
- C. 1/2
- D. 1

___________ **5.** A 20-amp AC rated snap switch can be used to control a motor load of _______ amps.
- A. 16
- B. 18
- C. 20
- D. 25

Problems

1. What minimum size running overload protection devices are required to protect a three-phase, 460-volt, 25-horsepower induction motor with a nameplate current of 25 amps?

2. What is the minimum size overloads required for a motor pulling 12 amps with a service factor of 1.40?

3. What is the minimum size overloads required to protect a motor with a nameplate current of 68 amps? The motor has a 75°C temperature rise.

4. What is the derated horsepower of a 30-horsepower motor used at an altitude of 9300′?

5. An induction motor has a nameplate rating of 22 amps with a service factor of 1.15. What is the amount of overload current that the motor draws without damage?

6. What size time-delay fuses are required to provide overload protection for a motor with a nameplate current rating of 32 amps?

7. How many amps will phases A and C pull on a three-phase motor drawing 38 amps if phase B is lost due to a blown fuse?

8. What is the derated horsepower of a 10-horsepower motor used to drive a load at an altitude of 4290′?

9. What size AC snap switch is required to disconnect a single-phase, 120-volt, 1-horsepower motor?

10. What size general-use switch (nonfusible) is required to disconnect a three-phase, 208-volt, 25-horsepower motor?

MOTOR CONTROL CIRCUITS

Chapter 6

Control circuit conductors are either tapped from the motor power supply circuit or supplied from the service equipment. Section 430-72 is applied where control circuits are tapped on the load side of the controllers. Section 725-12 is applied for control circuits supplied from a different source of power other than the motor circuit's source of power. Protection must be provided for control circuits no matter where they originate. Control transformers are installed to provide lower voltages for controlling motor systems.

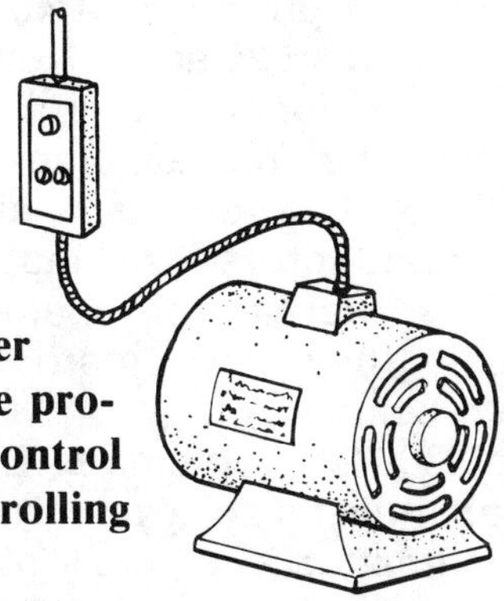

SIZING CONDUCTORS FOR CONTROL CIRCUIT—430-72 AND 725-12

When a motor control circuit derives its power from the power conductors supplying the motor and is tapped on the load side of fuses or circuit breakers, the protection for the control conductors may be either branch-circuit devices or supplementary protection. The method of protection is determined by the size of the control conductors and the rating of the motor power circuit device. If motor control circuits derive their power from a source other than the motor power circuit conductors, the control circuits are classified as *remote-control circuits*, and protection with fuses or circuit breakers must be provided for the conductors. See 725-12 and 725-35.

Section 430-74 requires that remote-control circuits have their disconnecting means mounted adjacent to the disconnecting means for the branch-circuit power conductors supplying the controller and motor. This rule identifies the two sources of power that make up the power and control circuits.

Table 430-72(b) lists the methods and procedures for selecting the protection device for a particular size control circuit conductor. Column A lists the conductor ampacity rating. Column B lists the maximum protection with the control conductors located in the control enclosure. Column C lists the maximum protection with the control conductors in a remote location relative to the control enclosure.

Protection for Control Circuits—430-72(b)

Motor control circuit conductors tapped from a motor power circuit larger than #14 AWG are protected according to the conductor ampacity in Tables 310-16 through 310-19. The overcurrent protection shall not exceed 10 amps for a #16 conductor and 7 amps for a #18 conductor. This requirement will no longer permit a 15-amp or 20-amp overcurrent protection device to protect #18 or #16 conductors used for control circuits.

Circuit loads may require overcurrent protection smaller than 15-amp; however, 15 amps is the smallest circuit breaker rating. Fuses are available in 1-amp, 3-amp, 6-amp, and 10-amp sizes per 240-6, Exception.

430-72(b), Ex. 1. Motor control circuit conductors that do not extend beyond the control equipment enclosure are considered to be protected by the motor power circuit fuses or circuit breaker, provided the devices do not exceed 400% of the ampacity rating of size #14 and larger conductors. Smaller size conductors, such as #16, may be protected at 40 amps, and size #18 conductors at 25 amps. Notice that the ampacity ratings for the control conductors are taken from the free air ampacities of Table 310-17 for 60°C wire. This is allowed because control conductors are installed in open enclosures instead of enclosed raceways. Therefore, they have more free space to dissipate heat.

Table 430-72(b), Column B, Exception 1 lists the maximum size overcurrent protection devices permitted for controller conductors rated #14 to #10. See Figure 6-1. *NOTE*: The overcurrent protection devices must be the next size rating below the values in Table 430-72(b), Column B, Exception 1.

Rating of OCPD's for Settings Listed in Table 430-72(b), Col. B, Ex. 1		
Size Wire	Size Device	
	Copper	Alum.
# 14	100 A	100 A
# 12	110 A	100 A
# 10	150 A	125 A

Table 430-72(b)

Figure 6-1. Overcurrent protection devices cannot exceed 400% of the control-circuit conductor's ampacity per Table 310-17.

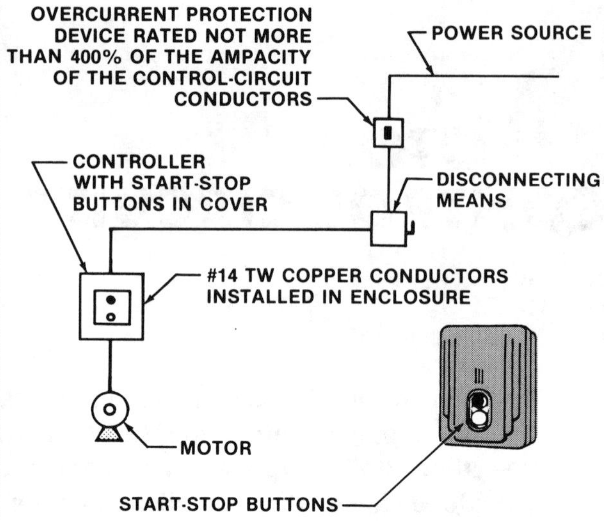

430-72(b), Ex. 1
Table 430-72(b), Col. B, Ex. 1

PROBLEM: What size overcurrent protection device is required to protect the control conductors inside the enclosure of the controller?

Step 1: *Table 310-17, Col. 1.*
#14 TW copper has a 25-A rating.

Step 2: *430-72(b), Ex. 1.*

25 A × 400% = 100 A

Answer: *240-6.*
100-A maximum fuse, cb, or fuse block

Figure 6-2. Sizing an overcurrent protection device for conductors inside the controller enclosure.

Start-and-stop stations are installed on the front covers of motor control centers or magnetic starters. Power conductors sized at 125% of the motor's full-load current are run to the various motors. See Figure 6-2.

If the 100-amp overcurrent protection device is increased to 150 amps to allow the motor to start and run, the control conductors would be selected as follows:

1. #12 would allow 110-amp device
   Violation
2. #10 would allow 150-amp device
   Permitted

430-72(b), Ex. 2. This exception applies to control conductors #14 AWG and larger that extend beyond the enclosure for the coil of the equipment being controlled. When the motor control circuit conductors are tapped from the line terminals of a magnetic starter and run to remote areas for connection to control switches and devices, the overcurrent protection devices cannot be larger than 300% of the ampacity of the control conductors used.

Table 430-72(b), Column C, Exception 2 lists

the maximum size overcurrent protection device permitted for controller conductors rated #14 through #10. Column C is applied for control conductors used for remote-control systems. See Figure 6-3. *NOTE:* The overcurrent protection device must be the next size rating below the values in Table 430-72(b), Column C, Exception 2.

Remote-control devices are located in various locations to start and stop a motor or motors. Remote-control conductors routed in conduit are sometimes run a great distance from the motor control center or magnetic starter. For this reason, the overcurrent protection device for the power circuit must be limited to 300% of the control conductor's ampacity rating per Table 310-16, Column 2.

Column C, Exception 2		
Size Wire	Size Device	
	Copper	Alum.
# 14	45 A	45 A
# 12	60 A	45 A
# 10	90 A	70 A

Table 430-72(b)

Figure 6-3. Overcurrent protection devices cannot exceed 300% of the control-circuit conductor's ampacity per Table 310-16.

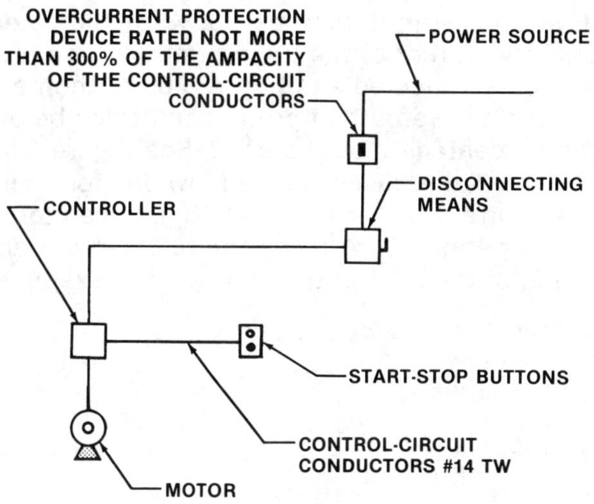

Table 430-72(b), Col. C, Ex. 2
430-72(b), Ex. 2

PROBLEM: What size overcurrent protection device is required to protect the control-circuit conductors of a remote-control circuit?

Step 1: *Table 310-16, Col. 1.*
#14 TW copper has a 20-A rating.

Step 2: *430-72(b), Ex. 2.*
15 A × 300% = 45 A

Answer: *240-6.*
45-A maximum fuse, cb, or fuse block with 15-A protection

Figure 6-4. Sizing an overcurrent protection device for conductors of a remote-control circuit.

If the overcurrent protection device for the power circuit exceeds 300% of the ampacity of the control conductors, the conductors must be protected at their individual ampacity ratings. For example, if the power circuit protection exceeds 300%, #14 control conductors must be protected with a 15-amp overcurrent protection device per the footnote to Table 310-16. See Figure 6-4. *NOTE*: Either individual overcurrent protection must be provided for the control conductors or the size of the conductors must be increased to permit a larger protection device.

430-72(b), Ex. 3. The protection on the primary side of the transformer may also protect the secondary conductors of the coil circuit. The protection for the transformer itself must be provided in accordance with 450-3. Section 240-3(i) allows only a two-wire secondary to use this type of protection. Generally, this rule applies where the transformer is installed within the starter enclosure. However, any installation of a transformer conforming to this rule meets the NEC requirement and may be utilized.

The protection must be in accordance with 450-3 and must not exceed the value determined by multiplying the secondary conductor ampacity by the secondary-to-primary voltage ratio. When the primary current is 9 amps or more, and 125% of the current does not correspond to a standard size breaker, the next higher size is permitted.

When the primary current is less than 9 amps but is 2 amps or more, the overcurrent protection device must not be more than 167% of the rated primary current. When the primary current is less than 2 amps, the overcurrent protection device is permitted to be 300% of the rated primary current of the transformer. Where protection is selected at 167% or 300%, the next lower size overcurrent protection device below the percentage must be used. See Figure 6-5.

If the 3-amp fuses should blow due to the inrush current of the coil, 430-72(c), Exception 2 would permit 500% times the full-load current rating of the control transformer. For example:

Step 1: *430-72(c), Ex. 2.*

$$1.67 \text{ A} \times 500\% = 8.35 \text{ A}$$

Step 2: *240-6, Ex.*
6-A fuses

Answer: **6-A fuses are permitted.**

CONTROL-CIRCUIT TRANSFORMERS

Protection for control-circuit transformers must be in accordance with Article 450. Section 430-72(c), Exception 3 or 4 permits a fuse or circuit breaker in the secondary circuit for the control transformer. Protection must be selected by the requirements of 450-3(b)(2). The overcurrent protection device supplying the motor circuit cannot exceed 250% of the motor's full-load current rating to apply this method of protection. Protection must be placed in each ungrounded hot leg. See Figure 6-6.

Control transformers rated less than 50 volt-amps that are located in the controller enclosure and an integral part of the controller require no further protection. The transformer is considered protected by the motor circuit's overcurrent protection device.

The full-load current rating of the control transformer is permitted by 725-11(a) to be increased to 167%. However, this percentage is limited to control transformers rated at 1000 volt-amps or less and 30 volts or less.

Control transformers with limited power sources are required by 725-11(a)(1) to be protected per 450-3(b)(2). Section 430-72(c), Ex. 3 refers to 725-11(a)(1), which refers to 450-3(b)(2) for selecting and placing overcurrent protection devices in the secondary of Class 1 control transformers.

DISCONNECTION OF MOTOR CONTROLS AND MOTOR POWER CIRCUIT CONDUCTORS—430-74

Motor control circuits must be disconnected from all sources of supply when the disconnecting means is in the open position. If the motor

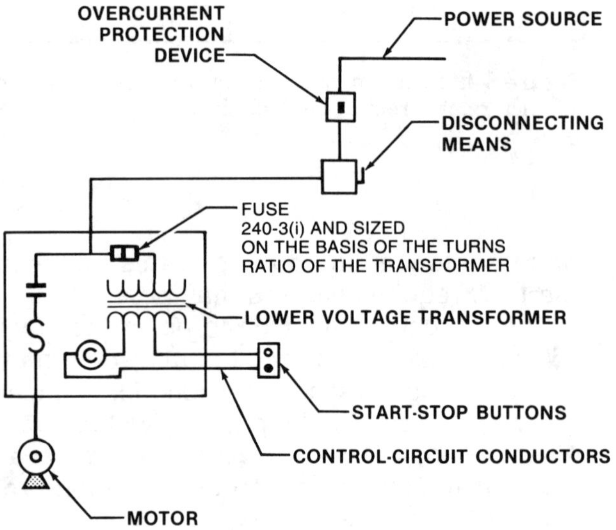

430-72(b), Ex. 3;
430-72(c), Ex. 2

PROBLEM: What size overcurrent protection device is required on the primary side of an 800-VA, two-wire, 480-V transformer?

Step 1: *430-72(b), Ex. 3; 450-3(b)(1), Ex. 1.*

$$I = \frac{VA}{V}$$

$$\frac{800 \text{ VA}}{480 \text{ V}} = 1.67 \text{ A}$$

Step 2: *450-3(b)(1), Ex. 1.*

$$1.67 \text{ A} \times 300\% = 5.01 \text{ A}$$

Answer: *240-6, Ex.*
3 A fuse

NOTE: 500% × primary FLC may be used per 430-72(c), Ex. 2.

Figure 6-5. Sizing an overcurrent protection device for the primary side of a transformer.

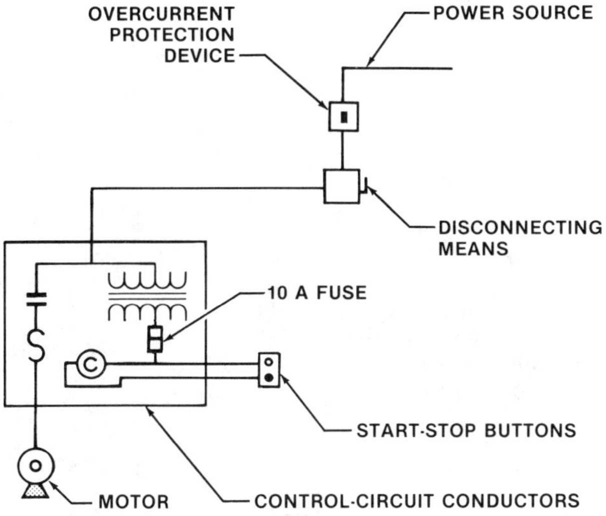

240-6; 430-72(c), Ex. 3 or 4
450-3(b)(2); 725-11(a)

PROBLEM: What size overcurrent protection device is required on the secondary side of a 200-VA, two-wire, 30-V transformer?

Step 1: *430-72(c), Ex. 3 or 4; 725-11(a).*

$$I = \frac{VA}{V}$$

$$\frac{200 \text{ VA}}{30 \text{ V}} = 6.67 \text{ A}$$

Step 2: *430-72(c), Ex. 3 or 4; 725-11(a).*

$$6.67 \text{ A} \times 167\% = 11 \text{ A}$$

Answer: *240-6, Ex.*
10 A fuse

Figure 6-6. Sizing an overcurrent protection device for the secondary side of a transformer.

control circuit is tapped from the line terminals of the magnetic starter, the disconnecting means for the starter may serve as the disconnecting means for both the motor power conductors and control circuit conductors.

If the motor control circuit is fed from another source and not tapped from the starter conductors, either an auxiliary contact must be installed in the disconnecting means for the controller or an additional disconnecting means must be mounted adjacent to that for the controller. See Figure 6-7. This requirement is to prevent shocking a worker servicing the controller. The two disconnects properly identifying power supply and power control should make the worker aware of the two power sources.

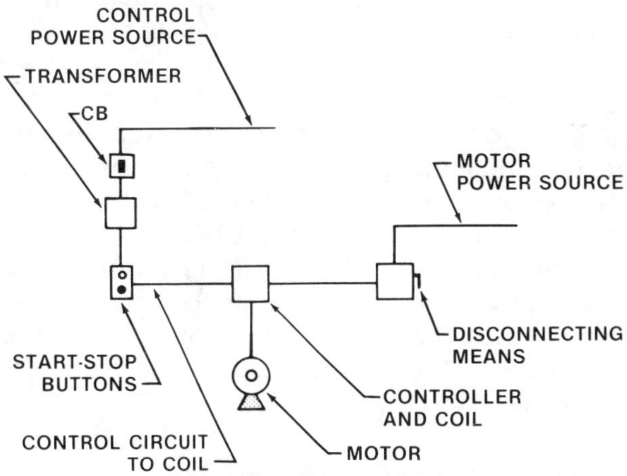

430-74

Figure 6-7. Motor control circuits must be disconnected from all supply sources when the disconnecting means is in the open position.

DESIGNING CONTROL CIRCUITS IN RACEWAYS—300-3(c) AND 725-15

Section 300-3(c) allows conductors of different systems to occupy the same raceway without regard to alternating current or direct current. All conductors in the raceway must be insulated for the maximum voltage of any one conductor, which must not exceed 600 volts. Section 725-15 allows the motor control conductors of a Class 1 circuit to occupy the raceway along with the power conductors supplying the magnetic starter and motor. If the power and control conductors are routed in the same raceway and supply the same motor, a short circuit in any conductor could damage the insulation of other conductors. However, only the one motor would be affected. See Figure 6-8.

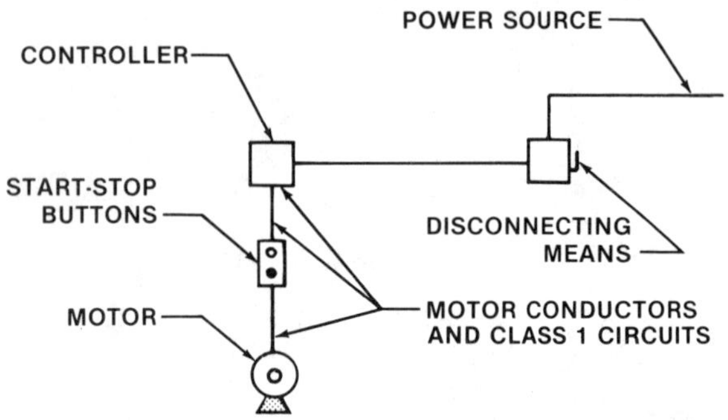

300-3(c); 725-15

Figure 6-8. Class 1 control circuits of 600 volts or less may occupy a raceway with power conductors.

Power and control conductors can occupy the same raceway system when supplying a number of controllers and motors. However, they must supply motors that are *functionally associated*. That is, the motors must work together to perform a job. If the motors are functionally associated and operate together, all the motors must operate to perform the work. If one of the motors were lost due to problems, the other motors would be shut down. If the motors do not operate together to drive loads to perform a certain job the power and control conductors must not occupy the same raceway system. See Figure 6-9.

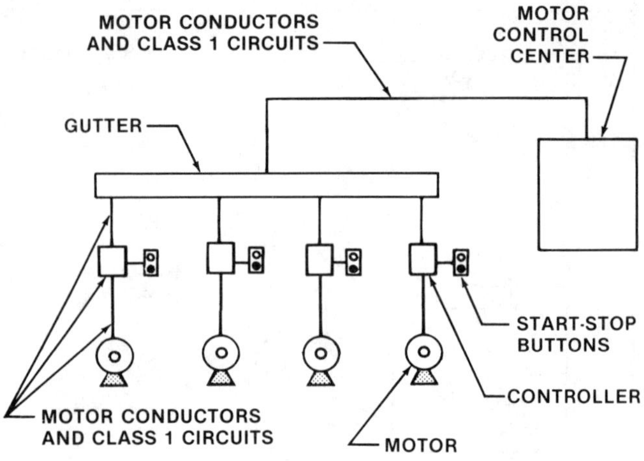

300-3(c)(1)

Figure 6-9. Class 1 control conductors and motor conductors, each of 600 volts or less, may occupy the same raceway to supply functionally associated motors.

CONDUCTORS OCCUPYING SAME ENCLOSURE—300-3(c)(1)(2) AND 300-32, EX. 1

Motor excitation, control relay, magnetic starter, or ammeter conductors rated at 600 volts or less may occupy the same enclosure. Also, conductors rated over 600 volts may occupy the same enclosure. However, a combination of conductors 600 volts or less with conductors over 600 volts must not occupy the same raceway. The motor enclosure and the starter enclosure may have excitation, control, relay, or ammeter conductors 600 volts or less together with power conductors over 600 volts. See Figure 6-10.

DESIGNING CONTROL CIRCUITS FOR MAGNETIC STARTER CONTACTOR AND ENCLOSURES—725-17

When the number of Class 1 remote-control, signal, and power-limited circuit conductors in a conduit exceeds three, and when they carry

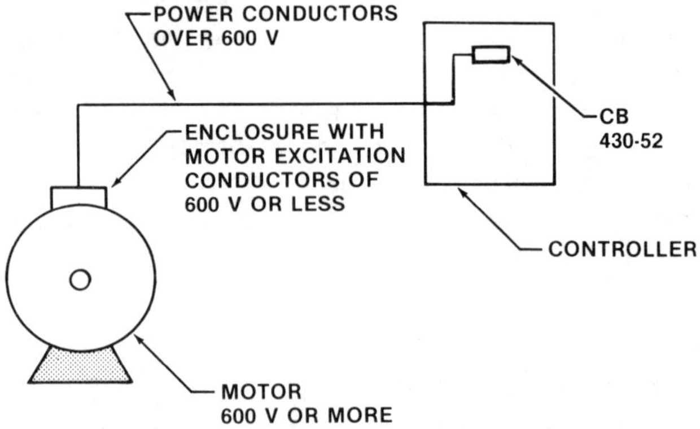

300-3(c)(2)
300-32, Ex. 1

Figure 6-10. Power circuits of 600 volts or more may occupy enclosures with control or monitoring circuits of 600 volts or less.

current continuously as permitted by 725-15, the derating factor of Note 8(a) of Notes to Ampacity Tables of 0 to 2000 Volts apply. The derating factors of Note 8(a) do not apply to conductors if the load is noncontinuous. If the number of conductors exceeds three, the load is considered continuous when the conductors carry current for three hours or more. If power and control conductors are in the same conduit, Note 8(a) applies to all current-carrying conductors that are of the continuous-duty type. Class 1 circuits may occupy the same enclosure as power conductors if they supply the same equipment and if that equipment is functionally associated.

Section 725-17 is a very controversial section of the Code. The authority having jurisdiction may not require the control conductors to be derated because they usually carry no more than 1 amp of current or less. Some inspectors feel that the heat generated by 1 amp of current is not enough to contribute excess heating to other conductors in the raceway. Other inspectors feel that 1 amp of current can produce heat ($I^2 \times R$) that will push the temperature above the limit for a particular insulation. Local interpretation must be checked before applying the rules of 725-17. See Figure 6-11. Derate conductors only if they carry continuous loads in excess of 10% of the ampacity of each conductor.

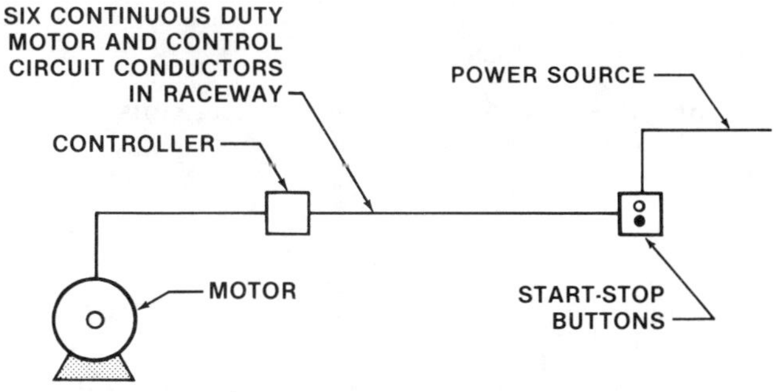

Figure 6-11. Note 8(a) of Notes to Ampacity Tables of 0 to 2000 Volts must be applied when continuous-duty conductors occupy the same raceway, and 10% of the ampacity of each control circuit conductor is exceeded.

Class 1 Circuits—725-11(a)(b)

Class 1 circuits are two types: *power-limited* and *power-unlimited*. In the power-limited type the voltage is limited to 30 volts or less at 1000 volt-amps. Class 1 circuits are used if the remote-control circuit is safety-control equipment that could fail to operate and create a hazard, even if the remote-control circuit would normally be classified as a Class 2 or 3 circuit. See Figure 6-12. Class 1 power-limited circuits are derived from the secondary of step-down transformers usually located in the enclosures of magnetic starters or motor control centers.

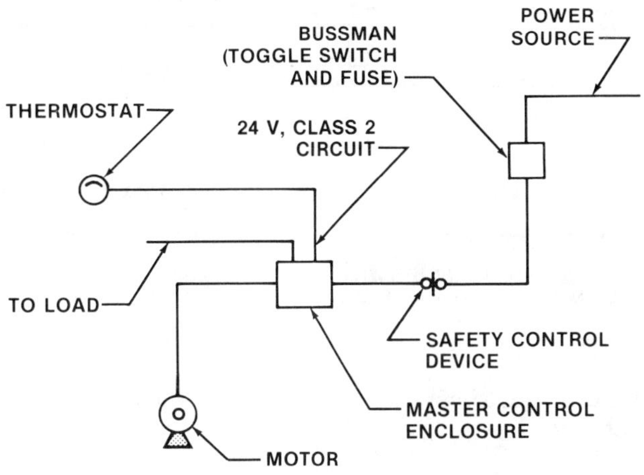

Figure 6-12. If the safety-control device fails to operate, creating a hazard, the circuit would be a Class 1 circuit instead of a Class 2 circuit.

Class 1 power-unlimited circuits are derived from the power circuit that supplies the controller and motor. The supply circuit voltage can be 120 volts to 600 volts. The operating voltages to the motor can be 600, 575, 550, 480, 240, 230, 220, 208, 120, 115, or 110 volts per 725-11(b). The coil control voltage to the start and stop buttons can be tapped from any of the supply voltages listed. See Figure 6-13.

Class 2 Circuits—Table 725-31(a)

Class 2 circuits are considered safe regarding both fire and electrical shock. In most cases, their currents of 5 milliamps or less would not cause damage should a ground fault occur because of the high-impedance winding of the transformer supplying their power. Class 2 circuits are limited to 30 volts or less and a maximum of 100 volt-amps. See Table 725-31(a) for power limitations of transformers for a Class 2 circuit. Class 2 circuits not inherently current-limited must have overcurrent protection. Class 2 circuits are not rigidly covered by the NEC and the wiring from the transformer to the controls may be bell wire or other suitable types of wiring. *NOTE:* Some Class 2 circuits operate at 5 amps or less.

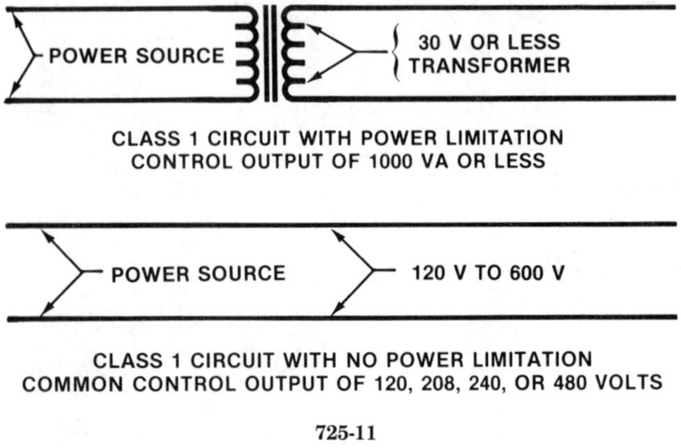

Figure 6-13. Class 1 circuits are power-limited or power-unlimited.

Class 2 circuits must not be installed in raceways with Class 1 circuits and power circuits because of the possibility of an inadvertent interconnection of the systems. Electricians and maintenance personnel could accidentally change a joint on conductors in an enclosure or junction box and put a high voltage on the low-voltage system, destroying components. Class 2 and Class 3 circuits may occupy the same raceway system provided the Class 2 circuit conductors are insulated to the extent required for Class 3 circuits. See Figure 6-14.

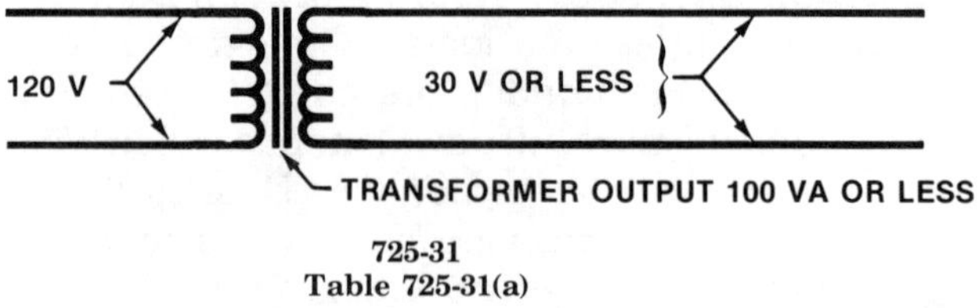

Figure 6-14. Class 2 power-limited circuits control low-voltage systems.

Class 3 Circuits—725-31

Class 3 circuits are considered hazardous regarding both fire and shock. They require additional safeguards because AC circuits from 30 volts to 150 volts limit the current to about 1 amp. See Table 725-31(a) for power limitations of transformers for Class 3 circuits. See Figure 6-15. *NOTE:* Class 3 circuits can operate with an ampacity exceeding 1 amp.

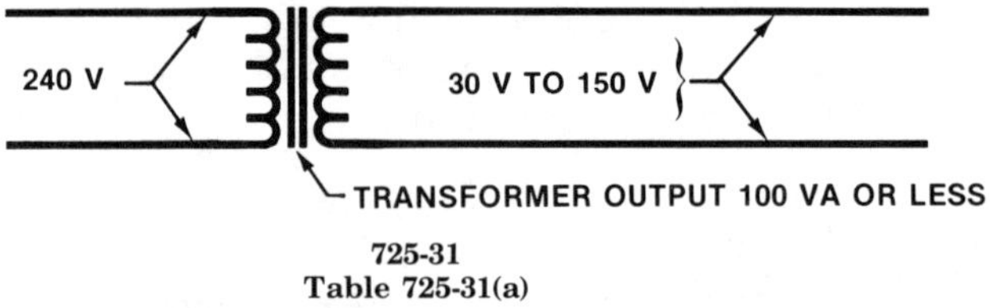

Figure 6-15. Class 3 power-limited circuits control 30-volt to 150-volt systems.

REVIEW—CHAPTER 6

Name Date

True-False

T F **1.** Control circuits tapped from the line side of the magnetic starter controlling a motor must comply with 430-72.

T F **2.** Remote-control circuits must have a disconnecting means installed adjacent to the disconnecting means for the power circuit to the motor.

T F **3.** Column B of Table 430-72(b) lists the ampacity of each conductor used for control circuits.

T F **4.** Column C of Table 430-72(b) lists the maximum protection for control conductors located remotely from the control enclosure.

T F **5.** The overcurrent protection device serving a motor circuit must not exceed 300% of the control conductor's rating where the buttons are mounted on the cover of the magnetic starter.

T F **6.** If the start and stop stations are mounted remotely from the starter, the ampacity of the conductors should be selected from Table 310-17.

T F **7.** Overcurrent protection devices protecting conductors inside magnetic starters that provide power to control circuits must not exceed 300% of the conductor's ampacity rating.

T F **8.** Overcurrent protection devices located on the primary side of transformers can be used to protect the secondary conductors where a two-wire to three-wire system is used.

T F **9.** Transformers having a primary current of 9 amps or more can have their overcurrent protection devices rated at 125% of the primary full-load current.

T F **10.** Transformers rated less than 50 volt-amps and used for control are not required to be protected by Article 450.

T F **11.** Transformers with a primary current less than 2 amps can have overcurrent protection devices selected at 550% of the primary full-load current.

T F **12.** Conductors derived from different voltage systems rated at 600 volts or less must never occupy the same raceway.

T F **13.** Conductors rated at 600 volts or less can be routed in the same raceway if they power and control the same motor and the control circuit is Class 1.

T F **14.** Class 1 circuits can occupy the same raceway as the power conductors if the motors served are functionally associated.

T F **15.** If Classes 2 and 3 circuits occupy the same raceway, they must be insulated for the highest voltage used.

Multiple Choice

_______ 1. No. 18 copper conductors used as control circuits have an ampacity of _______ amps.
A. 5
B. 7
C. 9
D. 10

_______ 2. No. 14 conductors used as control circuits to supply start-and-stop buttons mounted on the cover of a starter can be protected by an overcurrent protection device rated at _______ amps or less.
A. 100
B. 110
C. 125
D. 150

_______ 3. Remote-control circuit conductors of #12 copper can be protected at _______ amps or less.
A. 45
B. 50
C. 60
D. 70

_______ 4. The overcurrent protection devices must not exceed 167% of the primary full-load current for transformers rated less than _______ amps but _______ amps or more.
A. 9; 2
B. 10; 3
C. 12; 4
D. 15; 6

_______ 5. Conductors routed inside the enclosure of magnetic starters are figured from Table _______.
A. 310-16
B. 310-17
C. 310-19
D. 430-7(b)

_______ 6. No. 14 copper conductors used for remote-control circuits may be tapped from the load side of a starter supplied by a _______ amp or less device.
A. 45
B. 50
C. 60
D. 70

_______ 7. The maximum size overcurrent protection device that can be used to protect #10 aluminum conductors used for remote control is _______ amps or less.
A. 70
B. 80
C. 90
D. 100

__________ **8.** A Class 1 circuit operates at 30 volts or less and __________ volt-amps or less.
A. 50
B. 75
C. 100
D. 1000

__________ **9.** Overcurrent protection devices are computed at 300% of the values specified in Table 310-16 for __________ °C control conductor routed remotely.
A. 60
B. 75
C. 85
D. 90

__________ **10.** The ampacity value of conductors No. 18 or 16 is determined from Table _______ when used as control circuit conductors.
A. 430-7(b)
B. 430-72(b)
C. 430-148
D. 430-150

Problems

1. What is the maximum size overcurrent protection device permitted for three #14 copper conductors used for remote-control circuits?

2. What is the maximum size overcurrent protection device allowed for #12 copper conductors routed to start-and-stop stations in a motor control center?

3. What is the maximum size overcurrent protection device permitted for #16 copper conductors used for remote-control circuits?

4. What is the output, in amps, of a 50-VA control transformer used on 24 volts?

5. What is the output, in amps, of a 240-volt, 2.5-kVA control transformer mounted in a motor control center?

6. What size overcurrent protection device is required in the secondary side of a 1000-VA control transformer using 120-volt control circuits?

7. A 60-amp circuit breaker provides overcurrent protection for #12 copper conductors used for control circuits. If the 60-amp device trips open due to inrush current of the motor, what size copper conductors are required to increase the overcurrent protection device to 80 amps?

8. What size overcurrent protection device is required for #18 control circuits routed inside the enclosure of the starter?

9. What is the maximum size overcurrent protection device permitted to be installed on the primary side of a 480-volt, 2000-VA control transformer?

10. What is the maximum size overcurrent protection device permitted for a 100-VA control transformer with a 120-volt primary?

MOTOR CONNECTIONS AND TESTING

Chapter 7

Windings are connected in series for higher voltages and in parallel for lower voltages. Dual-voltage motors are available to match the supply voltage of the system to which they are connected. Winding leads are brought from the stator to the termination box on the motor to be connected to the supply voltage. Leads in the termination box are connected to the supply voltage in series or parallel to match the system voltage. Leads can be tested for continuity or the motor's rotation can be reversed by interchanging any two of the leads.

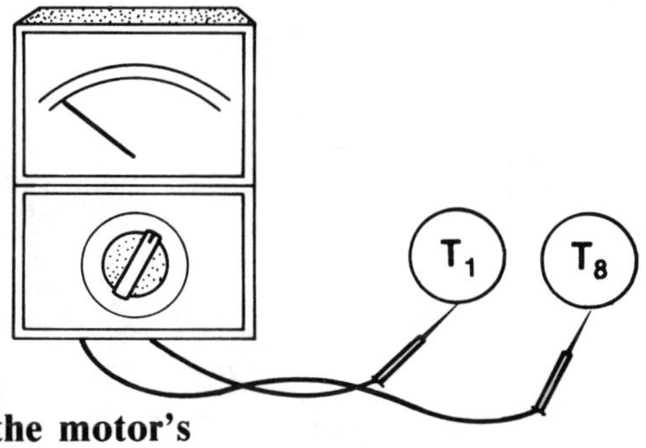

CONNECTING MOTORS

Motor windings can be connected for low or high voltage supply. Common windings are 120/208-volt, single-phase; 120/240-volt, single-phase; or 240/480-volt, three-phase.

Nameplates provide valuable information about the motor. The following information is listed on a motor's nameplate per 430-7(a).

- manufacturer's name
- voltage
- FLC
- frequency (if AC motor)
- number of phases
- rpm
- temperature rise
- class
- service factor (if assigned)
- hp (if ⅛ or more)
- type of motor
- plus additional information, depending upon type of motor

Single-phase Motors

Single-phase motors are connected to the power supply by connecting two stator lead wires to two supply wires. The electrician must make sure that the motor windings are connected in series for high voltage (240 V) or in parallel for low voltage (120 V) if the motor is dual-voltage. See Figure 7-1. The connections shown in Figure 7-1 are typical for the winding connections for single-phase operation on common voltages supplying motors.

The windings of dual-voltage (120/240-volt) motors are marked with one of the following methods:

1. The windings are labeled with an R, which represents the running winding.
2. The windings are labeled red for one set of windings and blue for the other set of running windings. However, the running windings can have any color to identify each set.

Windings. The windings of a single-phase dual-voltage motor can be connected in parallel and the motor will operate only on a 120-volt supply. The windings can also be connected in series and the motor will operate only on a 240-volt supply. See Figure 7-2.

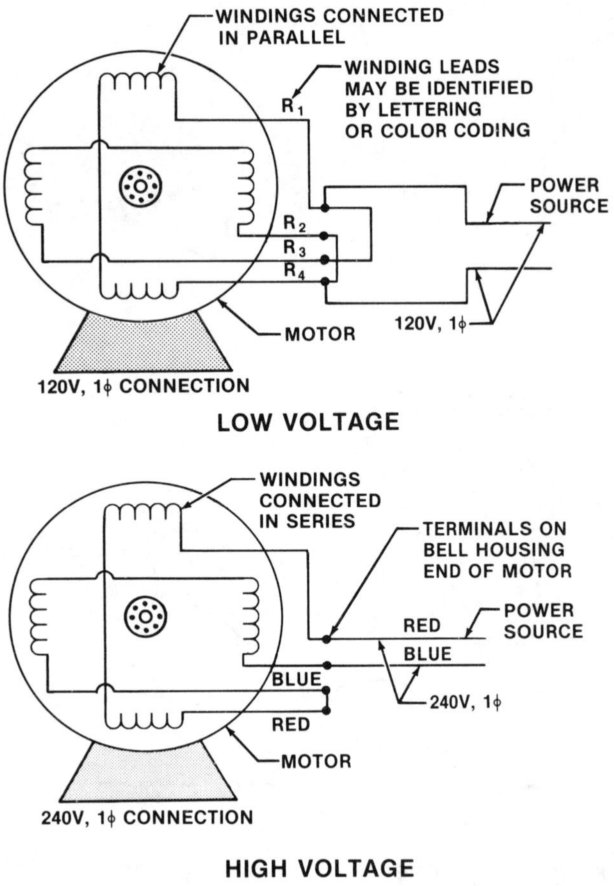

Figure 7-1. Connecting dual-voltage motor windings for low or high voltage operation.

The number of field poles determines the speeds of the motor. For a motor wound with six poles, the motor runs in the slow speed using six of the poles. The motor runs in the medium speed using four of the poles and in the fast speed when using only two of the six poles. See Figure 7-3.

Different speeds for a motor can also be obtained by using an auxiliary winding. The speeds are obtained by using parts of the auxiliary winding. All of the winding is used for the slow speed. Half of the winding is used for medium speed. The winding is not used at all for fast speed.

Starting Windings. Single-phase 120-volt and 240-volt motors must be provided with a means of starting the rotor in motion to catch the alternating magnetic fields of the field poles. A starter winding is connected in parallel with the running windings. A centrifugal switch disconnects the starting winding when the motor reaches its running speed. When a centrifugal switch disconnects the starter winding, the motor operates on the running windings only. Refer to Figure 7-3. Note that when the centrifugal switch is open, the starting winding is taken out of the circuit.

Three-Phase Motors

Three-phase motors that are designed for dual voltage are equipped with nine leads to connect for low or high voltage operation. The electrician connects the power leads to the stator leads by the schematic diagram on the motor's nameplate. If the nameplate is missing or unreadable, the electrician must identify the leads and connect them to match the supply voltage. Such leads may be identified through the use of an ohmmeter and a voltmeter. *NOTE*: A test light (continuity tester) may be utilized when testing the resistance of some windings.

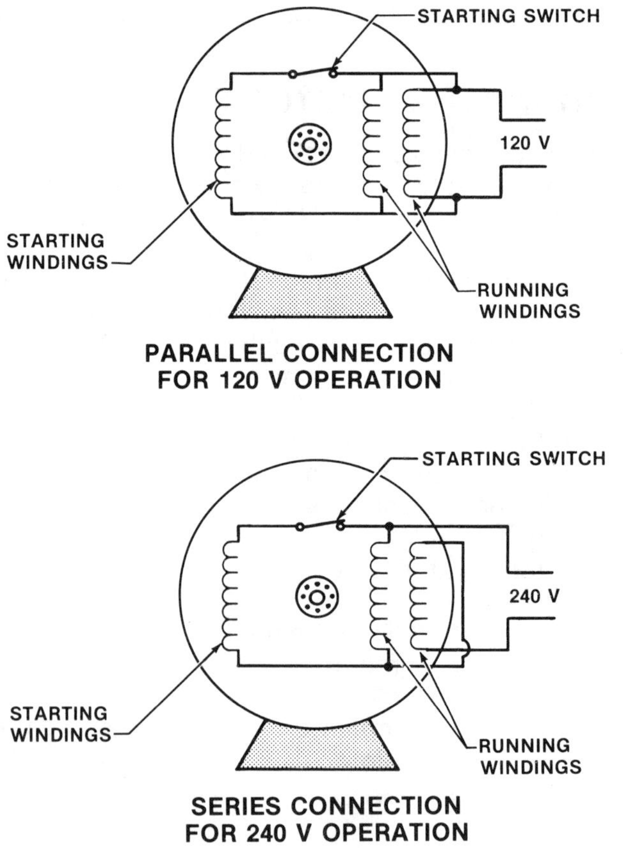

Figure 7-2. Connecting the windings of a motor for 120-volt or 240-volt operation.

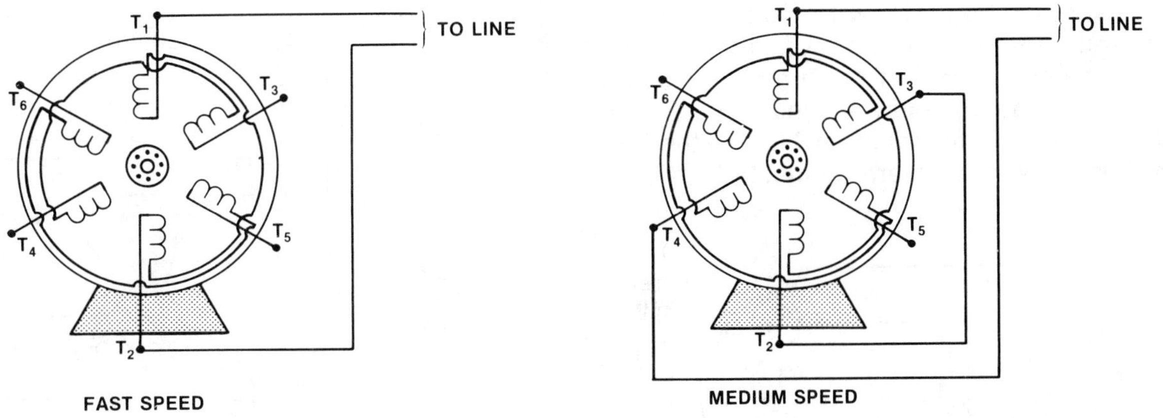

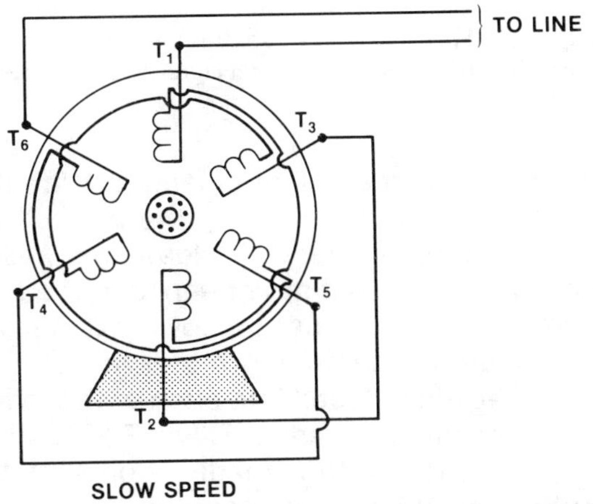

CONNECTING POLE WINDINGS FOR VARIOUS SPEEDS		
NO. OF POLES	SPEED	CONNECTIONS
Two	Fast	T_1 and T_2 to line
Four	Medium	T_2 to T_3 T_1 and T_4 to line
Six	Slow	T_2 to T_3 T_4 to T_5 T_1 and T_6 to line

VARYING MOTOR SPEED USING POLES

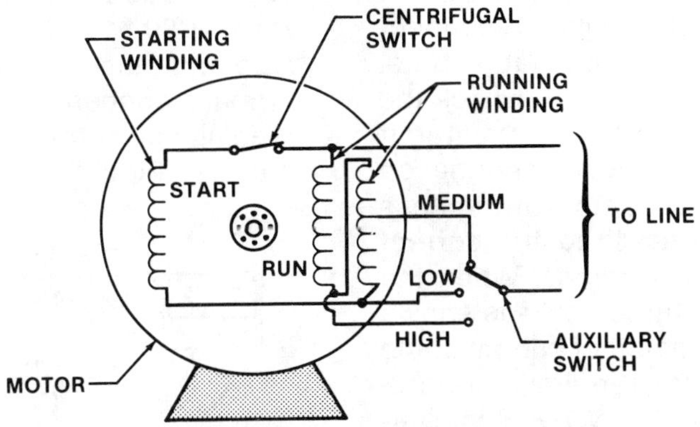

VARYING MOTOR SPEED BY TAPPING AN AUXILIARY WINDING

Figure 7-3. Motor speed is varied by the number of poles used or by tapping an auxiliary winding.

Wye-connected Windings. A wye-connected motor has one set of three leads that shows continuity when using an ohmmeter or test light. The wye-connected motor also has three sets of two leads that show continuity. Identify and mark the set of three leads with the numbers T_7, T_8, and T_9. The other motor leads with two leads each must be separately marked T_1 and T_4, T_2 and T_5, and T_3 and T_6. See Figure 7-4.

To connect the motor for high voltage, connect T_4 to T_7, T_5 to T_8, and T_6 to T_9. If the motor is wound for 240-volt or 480-volt operation, connect 480 volts to the motor terminals without a load being applied to the motor. Check the running current of the motor with a clamp-on ammeter. If the three hot legs are carrying approximately the same amount of current, the leads are marked properly and the windings are connected correctly. This will rarely happen the first time that the windings are connected.

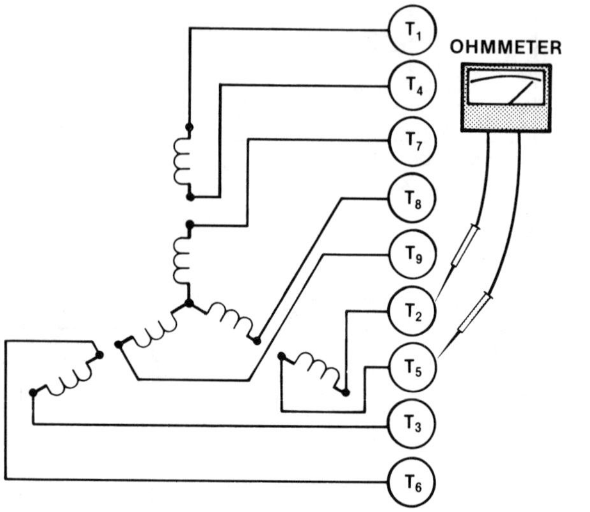

Figure 7-4. Locating each winding in a wye-connected motor with an ohmmeter. A test light could also be used.

If the motor does not run properly and current on L_1, L_2, and L_3 are not near normal, reconnect the three sets of two leads, one at a time. Reverse T_1 and T_4 and connect to T_7. Check the current once again with a clamp-on ammeter. If the current is not fairly even and near normal on L_1, L_2, and L_3, reverse T_2 and T_5, etc. This process of elimination ultimately leads to the correct connections of the motor windings. When the amperage on each of the three hot legs shows approximately the same amount of current in the normal operating range, the terminals are properly connected. See Figure 7-5. *NOTE:* Some terminals are usually marked and readable.

Applying Voltage and Identifying Motor Leads. If a three-phase motor is rated at 230/460 volts, a set of windings can be connected to a 230-volt supply and the leads from the windings of the motor will act like the primary and secondary of a transformer by mutual induction of the magnetic fields. The voltage readings can be used to properly mark the leads so that the motor's windings can be connected for operation. The voltage readings can be either additive, sub-

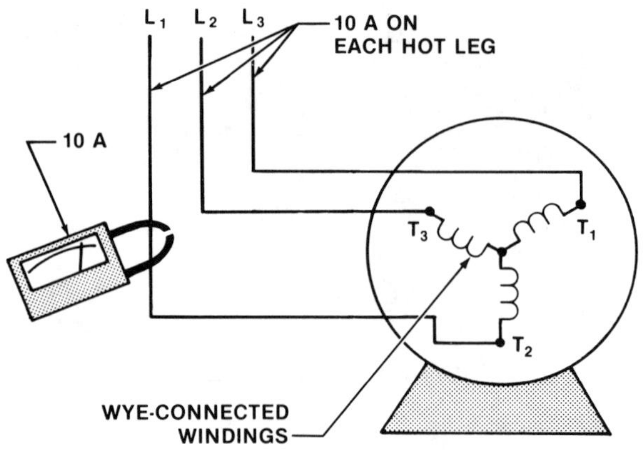

Figure 7-5. If current in each phase is about even and normal, the terminal connections are assumed to be correct.

tractive, balanced, or unbalanced. With the motor not loaded and running the 240-volt supply is connected to the set of leads marked T_7, T_8, and T_9. These leads have been identified with an ohmmeter. Always use the lower voltage for this procedure of identifying motor leads.

Voltage is induced into each of the remaining two-wire circuits marked T_1 and T_4, T_2 and T_5, and T_3 and T_6. The voltage readings on each two circuits should be approximately 125 to 130 volts. The voltage readings may vary from 80 to 125 volts or 90 to 130 volts, depending upon motor type and number of turns on the windings. The primary concern, however, is that the voltage readings be normal and as nearly equal as possible. See Figure 7-6.

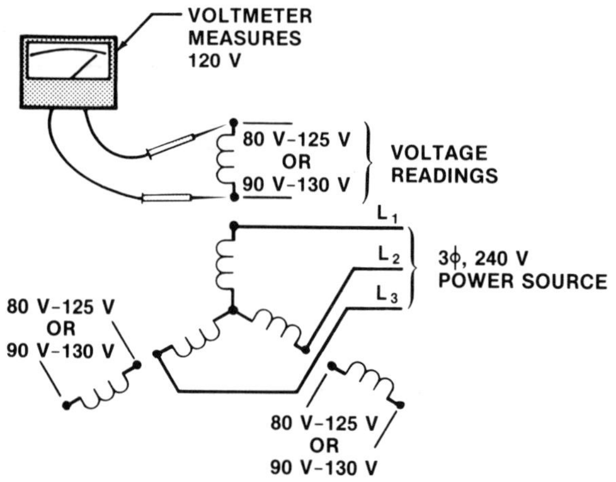

Figure 7-6. The voltage readings between the two-wire circuits of a wye-connected motor are taken with the leads of a voltmeter connected to the leads from each winding.

To determine if the leads are marked correctly, first connect T_4 to T_7 and take voltage readings between T_1 and T_8 and between T_1 and T_9. The voltage readings between the leads should be about 330 to 340 volts, respectively. If these values are the same for both windings, T_1, T_4, and T_7 can be permanently marked. See Figure 7-7.

If the two voltage readings are 125 to 130 volts each, interchange T_1 and T_4 by changing the original T_4 to T_1 and the original T_1 to T_4. The voltage readings will then read correctly and T_1, T_4, and T_7 can be permanently marked. If unequal voltage values are read between T_1 and T_8, and T_1 and T_9, then T_4 must be disconnected from T_7 and connected to T_8. The voltage values should read equal and about 330 to 340 volts between T_1 and T_7, and T_1 and T_9. Refer to Figure 7-7.

The second step is to take one of the remaining two-wire circuits marked T_5 and T_2 and connect T_5 to T_8. The voltage is then checked between T_2, T_7 and T_9. The voltage should read 330 to 340 volts from T_2 to T_7 and T_2 and T_9. The leads must be interchanged until both voltages on T_2 and T_7, and T_2 and T_9 are equal. Each voltage

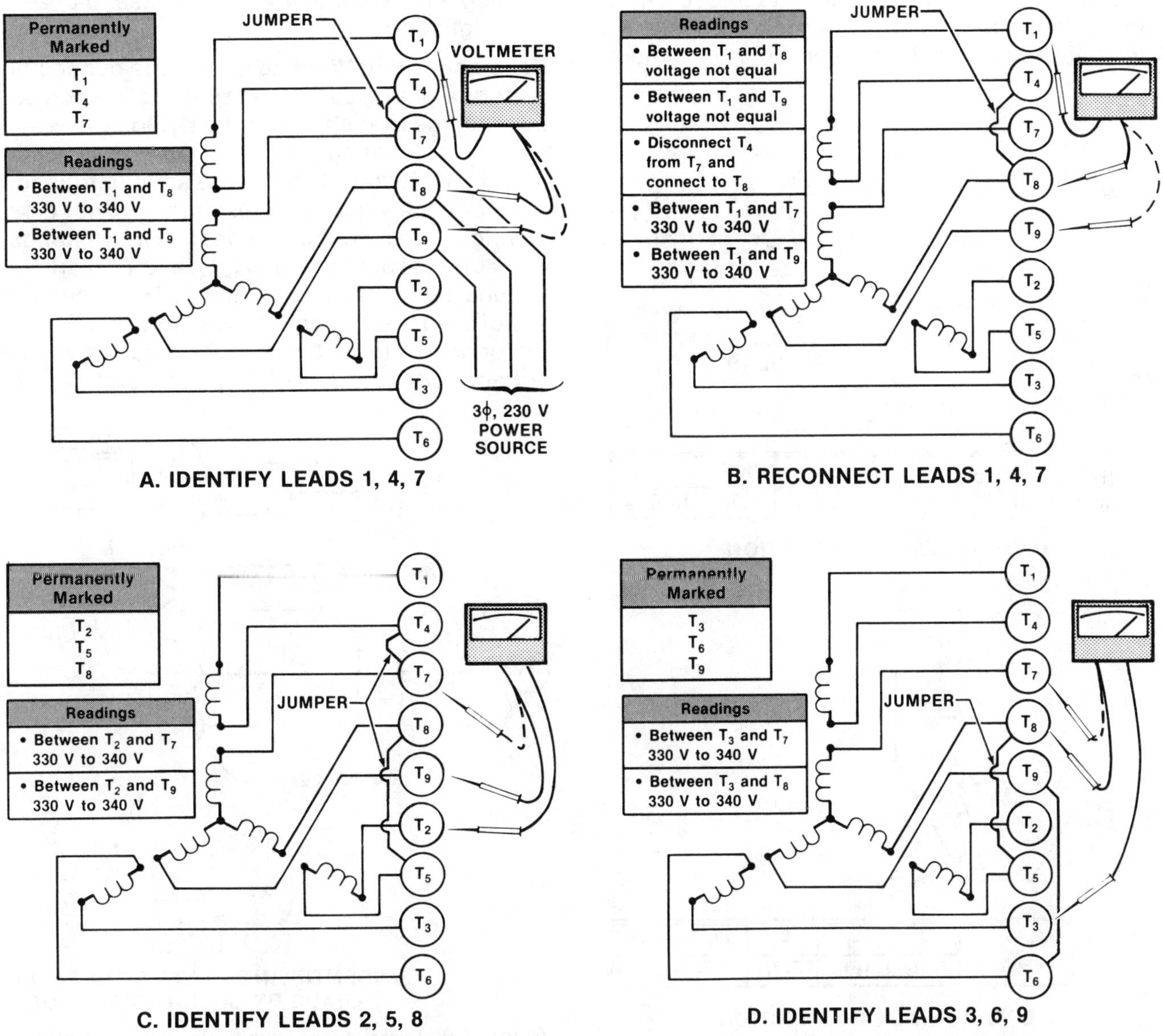

Figure 7-7. Three steps are followed to identify all leads of a wye-connected motor. If initial connections of leads 1, 4, and 7 do not produce proper voltage readings, reconnect these leads before proceeding to next steps.

reading should be equal at approximately 330 to 340 volts. With the voltages equal, T_2, T_5, and T_8 can be permanently marked. Refer to Figure 7-7. Notice that T_7, T_8, and T_9 are connected to the line.

The third step is to take the last two-wire circuit marked T_3 and T_6 and connect T_6 to T_9. Check the voltage reading between T_3 and T_7 and T_3 to T_8. The voltage should read 330 to 340 volts from T_3 to T_7 and T_3 to T_8. If the voltage readings are different than 330 to 340 volts, interchange T_3 and T_6. T_3 will become T_6 and T_6 will become T_3. The voltage readings should now be 330 to 340 volts. Refer to Figure 7-7.

Connect the motor for low voltage operation and with a clamp-on ammeter read the amperage phase-to-phase. If the flow of current (amps) on each phase (L_1, L_2, and L_3) is about even, the motor should be connected properly. Each lead must be marked permanently and each joint made for motor operation. Note that motors with higher voltage ratings must have lower voltage applied to T_7, T_8, and T_9 so the motor leads can be identified safely.

Motors with large horsepower ratings must be reduced. Connect the motor windings for either low or high voltage operation and read the voltage and current. The low voltage reading should be about 230 volts phase-to-phase. The current should be within 10% of the nameplate amperage listed on the motor. The higher voltage reading will be about 460 volts phase-to-phase. See Figure 7-8.

Delta-connected Windings. A delta-connected motor has three sets of three leads in each set of windings that show continuity. Identify each set by first locating the center tap of each winding. The resistance is twice the amount from the ends of each winding. A test light will not glow as bright when connected on the two windings. Mark one set of the leads T_1, T_4, and T_9. Mark the second set of leads T_2, T_5, and T_7. Mark the third set of leads T_3, T_6, and T_8. See Figure 7-9.

Connect T_4 to T_7, T_5 to T_8, T_6 to T_9 for the higher voltage. Connect T_1 to L_1, T_2 to L_2, and T_3 to L_3

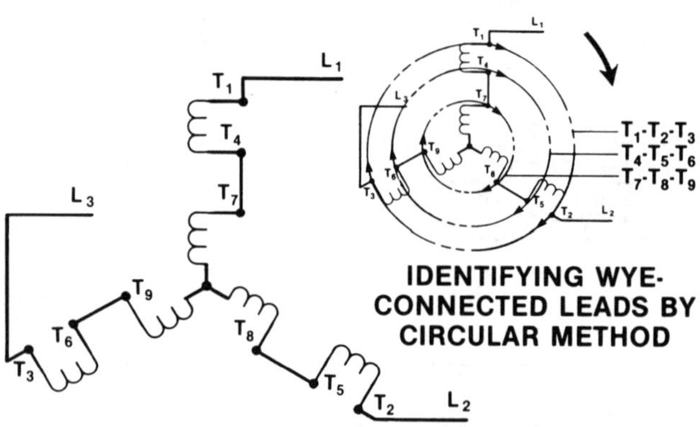

High Voltage	L_1	L_2	L_3	Tie Together
	T_1	T_2	T_3	T_4-T_7, T_5-T_8, T_6-T_9

HIGH VOLTAGE CONNECTION

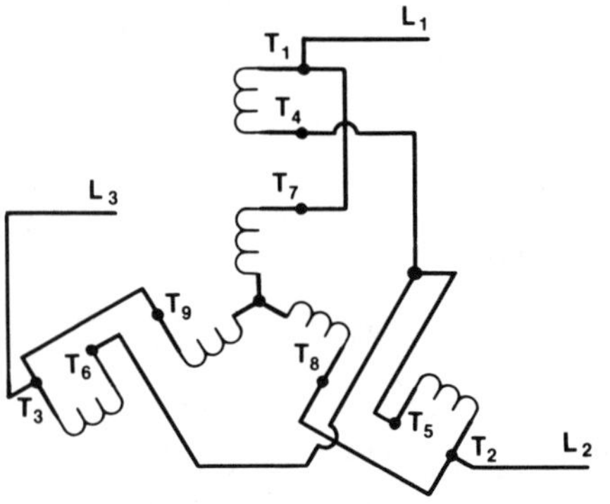

Low Voltage	L_1	L_2	L_3	Tie Together
	T_1-T_7	T_2-T_8	T_3-T_9	T_4-T_5-T_6

LOW VOLTAGE CONNECTION

Figure 7-8. Connecting the leads and windings for low or high voltage operation.

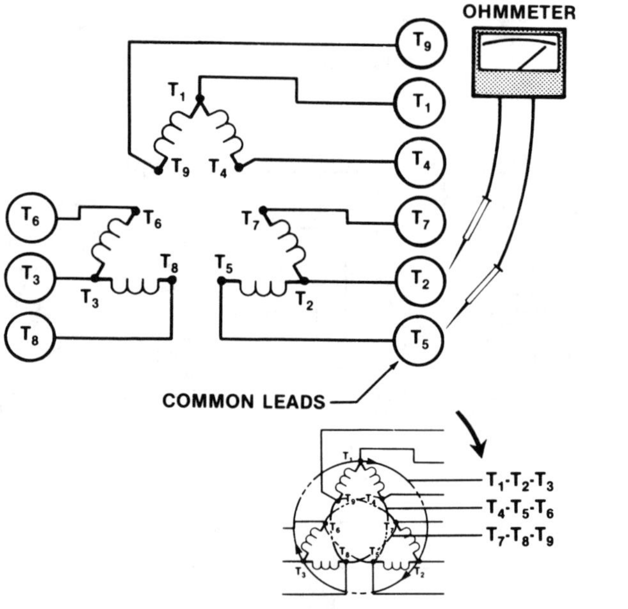

IDENTIFYING DELTA-CONNECTED LEADS BY CIRCULAR METHOD

Figure 7-9. Locating each winding in a delta-connected motor using an ohmmeter. Some of the leads are readable.

to bring the supply voltage to the motor. Apply 480 volts to the motor leads without a motor load. If the current is about even and normal the motor leads are connected properly. This rarely happens the first time the leads are connected. If the current is not even and normal, reverse T_9 and T_4 and reconnect. If the current flow is not even, reverse T_7 and T_5 and reconnect. Check current flow and, if the current is not even, reverse T_6 and T_8. The motor windings should now be connected properly for 480-volt operation. Check the current flow and if it is about even and normal connect the power supply and load permanently. See Figure 7-10. *NOTE:* Some of the leads are usually marked.

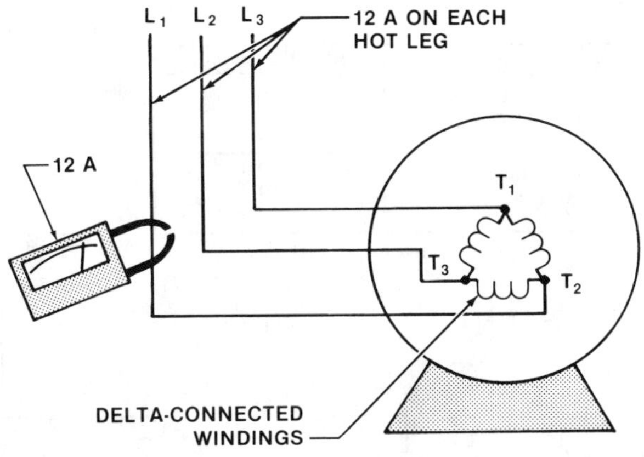

Figure 7-10. If current is about even and normal in each phase, the terminal connections are assumed to be correct.

Applying Voltage and Identifying Motor Leads. The common lead of the first winding is marked T_1. The other two leads are marked T_4 and T_9. The common lead of the second winding is marked T_2. The other two leads are marked T_5 and T_7. The common lead of the third winding is marked T_3. The remaining leads are marked T_6 and T_8. After all the leads of the motor windings have been temporarily identified and marked, connect T_1, T_4, and T_9 to a 240-volt, three-phase supply. The windings of T_1, T_4, and T_9 will induce a voltage into the other two windings by mutual induction. See Figure 7-11.

With the 240-volt, three-phase power supply applied to T_1, T_4, and T_9, the motor can be turned on and run without a load. T_4 is connected to T_7 and the voltage values read with a voltmeter. The voltage reading should measure approximately 460 volts. If the voltage reading is about 460

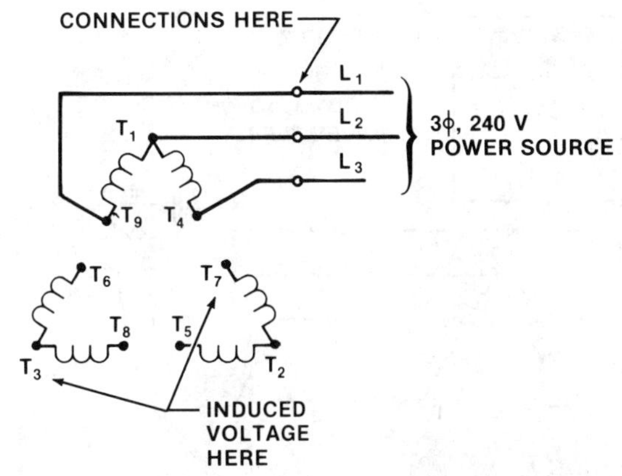

Figure 7-11. When the power supply is connected to T_1, T_4, and T_9, voltage is induced into the windings of T_2, T_5, and T_7 and the windings of T_3, T_6, and T_8.

volts, the lead markings for T_4, T_9, T_7, and T_5 can be permanently marked. See Figure 7-12.

If the voltage readings between T_1 and T_2 measure about 400 volts or less, interchange T_5 with T_7 or T_4 with T_9, and read the voltage values. If the voltage reading is approximately 460 volts between T_1 and T_2, the leads are properly identified. However, should the voltage between T_1 and T_2 read about 230 volts or less, interchange both T_5 with T_7 and T_4 with T_9. The new voltage reading should be approximately 460 volts between T_1 and T_2. Refer to Figure 7-12. The leads T_4, T_9, T_7, and T_5 can be permanently marked.

The next step is to connect T_6 to T_9 and measure the voltage between T_1 and T_3. The voltage should measure about 460 volts. T_6 and T_8 can be permanently marked. If the voltage between T_1 and T_3 does not equal 460 volts, interchange leads T_6 and T_8. With the voltage between I_1 and T_3 reading 460 volts, leads T_6 and T_8 can be permanently marked. Refer to Figure 7-12.

Shut the power off to the motor and connect L_1 to T_2, L_2 to T_5, and L_3 to T_7. Note the direction of rotation. The rotor should turn in the same direction as before. Now reconnect L_1 to T_3, L_2 to T_6, and L_3 to T_8. The rotor should rotate in the same direction. If the rotor does not rotate in the same direction as the previous test, the leads must be carefully checked and the mismarked lead properly marked. All the motor leads are now permanently marked and ready to be connected to the 240-volt, three-phase supply circuit.

The nine leads of the delta-connected motor must be connected for the lower voltage operation. Shut the motor off and reconnect the power circuit and motor leads as follows:

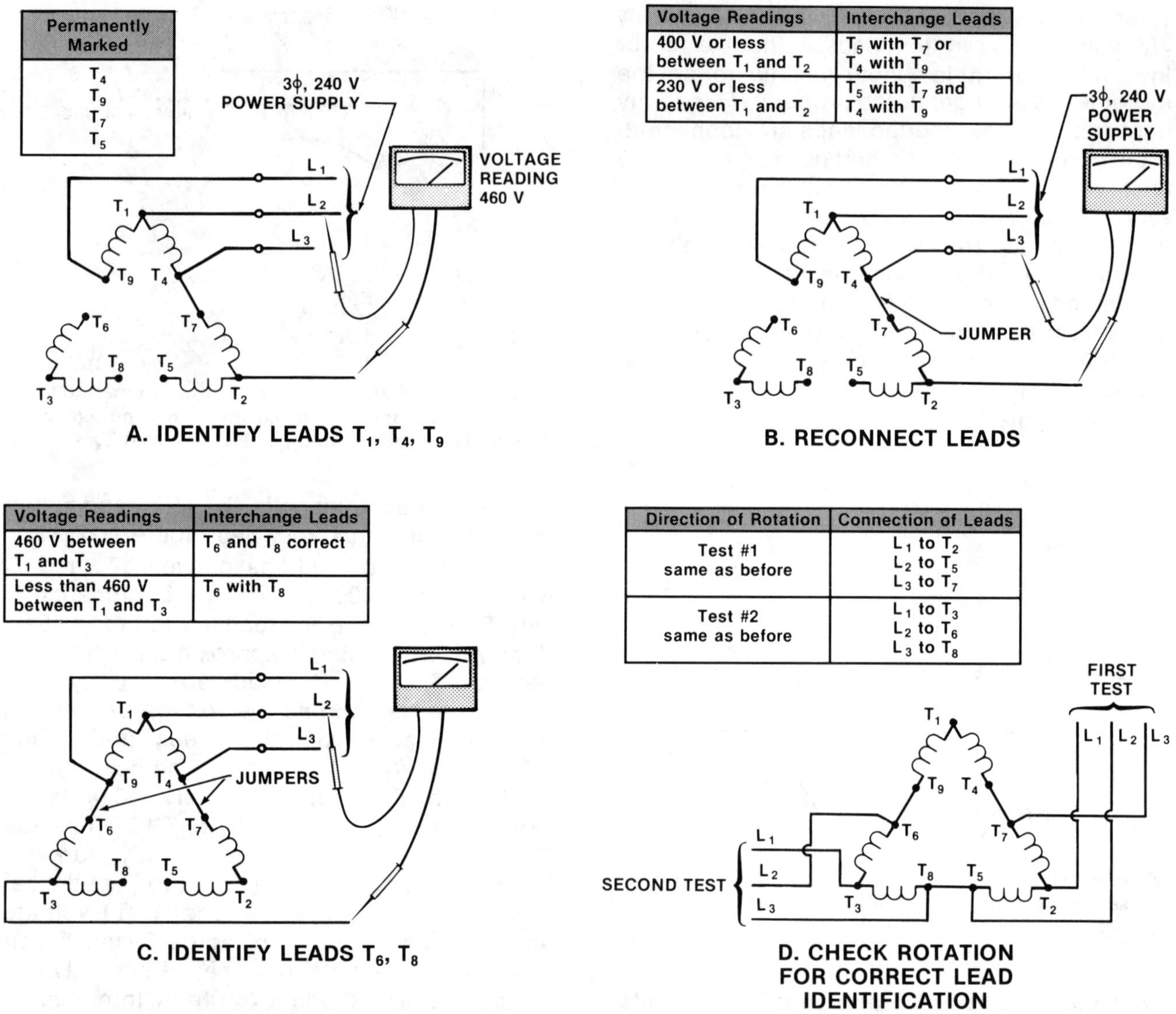

Figure 7-12. Two steps are followed to identify all leads of a delta-connected motor. If initial connections of T_1, T_4, and T_9 do not produce proper voltage readings, connect T_5 with T_7 or T_4 with T_9 before proceeding to step 2. Direction of rotation test verifies proper connection of leads.

1. connect L_1 to T_1, T_6, T_7
2. connect L_2 to T_2, T_4, T_8
3. connect L_3 to T_3, T_5, T_9

See Figure 7-13 for the proper procedure for connecting the windings and leads for low- or high-voltage operation. With the motor leads connected for low-voltage operation, turn on the power to the motor windings. Check the voltage from phase-to-phase and a 240-volt reading should be obtained. Check the amperage phase leg (A, B, and C) and current should measure about equal. Connect the motor to the driven load. The motor current pulled should be about even and normal on all three phases.

REVERSING MOTOR ROTATION

The direction of rotation of single-phase motors is reversed by changing the direction of the current flow in relation to the starting or running windings. The end plate of the motor must be removed and the four lead wires reconnected in order to reverse rotation. Standard rotation of the motor is counterclockwise viewing the motor from the front end. The direction of three-phase motors is changed by interchanging any one of the three leads from the stator of the motor to the power supply leads. This interchanging of leads can be accomplished either in the magnetic starter or disconnecting means in the motor circuit conductors. See Figure 7-14.

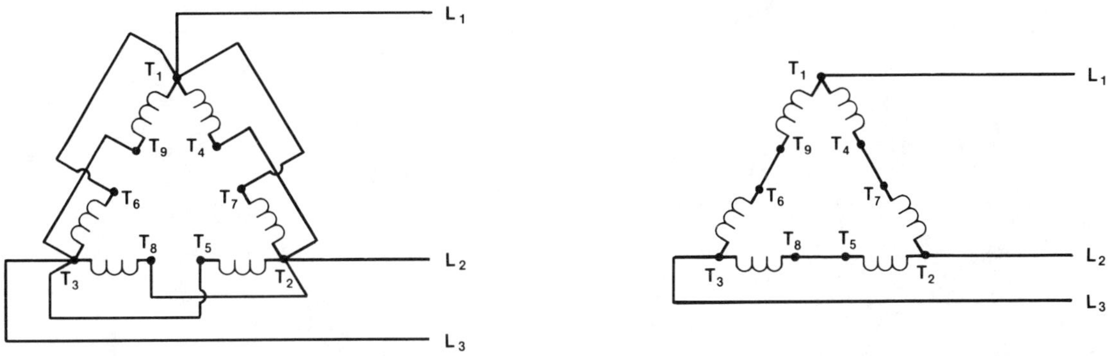

Low Voltage	L_1	L_2	L_3	Tie Together
	T_1-T_6-T_7	T_2-T_4-T_8	T_3-T_5-T_9	———

High Voltage	L_1	L_2	L_3	Tie Together
	T_1	T_2	T_3	T_4-T_7, T_5-T_8, T_6-T_9

LOW VOLTAGE   HIGH VOLTAGE

Figure 7-13. Connecting the windings of a three-phase delta motor for low- or high-voltage operation.

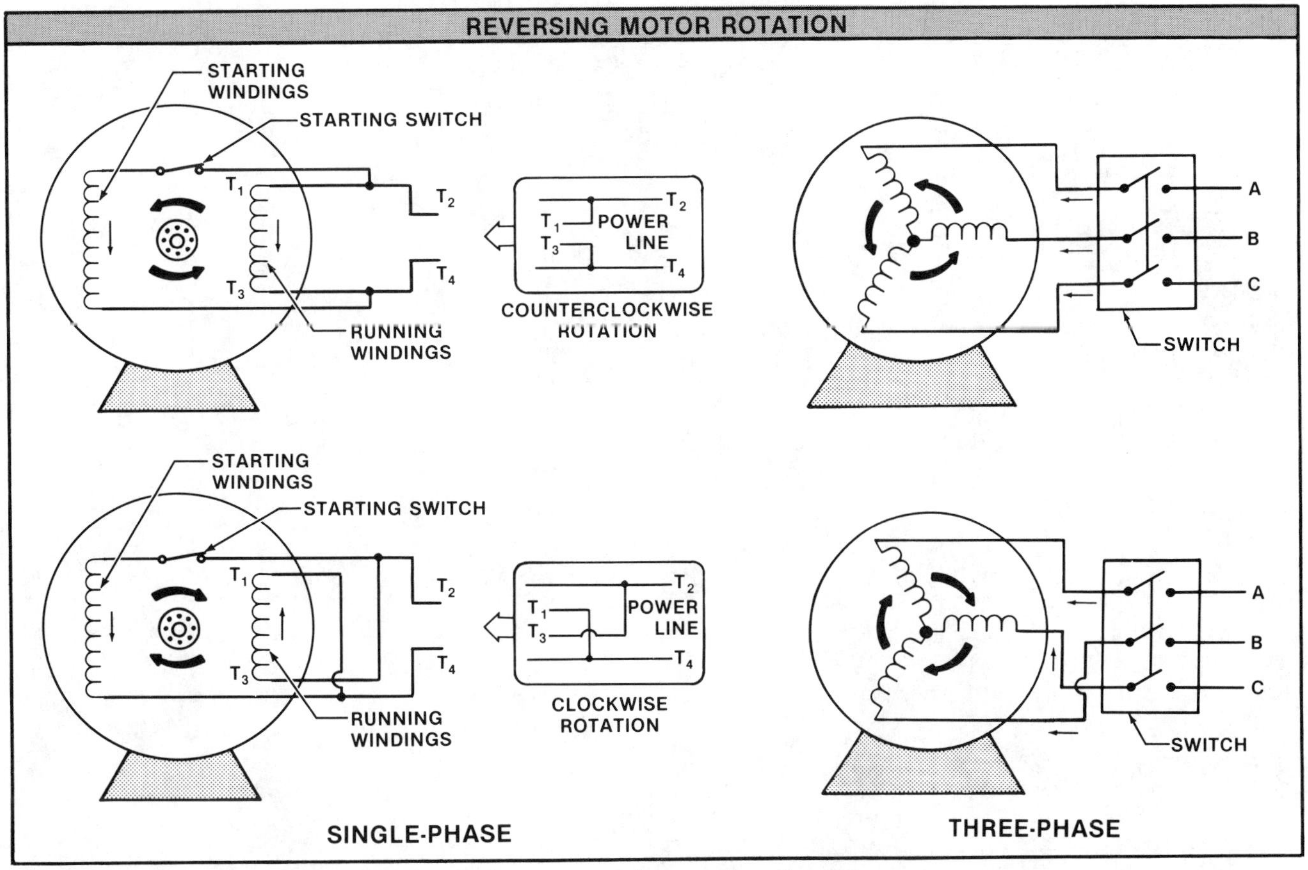

Figure 7-14. Leads must be interchanged to reverse the rotation of motors.

REVIEW—CHAPTER 7

Name ______________________________ Date ______________

True-False

T F **1.** Motor windings must be connected in parallel when served by high voltage systems.

T F **2.** The windings of dual-voltage motors may be connected to low or high voltage supply circuits.

T F **3.** Where leads to motor windings are labeled with an S, the markings show that the leads are connected to the running windings.

T F **4.** The leads to windings that are connected to a 120-volt supply are connected in parallel.

T F **5.** Motors with nine leads are more difficult to connect than motors with six leads where the marking of the leads are unreadable.

T F **6.** The amperage a motor pulls can be measured with an ohmmeter.

T F **7.** Each ungrounded conductor that supplies a motor should read about the same amount of amps.

T F **8.** The voltage to motors should never be below 10% of the supply voltage to the motor.

T F **9.** To obtain higher voltage in a wye-connected motor, T_1 is connected to L_1, T_2 to L_2, and T_3 to L_3.

T F **10.** The starting windings of split-phase motors are connected in series with the running winding.

T F **11.** The number of poles to a motor determines the number of speeds that can be used.

T F **12.** All the poles of a motor are used to obtain the fastest speed.

T F **13.** A centrifugal switch mounted inside the housing of motors is used to disconnect the motor.

T F **14.** Viewing the motor from the front end, standard rotation of motors is counterclockwise.

T F **15.** The rotation of three-phase motors can be reversed by interchanging any two phase leads.

Completion

__________ **1.** The rotation of single-phase motors can be reversed by changing the direction of the current flow to the starting or __________________ winding.

__________ **2.** The interchanging of the conductor leads can be accomplished either in the magnetic starter or ______________________.

______ **3.** The starting winding is used to ______ the motor.

______ **4.** The running winding cannot be used to ______ and run split-phase motors without an additional winding.

______ **5.** The running winding has ______ resistance than the starting winding.

______ **6.** A(n) ______ winding can be used to obtain more than one speed in split-phase motors.

______ **7.** The centrifugal switch opens the circuit to the ______ winding at a predetermined time.

______ **8.** All the poles of a motor are used to obtain ______ speed.

______ **9.** The running winding has ______ wire than the starting winding.

______ **10.** A(n) ______ can be used to identify the leads of motors.

______ **11.** A(n) ______ can be used to test the voltage supply to motors.

______ **12.** A(n) ______ can be used to measure the current in amps in each leg of circuits feeding motors.

______ **13.** The running current of motors should be within ______% of the motor's nameplate current.

______ **14.** When connecting the nine leads of three-phase, wye-connected motors to operate on lower voltage, the leads marked ______ are tied together.

______ **15.** The starting winding of split-phase motors has ______ turns than the running winding.

Multiple Choice

______ **1.** The fast speed of six-pole motors operates on ______ pole(s).
   A. one
   B. two
   C. three
   D. four

______ **2.** Four-pole motors provide ______ speeds.
   A. two
   B. three
   C. four
   D. six

______ **3.** Split-phase motors with dual windings connected in series must be connected to ______ volts.
   A. 120
   B. 240
   C. 480
   D. 600

_____ **4.** The starting windings are wound on the top of the _____.
A. armature
B. rotor
C. stator
D. running windings

_____ **5.** The leads of a three-phase, wye-connected motor operating on 480 volts must be connected in _____.
A. series
B. parallel
C. series and parallel
D. parallel with series taps

_____ **6.** The direction of rotation of three-phase motors can be reversed by interchanging leads _____.
A. L_2 and L_3
B. T_4 and T_5
C. T_6 and T_7
D. T_8 and T_9

_____ **7.** The leads of a six-pole motor operating at medium speed are connected to operate on _____ poles.
A. two
B. four
C. five
D. six

_____ **8.** Three-phase, nine-lead motors are equipped with _____ windings.
A. four
B. five
C. six
D. nine

_____ **9.** Three-phase, nine-lead, delta-connected motors have L_1 connected to leads _____ when connected for low-voltage operation.
A. T_1-T_6-T_7
B. T_2-T_4-T_8
C. T_3-T_5-T_9
D. T_7-T_8-T_9

_____ **10.** The windings of split-phase motors connected for 120-volt operation are connected in _____.
A. series
B. parallel
C. series and parallel
D. parallel with series taps

Problems

1. Number the leads of each winding for the following motor.

   A. ____________
   B. ____________
   C. ____________
   D. ____________
   E. ____________
   F. ____________
   G. ____________
   H. ____________
   I. ____________

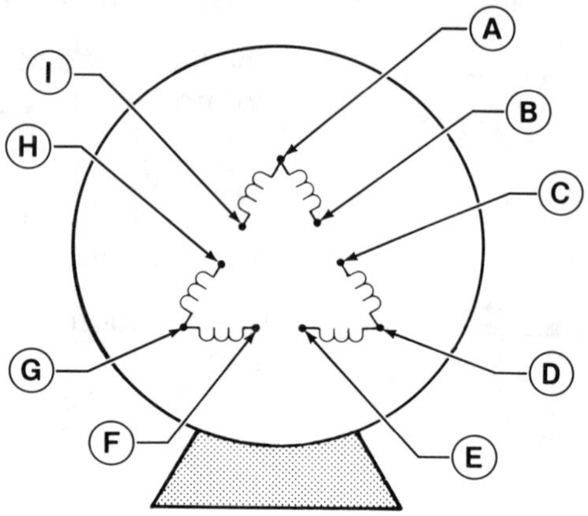

2. Number the leads of each winding for the following motor.

   A. ____________
   B. ____________
   C. ____________
   D. ____________
   E. ____________
   F. ____________
   G. ____________
   H. ____________
   I. ____________

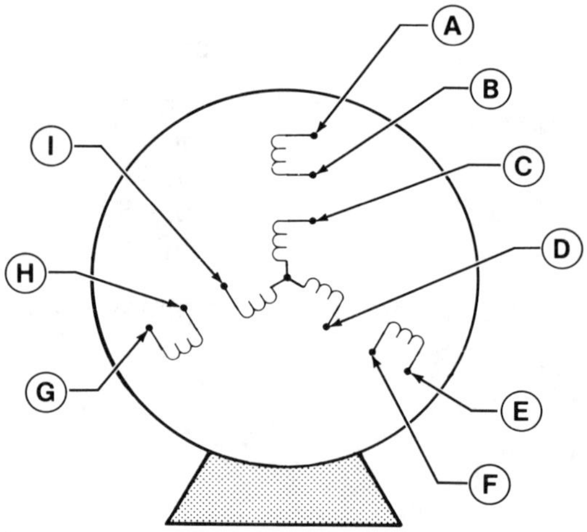

3. Connect the leads in the six-pole motor for fast speed operation.

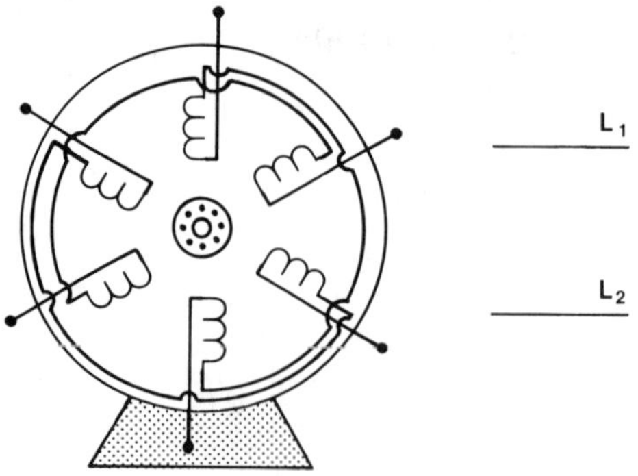

4. Connect the leads in the motor for high-voltage operation.

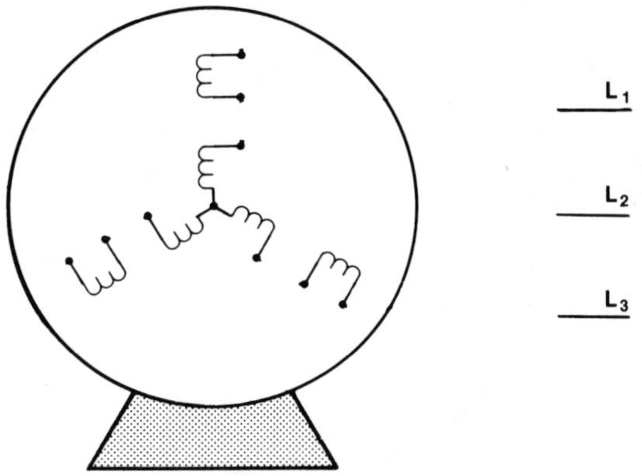

5. Draw a two-pole motor and show its windings.

6. Draw a split-phase motor and show its poles, windings, and starting switch.

7. Draw a split-phase motor and show the rotation being reversed for clockwise rotation.

8. Connect the leads in the motor for low-voltage operation.

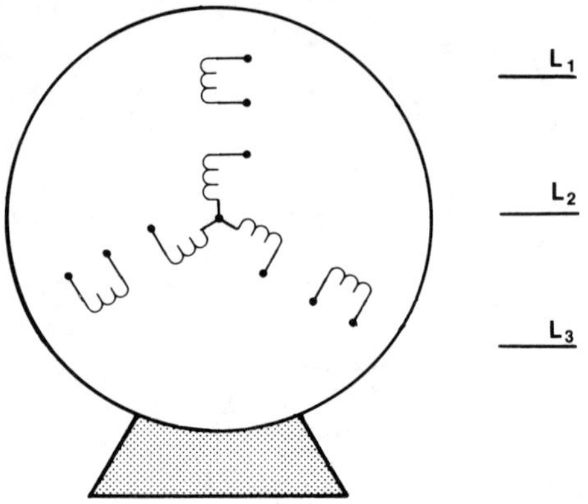

9. Connect the leads in the motor for high-voltage operation.

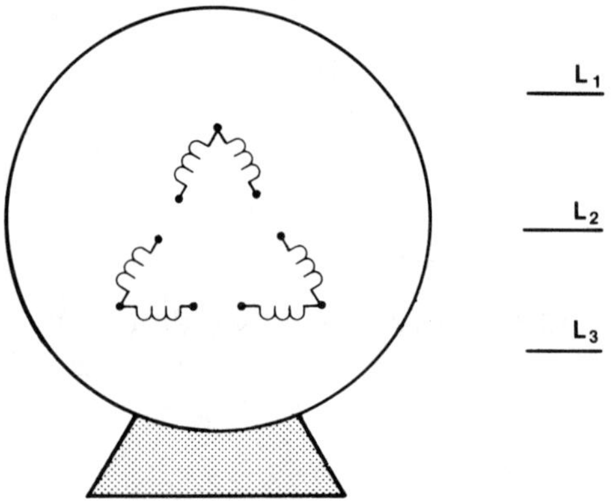

10. Connect the six-pole motor for low-speed operation.

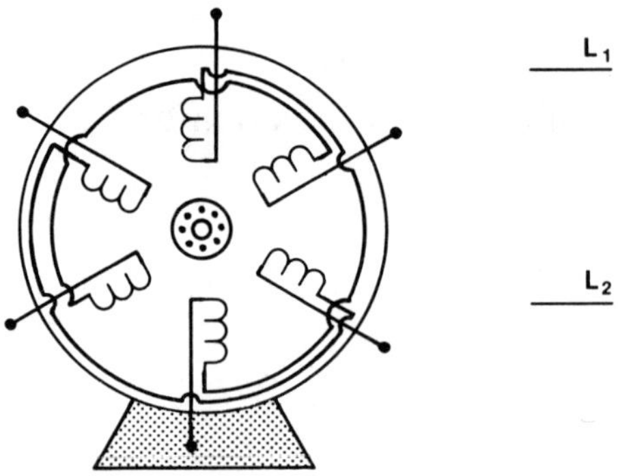

MOTOR CONTROL HOOKUPS

Chapter 8

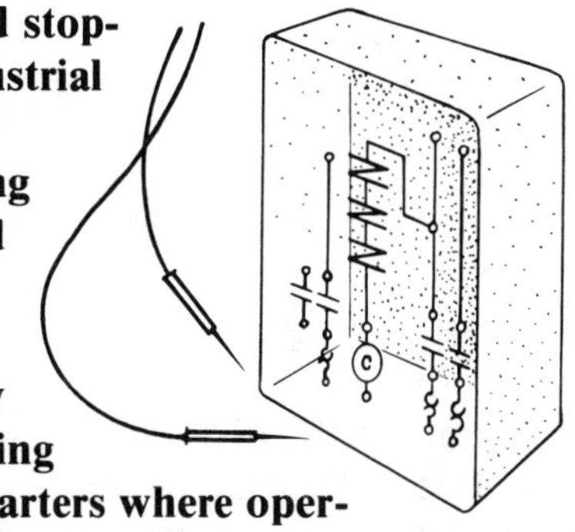

All electrical motors must have a means of starting and stopping the motor and the driven load. Controllers for industrial motors are either *manual starters* or *magnetic starters*.

Using manual starters is the simplest method of starting and stopping a motor load. The power supply is connected to the starter in series through the contacts to the motor leads. Magnetic starters can be controlled by pressure, temperature, light, and start-and-stop buttons. This variety of control devices provides automatic starting and stopping of motors. Magnetic starters are preferred over manual starters where operators are not required to be present.

MAGNETIC STARTERS

Magnetic starters are the most commonly used controllers in the electrical industry for starting and stopping motors. Magnetic starters are equipped with normally open (NO) contacts, which are energized by applying voltage to a coil. The control voltage used to energize the coil and contacts can range from 24 volts to 480 volts. The device used to control the coil may be automatic or manually operated. Each method requires a different wiring procedure. An *overload relay unit* is provided in the magnetic starter circuitry to protect the windings from overload conditions. See Figure 8-1.

Magnetic Starter Components

L_1, L_2, and L_3 are the terminals that connect the branch-circuit conductors from the power line to the magnetic starter. Branch-circuit conductors are sized and selected per 430-22(a) and Table 310-16.

T_1, T_2, and T_3 are the terminals on the magnetic starter that connect the branch-circuit conductors to the motor lines. The wire-bending space at the terminals in the enclosure housing the magnetic starter must comply with Table 430-10(b). For example, to determine the wire bending space required for one #4 conductor per terminal inside the enclosure housing a magnetic starter, use the following procedure:

Step 1: *Table 430-10(b).*
one #4 conductor

Answer: **2″**

The space required in the terminal housing on the motor must comply with 430-12. The stationary contacts connect the branch-circuit conductors from the power supply to the motor leads. Contacts making and breaking or starting and stopping motors that are driving loads will eventually become tarnished and worn. If they become pitted or burned they can be replaced by using a handy contact kit. The coil is the device that causes the contacts to energize and de-energize the power circuit to the motor. The control circuits operating the coil are selected

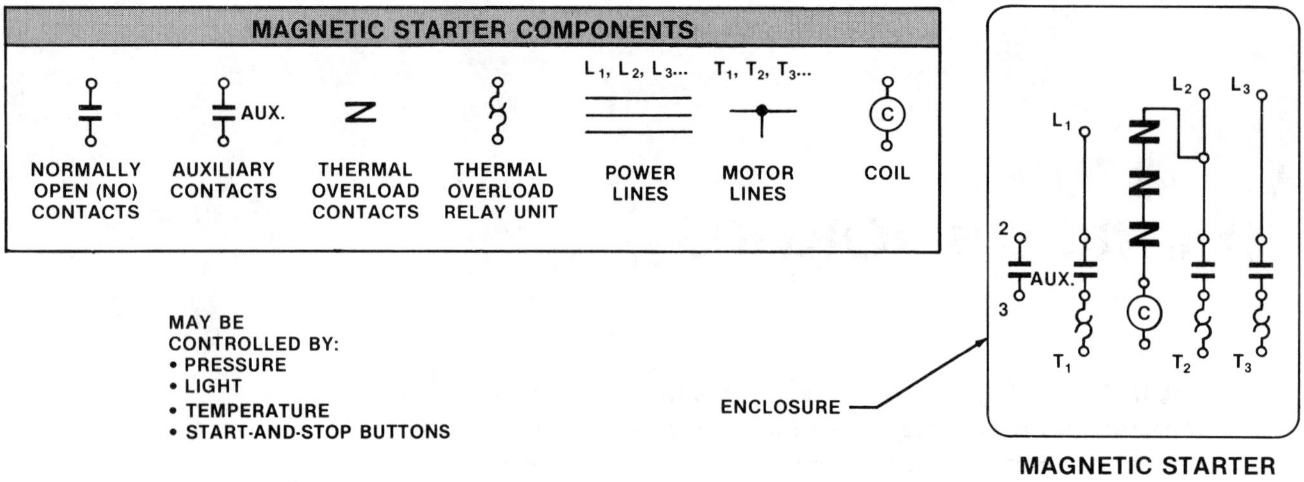

Figure 8-1. Magnetic starters are used to start and stop motors automatically.

and sized from 430-72 and 725-12. Auxiliary 2 and 3 are the points of connection bridging the circuit conductors to the coil. The thermal overload relay unit senses the rise in temperature in the motor windings. If the temperature rise in the motor windings exceeds the setting of the overload relay, the overload contacts open the coil circuit, dropping out the power circuit to the motor. Overload contacts are connected in series from L_2 to the coil and from the coil to the controlling devices supplied by L_1. The overload contacts actually open the coil control circuit due to an overload.

Two-wire Control Circuit

Two-wire control circuits will not provide a voltage release during a power failure. The contacts of the pilot device controlling the circuit to the coil usually remain closed and connect power immediately to the coil when power to the circuit is restored. No voltage release means that the coil circuit is maintained through the contacts of the pilot device until it is disconnected. Two-wire pilot devices are devices such as single-pole switches, float switches, pressure switches, thermostats, and limit switches. See Figure 8-2.

Three-wire Control Circuit

A three-wire control circuit provides a voltage release during a power failure. The start button, which is normally open (NO), must be pushed to energize the power to the coil, causing the contacts to close and start the motor.

No voltage protection means that the coil circuit is maintained through the normally closed (NC) contacts of the stop button. A three-wire control circuit has a start button with normally open contacts and a stop button with normally closed contacts. The contacts of the start button are connected in parallel. The contacts of the stop button are connected in series. An extra set of contacts (auxiliary 2-3) completes the control circuit through the stop button and holds power to the coil until the stop button is pressed. Pressing the normally closed series stop button de-energizes the circuit to the coil and drops out the control circuit. The auxiliary contacts (2-3)

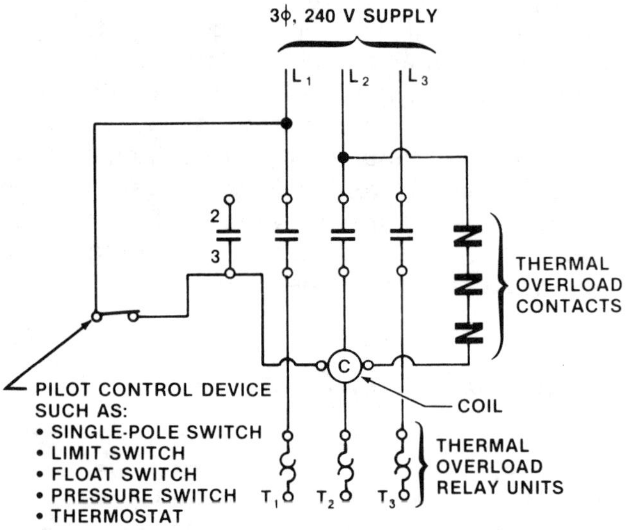

Figure 8-2. Magnetic starter controlled by a two-wire circuit.

will not close until the coil is energized by the start button. See Figure 8-3.

In a three-wire control circuit, pressing the normally open start button energizes the power to the coil and closes the motor starter contacts. Pressing the normally closed stop button de-energizes the holding circuit to the coil, dropping out the motor starter contacts. See Figure 8-4.

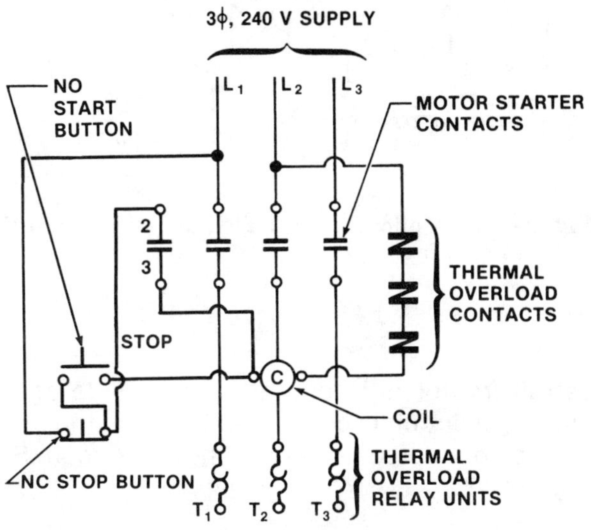

Figure 8-3. Magnetic starter controlled by a three-wire circuit.

ADDING CONTROL DEVICES

Control devices may be added to motor control circuits to perform various operations. Start buttons, stop buttons, pilot lights, auxiliary contacts, jog buttons, master stop buttons, hand-off automatic switches, and forward-reverse-stop pushbutton stations are examples of control devices that may be added to motor control circuits.

Adding Start Buttons

Extra start buttons can be added to energize the control circuit to the coil, which will close contacts to start the motor. As many start buttons can be added as needed for control purposes. All start buttons must be connected in parallel. See Figure 8-5.

Adding Stop Buttons

Extra stop buttons can be added at various locations to stop the motor by disconnecting power

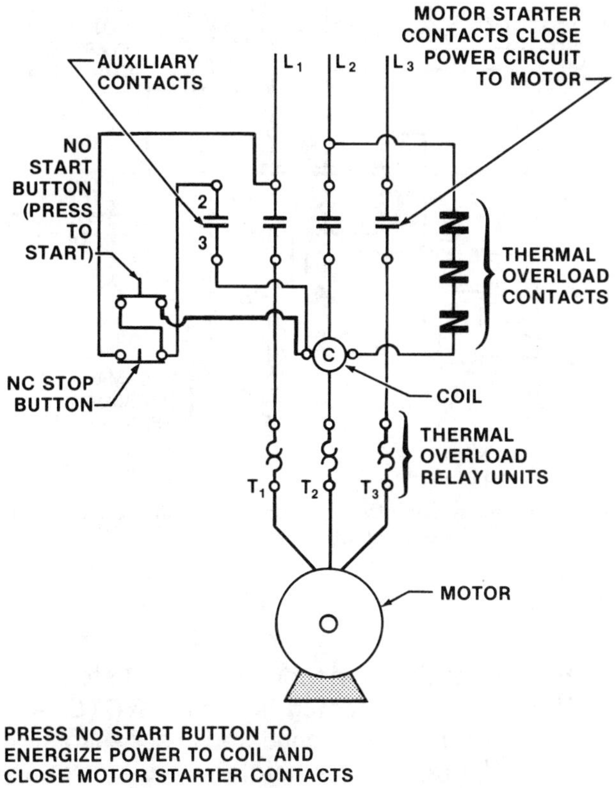

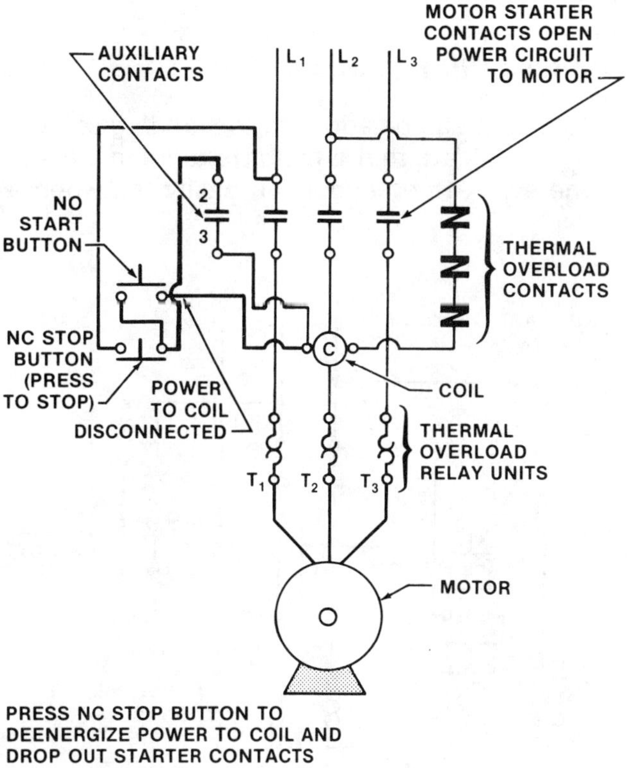

Figure 8-4. The NO START button energizes and the NC stop button de-energizes the coil of a three-wire control circuit.

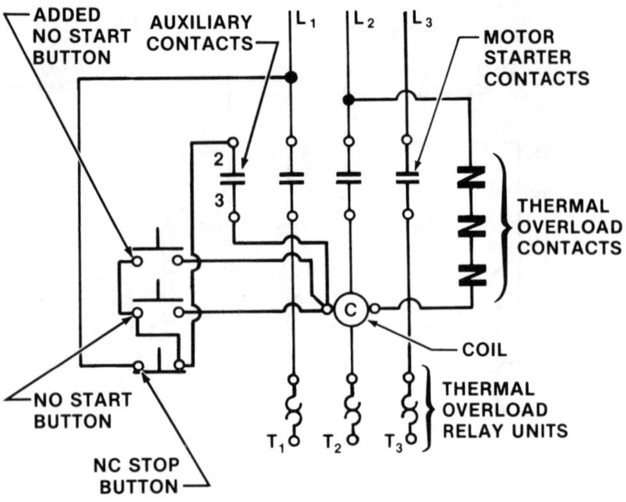

Figure 8-5. A variety of motor control devices may be added to a motor control circuit.

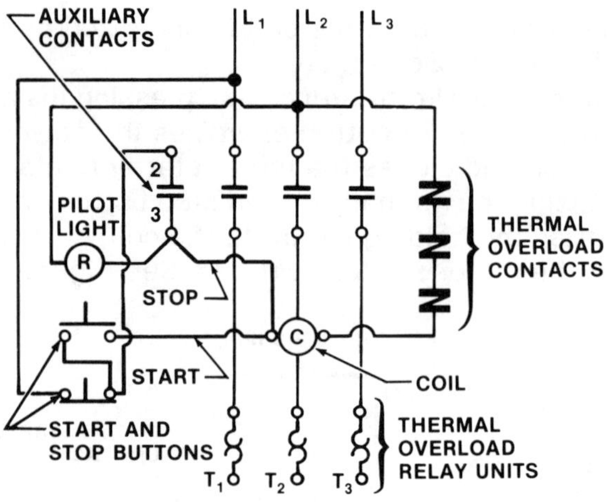

Figure 8-7. A pilot light is added to indicate that the motor is running.

to the coil of the magnetic starter. All stop buttons must be connected in series. *NOTE*: Any type of switch can be connected in series to stop a motor by breaking the holding circuit (auxiliary contacts 2-3) to the coil. See Figure 8-6.

Adding Pilot Lights

Pilot lights can be added to the holding coil circuit to indicate that a motor is running. This is done so that personnel can easily tell when a certain motor is in operation. One side of the pilot light circuit is connected to L_2 and the other side is connected to the holding coil circuit. See Figure 8-7.

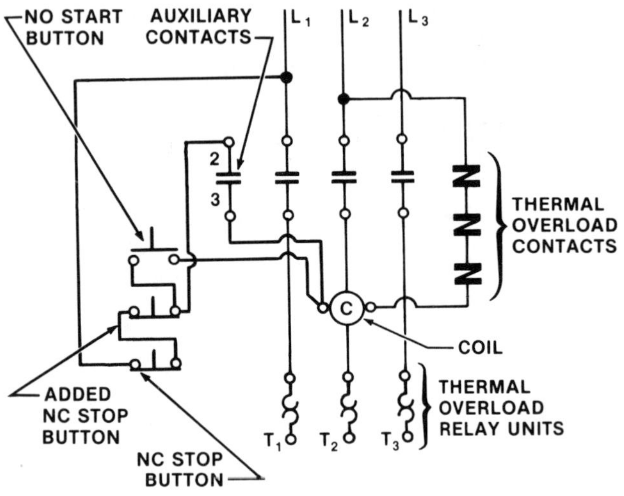

Figure 8-6. Series stop buttons can be placed in various locations to stop a motor.

Adding Auxiliary Contacts To Control Other Starters

An extra auxiliary contact can be added to control a circuit to another coil or device. The auxiliary contacts are added to one side of a magnetic starter. They may be normally open or closed. If they are closed, they are opening a control circuit. If they are open, they are closing a control circuit. One side of the auxiliary contact is connected to L_1 and the other side is routed and connected to the coil to be controlled. See Figure 8-8.

Adding Jog Buttons

The purpose of jogging is to have the motor run as long as the jog button is depressed. The magnetic starter must be wired so that there is no chance of locking during the jogging period. This is accomplished by adding a jog relay in the control circuit. For example, jogging is required to slowly move the drum of a printing press so that ink may be cleaned off the drum after a printing job. The normally open contacts of the

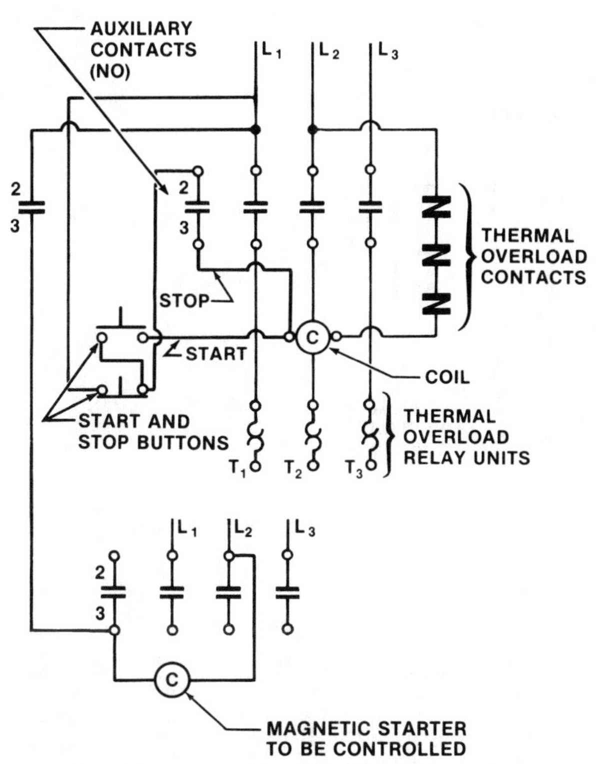

Figure 8-8. A magnetic starter with NO auxiliary contacts controls another magnetic starter.

jog button are connected in parallel with the start button. Two normally open contacts (CR-M) are connected in series with the stop button and auxiliary terminal no. 3. The motor can be jogged by holding down the jog button, which connects power to the main coil. The normally open contacts will prevent the holding coil from locking in and running the motor. See Figure 8-9.

Adding a Master Stop Button

A master stop button (safety) can be connected in series with wire from L_1 to the first stop button in the control circuit. The master stop button can be manually turned OFF and its contacts remain in the open position. With the contacts of the stop button in the open position, the power of L_1 is disconnected from the components of the control circuit and the coil of the magnetic starter cannot be energized. See Figure 8-10.

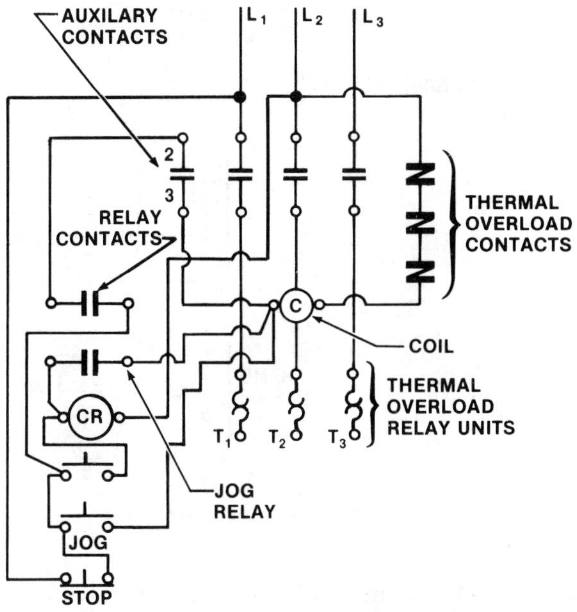

Figure 8-9. Jogging a motor using a jog button energizes the coil of a magnetic starter.

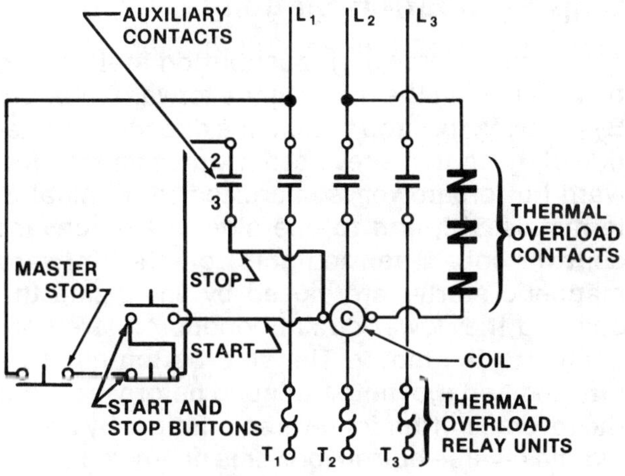

Figure 8-10. A master stop button is added to the control circuit of a magnetic starter.

Adding a Hand-off Automatic Switch

A hand-off automatic switch can be used to start a motor manually or automatically. A remote-control device may be used when starting the motor automatically. Pilot devices such as limit switches, pressure switches, and float switches can be used to energize the coil and start the operation of the motor. See Figure 8-11.

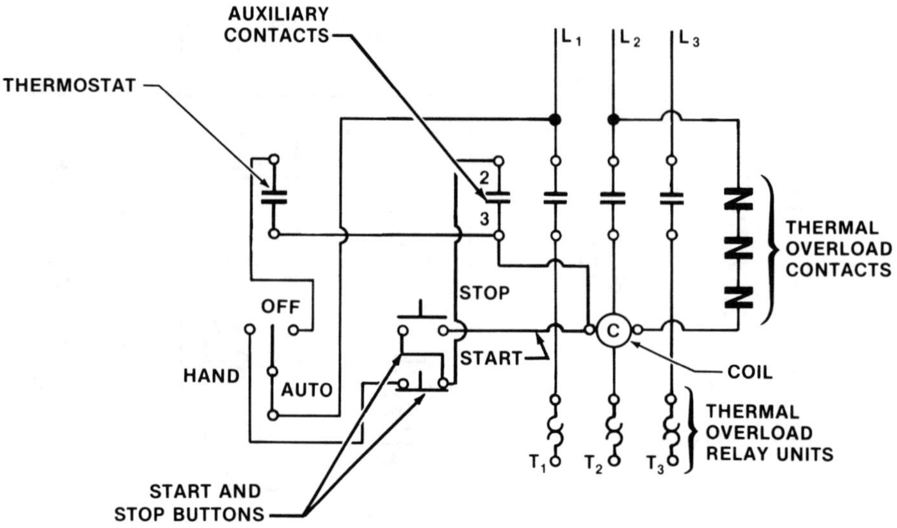

Figure 8-11. Adding a hand-off automatic switch in the control circuit.

Adding a Forward-Reverse-Stop Pushbutton Station

A forward-reverse-stop pushbutton station can be wired to run the motor in the forward rotation. By pressing the stop button, the direction of rotation of the motor is reversed. By pressing the forward button, power is energized to terminal 3, terminals 3–6, and to one side of the forward control coil. The contacts of the forward magnetic starter are closed by energizing the coil and the power circuit conductors are connected to the motor. The stop button must be pressed and the motor stopped before pressing the reverse button for reverse rotation. By pressing the reverse button, power is energized to terminals 5–7 and to one side of the reverse control coil. The contacts of the reverse magnetic starter are closed by energizing the coil and the power circuit conductors are connected to the motor. See Figure 8-12.

TROUBLESHOOTING

Troubleshooting a motor control circuit should always be done beginning with the most obvious components. By checking the most obvious components first, the trouble is often quickly found and may be easily rectified. A basic procedure to follow when troubleshooting a motor control circuit is:

(1) Check supply voltage.
(2) Check all terminal connections.
(3) Check fuses.
(4) Check magnetic starter.
(5) Check push buttons.
(6) Check coil.

Checking Supply Voltage

Supply voltage is checked by using a voltmeter. See Figure 8-13. Measure the phase-to-phase voltage from L_1 to L_2, L_1 to L_3, and L_2 to L_3. Read the phase-to-phase voltage. For example, the phase-to-phase voltage may be 48 volts or 240 volts. *NOTE*: Phase-to-phase voltage shown in Figure 8-13 is 240 volts.

Checking Fuses

After verifying the supply voltage, measure voltage from the line side of L_1 to the load side of L_2. See Figure 8-14. If 240 volts is read, the fuse in L_2 is good. If no voltage is measured, the fuse in L_2 is bad. Measure from the line side of L_2 to the load side of L_1. If 240 volts is read, the fuse in L_1 is good. Next, measure from the line side of L_1 to the load side of L_3. The voltmeter will measure 240 volts if the fuse in L_3 is good. There will be no measurement of voltage if the fuse is bad. *NOTE*: Alternate probe points may also be

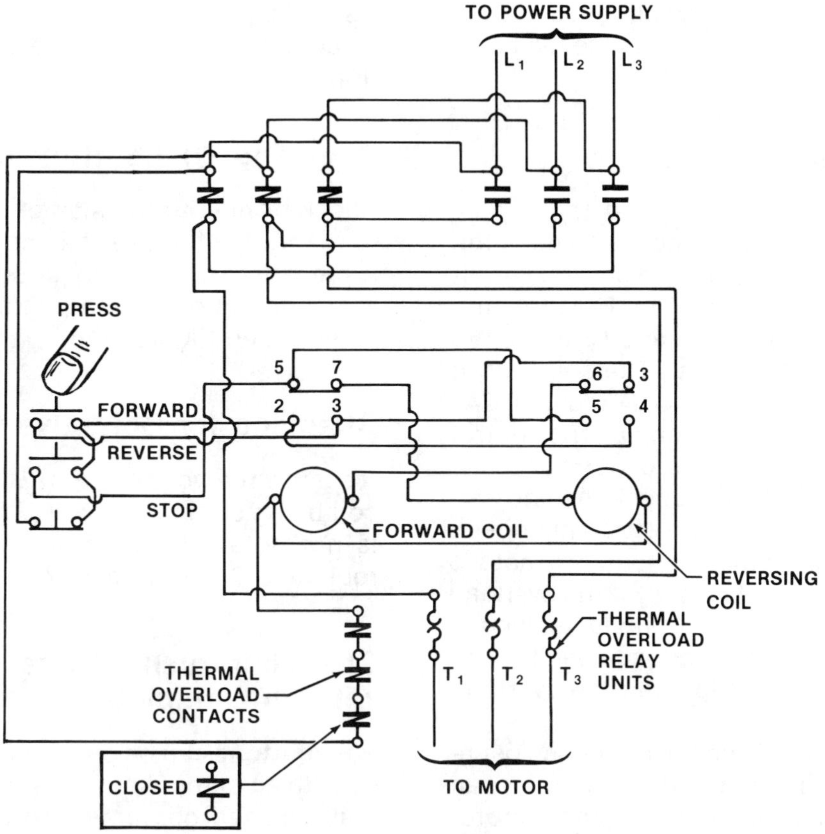

Figure 8-12. Pressing the forward button energizes the coil for forward rotation using a forward-reverse-stop pushbutton station.

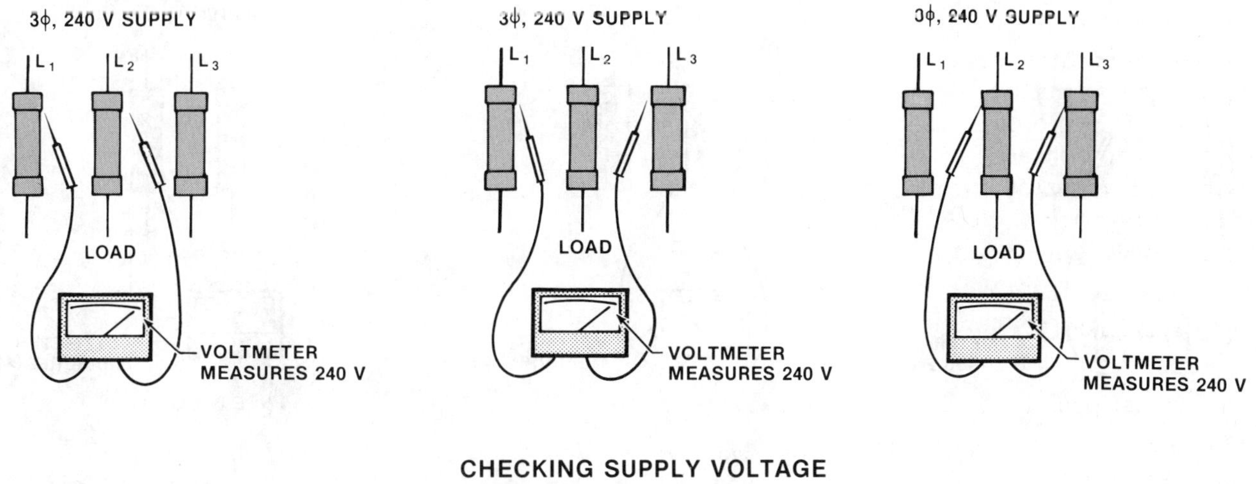

Figure 8-13. A voltmeter may be used to check supply voltage and fuses.

used to check fuses. For example, fuse L_1 may also be checked by measuring from the load side of L_1 to the line side of L_3.

Checking the Magnetic Starter

When it has been determined that the supply power is available to the line and load side of the fuses and the fuses are good, the control circuit must be checked. See Figure 8-15. The first step is to measure the voltage to L_1 and L_2 on the terminals of the magnetic starter. The coil cannot be energized without voltage to L_1 and L_2. If 240 volts is measured between L_1 and L_2, the control circuit voltage to the controls is correct.

Leave the probe on L_1 and check the line and load side of the overload contacts. This check will verify if any one of the overload contacts is open in the control circuit from L_2. If the overload contacts are open, reset the overload and check for the correct voltage. An overload condition on the motor windings can trip open the overload contacts.

If the inlets and outlets on the motor enclosure are blanketed with lint or dirt, this condition can create an overload. A stuck bearing on a motor or driven load can cause the motor windings to be overloaded. *NOTE*: If the overload contacts are not open, check for a loose connection or broken wire. Loose connections are common causes of trouble in motor control circuits. A visual inspection will sometimes detect loose, discolored, or burned wires, or other malfunctions.

Checking Pushbuttons (Start and Stop)

Check from L_1 to the side of the coil connected to L_2 through the overload contacts. If 240 volts is measured, the control circuit is complete and the contacts of the magnetic starter should be closed. See Figure 8-16.

Checking for Defective Coil

Measure the voltage from L_1 to the side of the coil being supplied from L_1. If a voltage reading is not obtained, the coil is bad and must be replaced. See Figure 8-17.

Checking Pushbuttons with an Ohmmeter

Pushbuttons can be checked for proper continuity with an ohmmeter. The stop buttons have normally closed contacts and a resistance when read. Place one of the ohmmeter probes on one side of the stop button and the other probe on the other side of the stop button. If the contacts are closed, the ohmmeter needle will peg. See Figure 8-18.

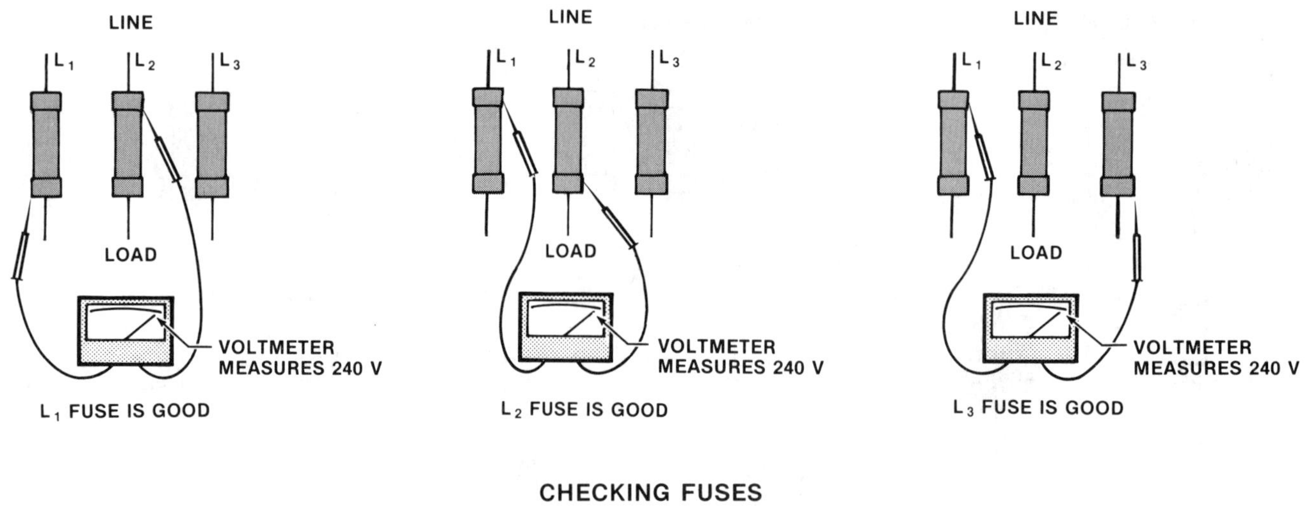

Figure 8-14. A voltmeter may be used to check fuses.

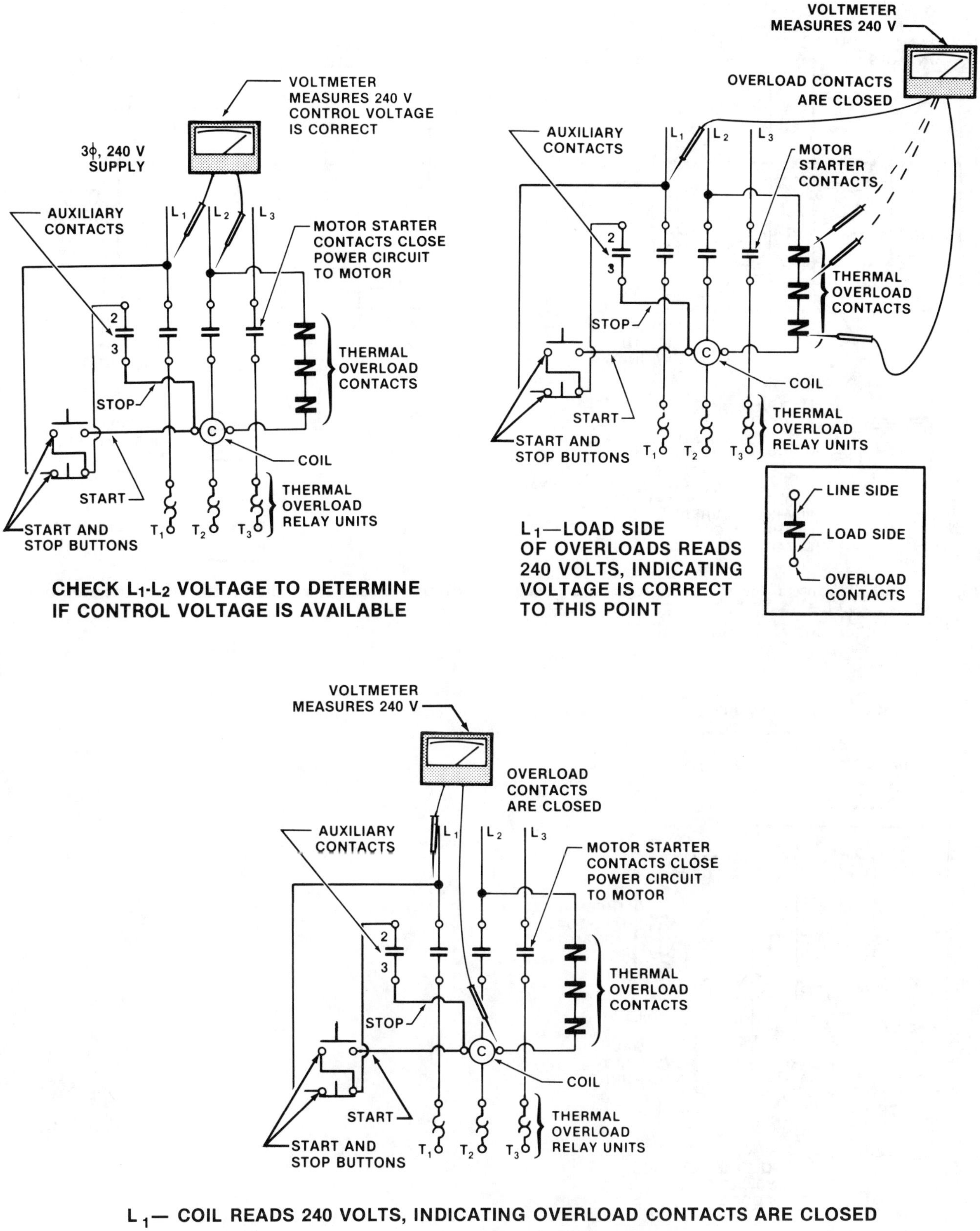

Figure 8-15. A voltmeter may be used to check the control voltage.

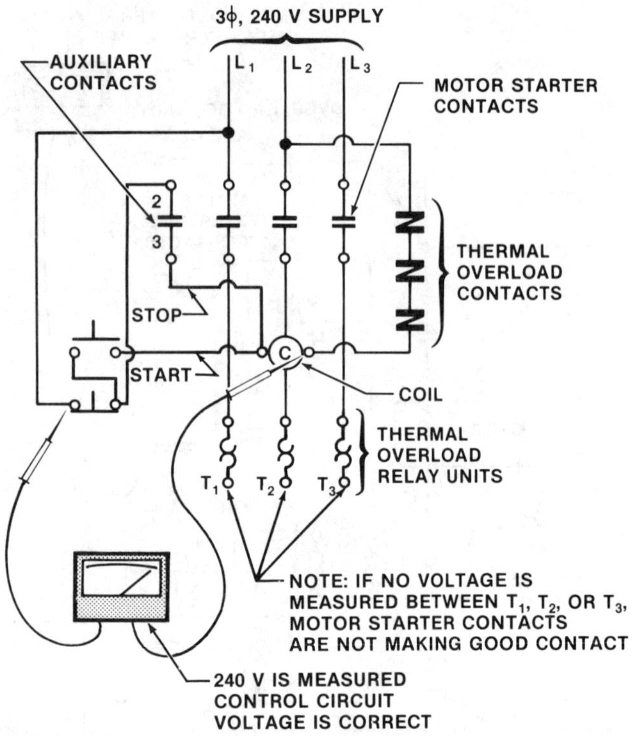

Figure 8-16. To determine if pushbuttons are good, check the control voltage from L_1 through the pushbuttons to the other side of the coil circuit.

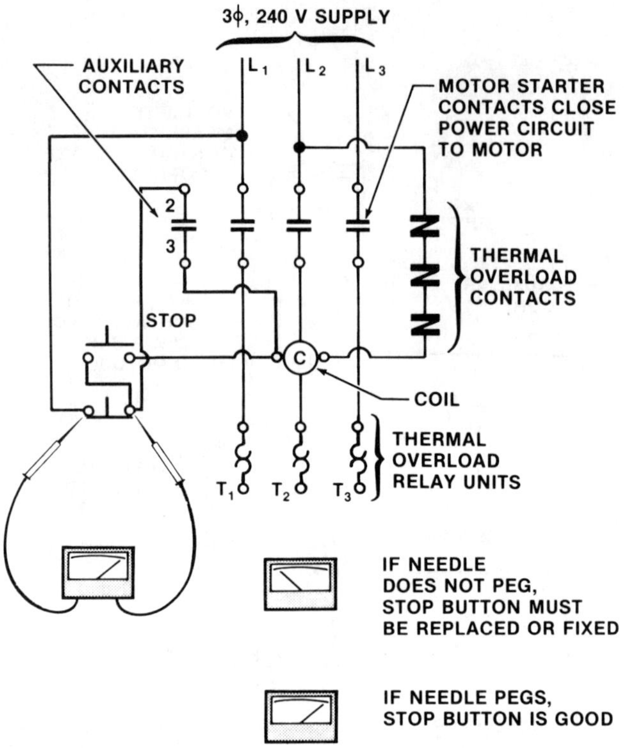

Figure 8-18. Check the continuity of the stop button with an ohmmeter.

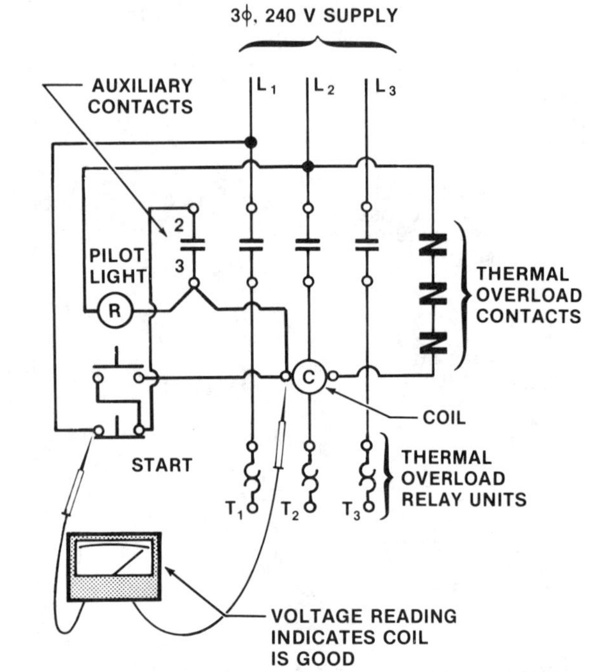

Figure 8-17. Check the control voltage of a magnetic starter by measuring voltage from L_1 to the L_1 side of the coil to determine if the coil is good. Disconnect the pilot light before testing.

REVIEW—CHAPTER 8

Name ______________________ Date ______________________

True-False

T F 1. Magnetic starters are the most commonly used types of controllers to start and stop motors.

T F 2. Magnetic starters are equipped with coils that can be used on different control voltages for safer operation.

T F 3. The voltage used to energize the coils that make the contacts to connect the power supply ranges from 24 volts to 4160 volts.

T F 4. A coil is connected in series with the overloads to protect the windings of motors.

T F 5. Conductors supplying motors must be at least 135% of the motor's full-load current rating.

T F 6. Overloads used to trip open contacts holding the control circuit are selected to respond to the heating of the motor windings.

T F 7. Three-phase motor starters are usually equipped with three overload units.

T F 8. Two-wire control circuits must have a holding circuit in series with the auxiliary contacts.

T F 9. A three-wire control circuit will provide a voltage release if there is a power failure.

T F 10. Float switches are used where pressure causes the control circuit to be energized.

T F 11. Start buttons are connected in series with the common line.

T F 12. Stop buttons are equipped with normally closed contacts.

T F 13. L_2 is connected in series with the overload contacts.

T F 14. Pressing the stop button will provide voltage to the coil and energize the power to the motor.

T F 15. T_1, T_2, and T_3 connect the power supply lines to the motor leads and windings.

Completion

__________ 1. Start buttons are provided with normally __________ contacts.

__________ 2. Any type of switch can be connected in __________ to break the power to the holding coil circuit.

__________ 3. Limit switches are used to __________ a motor due to the operation of the machine.

__________ 4. Start buttons can be placed at various locations to __________ the coil and contacts.

__________ 5. __________ can be connected to the holding coil to indicate that a motor is running.

__________ 6. Master (safe) buttons can be connected in series with L ________ and all stop buttons to prevent the control circuit from being energized.

__________ 7. Auxiliary __________ can be added to a magnetic starter to energize the coil of a second starter.

__________ 8. __________ buttons can be used to slowly move the drum of a printing press.

__________ 9. A(n) __________ hand-off switch can be used to start the motor manually or automatically.

__________ 10. A bad fuse can be checked by measuring the voltage from the ________ of one phase to the load side of another.

__________ 11. Overloads that have tripped open usually have to be __________ before the contacts will make.

__________ 12. Contacts of defected relays will arc, spark, or glow __________.

__________ 13. The ohmmeter needle will peg and fall back to zero if the capacitor being checked is __________.

__________ 14. The continuity of contacts in start-and-stop buttons can be measured using a(n) __________ with the power off.

__________ 15. No voltage protection in the coil circuit means that the coil circuit is maintained through the normally closed contacts of the __________ button.

Multiple Choice

__________ 1. Overload units trip open due to an increase in __________.
   A. conduit size
   B. water level
   C. copper size
   D. heat

__________ 2. Auxiliary contacts are usually identified with the markings __________.
   A. L_1-L_2
   B. T_2-T_3
   C. 2-3
   D. L_1-T_2

__________ 3. The power supply to the line side of the magnetic starter is usually labeled __________.
   A. T_1-T_2-T_3
   B. L_1-L_2-L_3
   C. L_1-T_2
   D. L_3-T_3

__________ 4. The leads to the windings of the motor are usually classified by the markings __________.
   A. L_1-L_2-L_3
   B. T_1-T_2-T_3
   C. L_1-L_2-T_3
   D. A_1-B_2-C_3

_______ **5.** The power conductors from the service equipment connect to terminals L_1-L_2-L_3 in the _______.
A. magnetic starter
B. limit switch
C. start button
D. stop button

_______ **6.** The coil is represented by the letter _______.
A. A
B. B
C. C
D. D

_______ **7.** No. 1/0 conductors terminating one per lug in the enclosure of a magnetic starter must have a minimum clearance of _______″.
A. $1\frac{1}{2}$
B. 2
C. 3
D. 5

_______ **8.** A three-phase magnetic starter requires at least _______ overload unit(s).
A. one
B. two
C. three
D. four

_______ **9.** Start buttons are equipped with _______.
A. normally open contacts
B. normally closed contacts
C. normally open and normally closed contacts
D. series buttons

_______ **10.** A magnetic starter with a start-and-stop station normally has _______ contact terminals on the line side.
A. two
B. three
C. four
D. five

Problems

1. Identify the components marked A through K.

A. _______ G. _______
B. _______ H. _______
C. _______ I. _______
D. _______ J. _______
E. _______ K. _______
F. _______

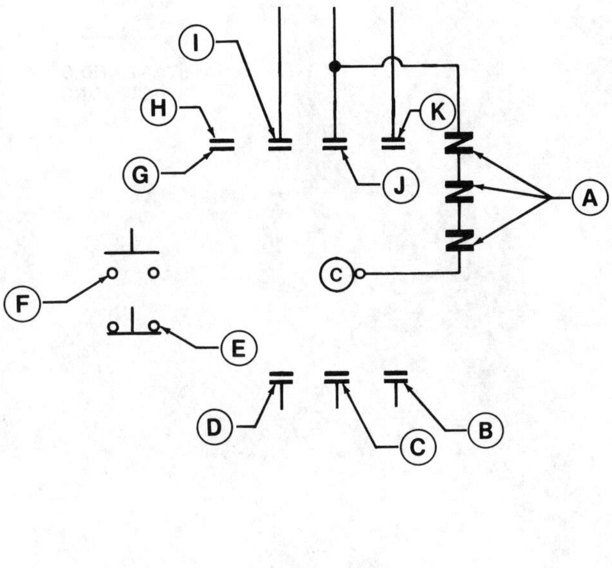

2. Connect the following magnetic starter for two-wire operation.

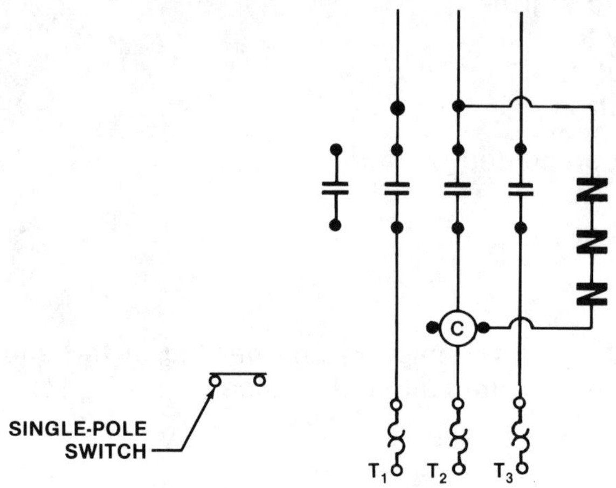

3. Connect the following control circuit for three-wire operation.

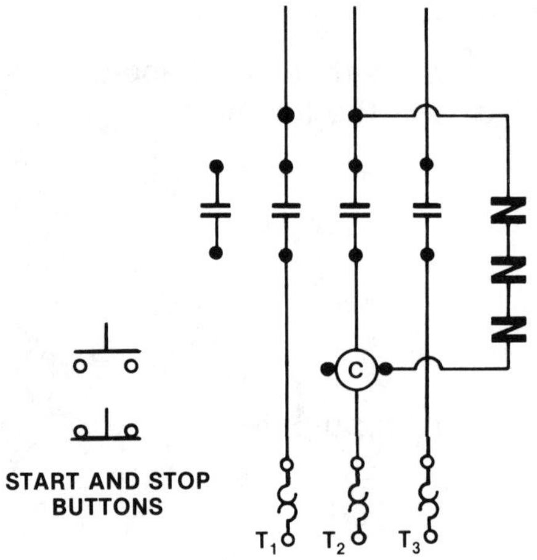

4. Connect the additional stop button to stop the motor from a different location on the machine.

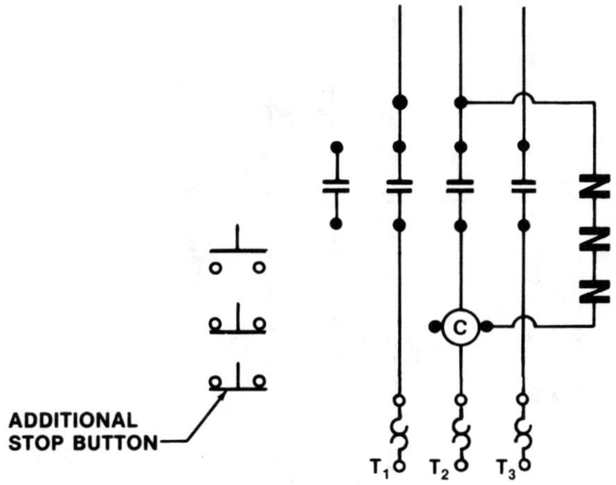

5. Connect the additional start button to start a motor from a different location on the machine.

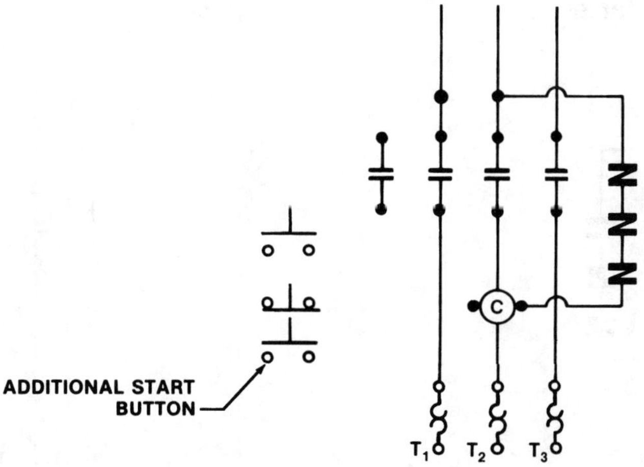

6. Connect the following control circuit with the master (safe) switch disconnecting the control circuit.

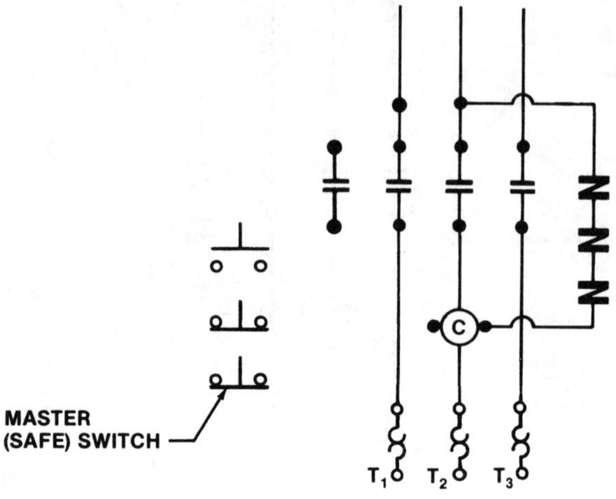

7. Connect the following control circuit to energize the magnetic starter whose extra set of contacts will call an additional starter.

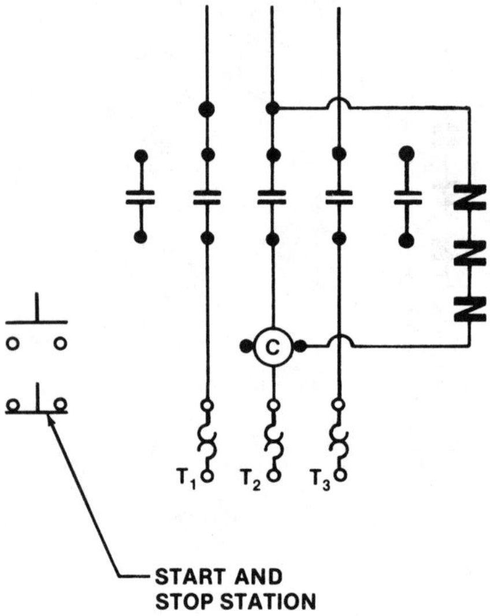

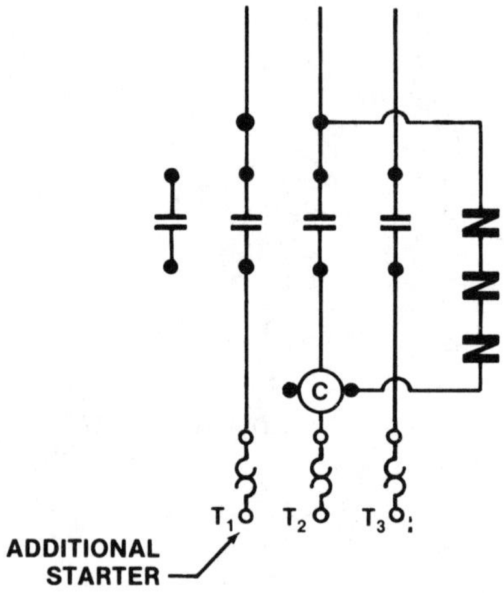

8. Connect the following control circuit to energize the coil with a thermostat.

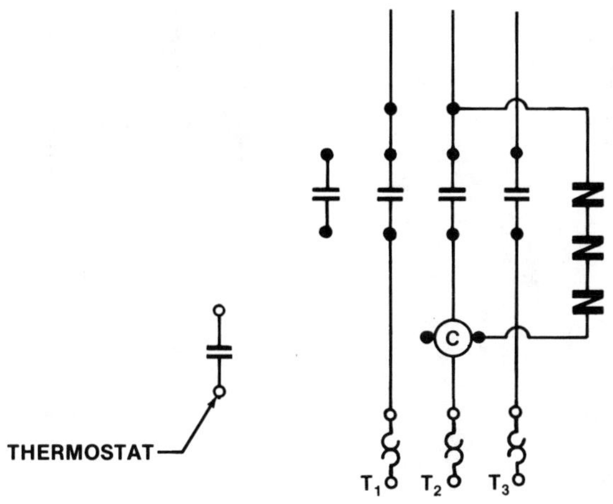

9. Connect the leads of the voltage tester to the fuse that is blown, and show the testing procedure.

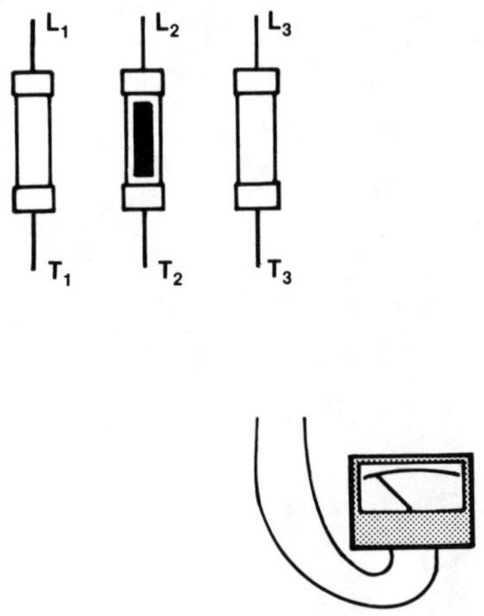

10. Connect the control circuit to jog the motor.

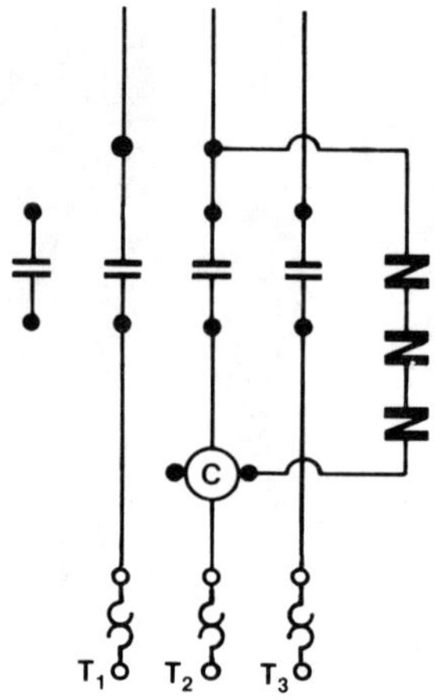

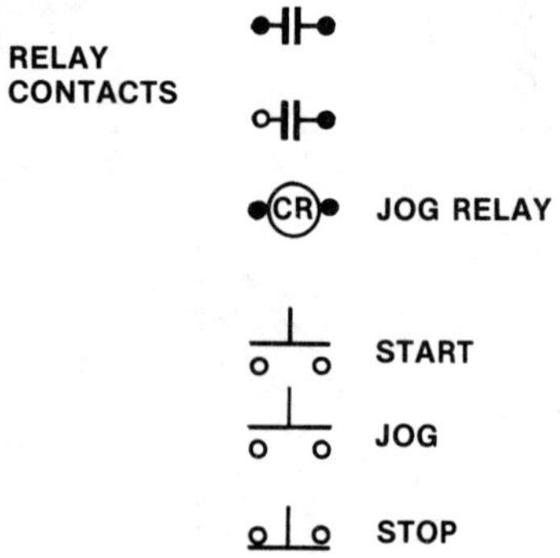

MOTOR EXAM

Chapter 9

The four tests in this chapter provide a comprehensive review of the material presented in chapters 1 through 8. Each test contains a variety of True-False, Completion, and Multiple Choice questions. Record your answers in the spaces provided. Each test also contains Problems. Show your work in the spaces provided. Your instructor may require Code validation for specific questions and/or problems.

Subject matter covered in these tests includes:

- Motor Theory
- Motor Identification
- Motor Nameplates
- Troubleshooting Procedures
- Testing Connections
- Testing Windings
- Finding Starting Torque
- Finding Full-Load Torque
- Finding Locked-Rotor Current
- Finding Horsepower
- Finding Full-Load Current
- Sizing Conductors
- Sizing Overcurrent Protection Devices
- Sizing Overload Protection Devices
- Sizing Disconnects
- Sizing Reduced Starters
- Sizing Controllers
- Selecting Control Circuits

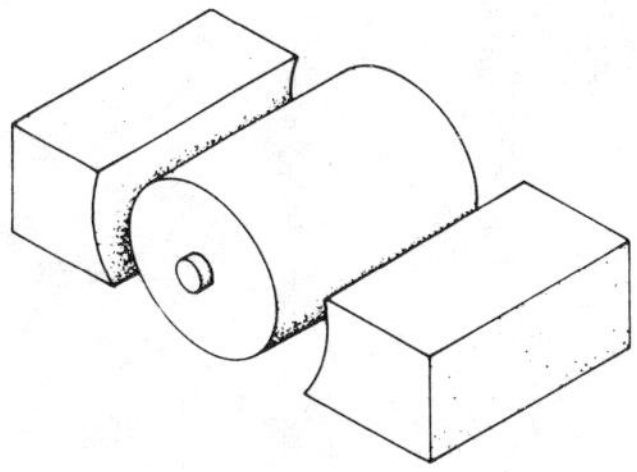

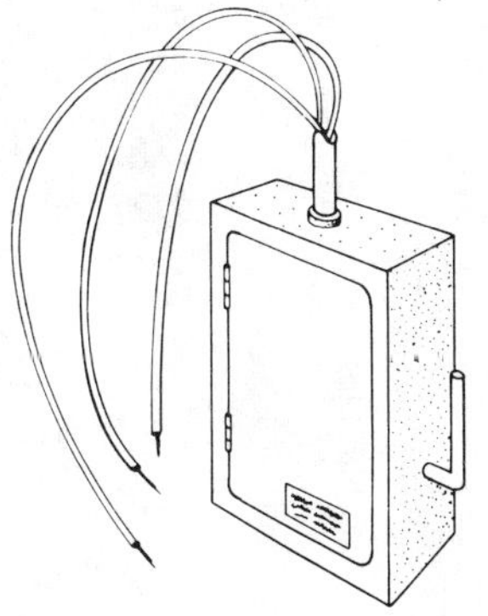

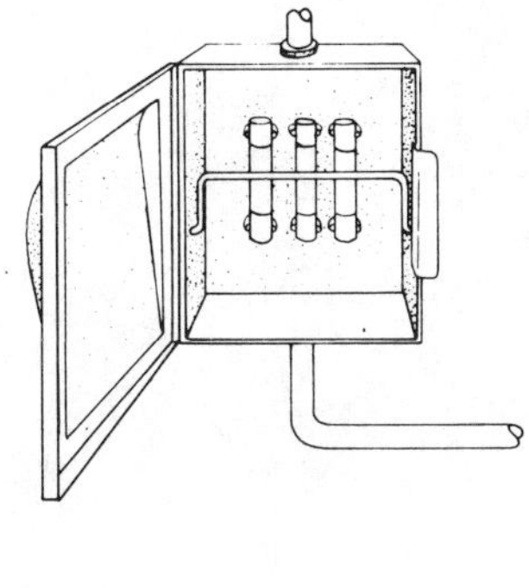

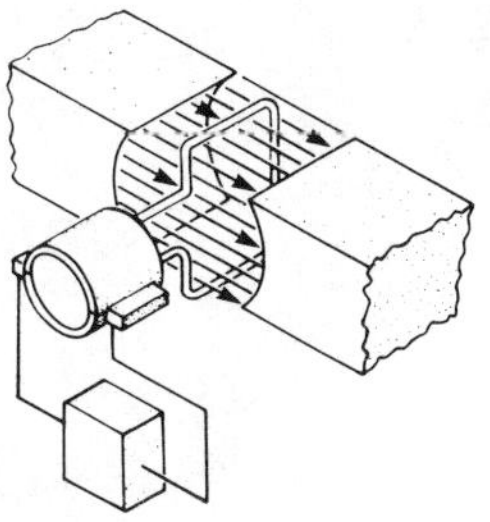

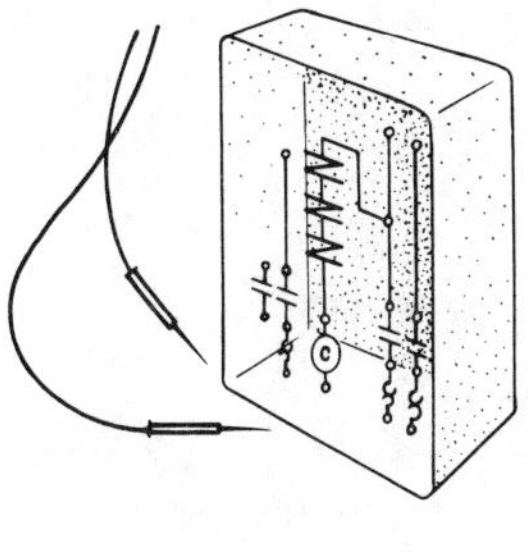

TEST 1—CHAPTER 9

Name ______________________________ Date ______________

True-False

T F **1.** The most popular method of starting and stopping motors is with magnetic starters.

T F **2.** Contacts used for start buttons are normally closed.

T F **3.** Overload contacts are connected in series so that any one set of contacts can trip open and disconnect the control circuit.

T F **4.** Single-phase, split-phase motors can be wound for 120-volt operation only.

T F **5.** The windings of single-phase induction motors are connected in series for the lower voltage.

T F **6.** On wye-connected motors, T_7, T_8, and T_9 tie together to form the wye.

Completion

__________ **7.** An extra set of __________ can be added to a magnetic starter to energize the coil of another starter.

__________ **8.** Terminals T_1, T_2, and T_3 on a magnetic starter connect the branch circuit to the __________ leads.

__________ **9.** Two-wire control circuits do not have to depend on a(n) __________ coil circuit.

__________ **10.** Three speeds can be obtained from motors equipped with __________ pole(s).

__________ **11.** Single-phase, split-phase motors must be provided with a(n) __________ winding to start and run the motor.

__________ **12.** The current flow in each phase leg of delta connected motors should measure even and __________.

Multiple Choice

__________ **13.** When the leads of wye-connected motors are identified, it makes no difference which leads are marked __________.

A. T_1-T_4
B. T_2-T_5
C. T_3-T_6
D. T_7-T_8-T_9

__________ **14.** An autotransformer with a 65% tap reduces the starting current to a value of __________%.

A. 25
B. 35
C. 42
D. 64

______ **15.** Part-winding starting reduces the starting current and torque to a value that is ______% of normal.
A. 25
B. 35
C. 42
D. 65

______ **16.** The phase current of delta-connected motors is equal to ______% of the line current.
A. 42
B. 58
C. 65
D. 80

______ **17.** Stop contacts are connected in series with the coil and contacts ______.
A. 1 and 2
B. 1 and 3
C. 2 and 3
D. none of the above

______ **18.** A ______-amp device or smaller can be used to protect #14 control circuits used on the enclosure of a magnetic starter.
A. 100
B. 110
C. 125
D. 150

Problems

19. Connect the terminals of the magnetic starter for the power and control circuits.

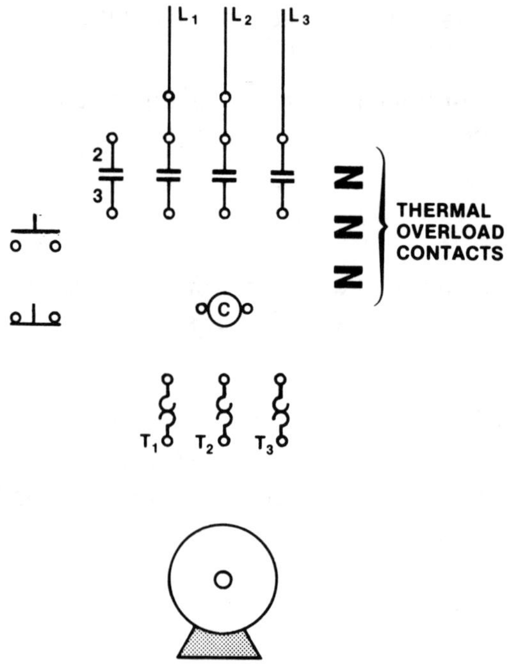

20. Place the leads on the proper terminals to measure the coil circuit voltage.

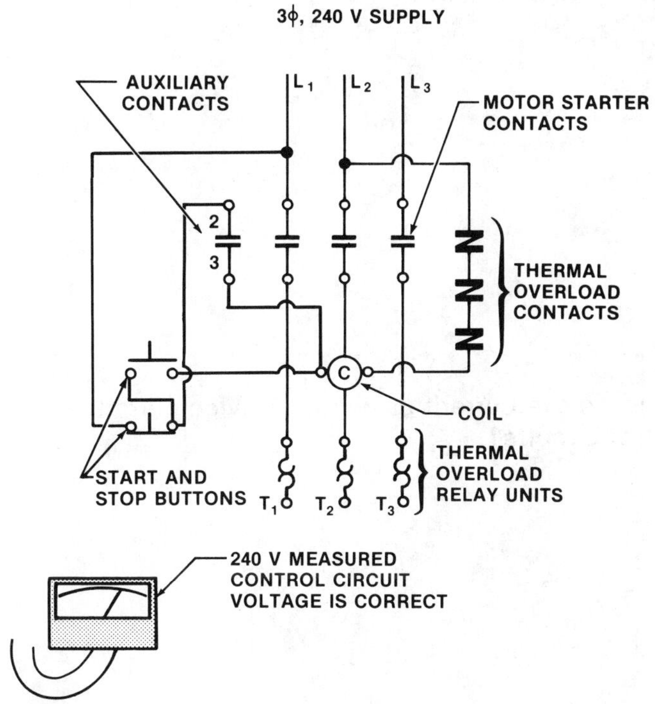

21. Connect power to leads 1, 4, and 9 to identify and check defaced leads.

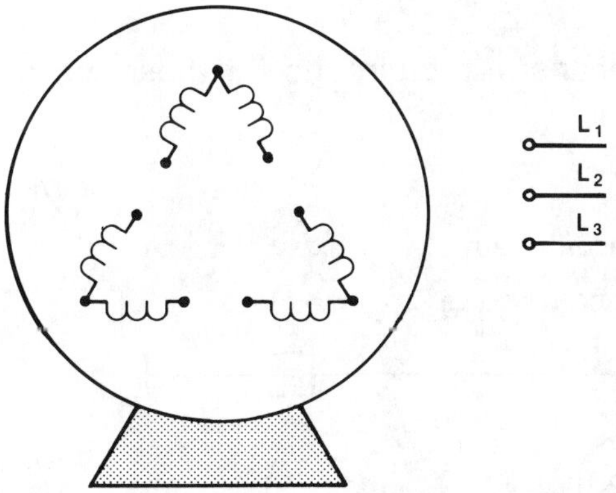

22. A 90-amp overcurrent protection device is increased to 110 amps to allow a motor to start and run. What size copper conductors are needed for the remote control circuit?

23. What size overcurrent protection device is required to protect the primary side of a 900 VA control transformer supplied by a 480-volt, single-phase system?

24. What is the maximum size overcurrent protection device permitted for #10 copper conductors used for remote control circuits?

25. Place a disconnecting means in the circuits for these motors.

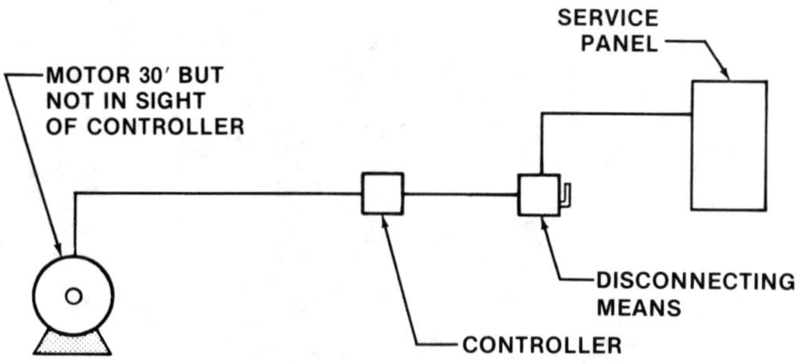

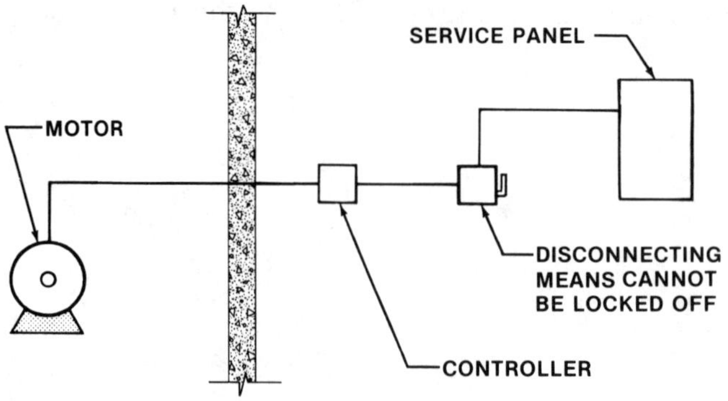

TEST 2—CHAPTER 9

Name ______________________________ Date ______________

True-False

T F **1.** Overcurrent protection devices rated 15-amp and 20-amp can no longer protect #18 and #16 conductors when used for control circuits.

T F **2.** The percentages of protection for control circuits listed in Table 430-72(b) can be exceeded if the overcurrent devices trip open.

T F **3.** A control circuit can be supplied by a transformer that reduces the voltage to a safer operating value for personnel.

T F **4.** The operation of electric motors is based on the principles of magnetism.

T F **5.** Unlike poles of electromagnets attract, and like poles repel with AC power connected to the windings.

T F **6.** The magnetic lines of force around a coil carry electricity.

Completion

__________ **7.** Class 1 circuits can operate at power outputs that are __________ or non-limited.

__________ **8.** Class __________ circuits can be wired with bell wire because their output is limited to a nonhazardous value.

__________ **9.** The circuit breaker in the service equipment can be used as the disconnecting means for motors rated at __________ horsepower or less.

__________ **10.** It is mandatory that a disconnecting means be placed within sight of the ______.

__________ **11.** Where the power for control circuits is brought in by a separate source, disconnects must be provided for the control conductors and the __________ conductors.

__________ **12.** Motors installed in altitudes above __________′ must not be loaded to full capacity.

__________ **13.** Wye-connected motors have three sets of windings with two leads and one set with __________ leads.

__________ **14.** The rotation of single-phase motors can be reversed by reversing the flow of current through the __________ winding.

__________ **15.** When all overload contacts are closed, voltage can be measured from the coil to __________.

__________ **16.** Running overload units respond to temperature and __________.

__________ **17.** Remote control circuits using #10 aluminum conductors can be protected by a(n) __________-amp or less device.

__________ **18.** Control circuit conductors can be increased in size to match the proper ______.

Problems

19. Connect the overload contacts from L_2 to one side of the coil.

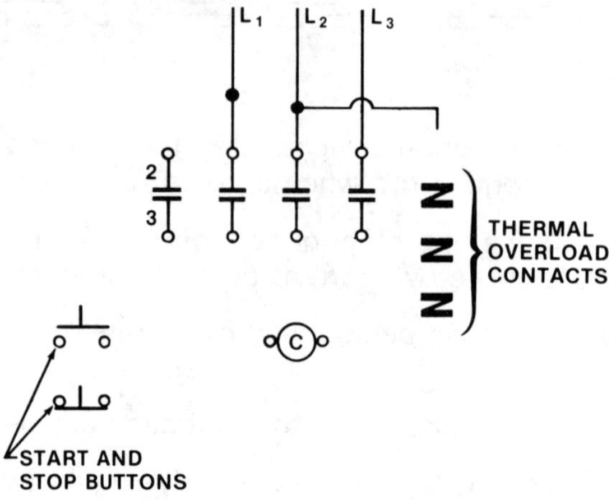

20. What size time-delay fuses used as a single unit are required to start and run a 30-horsepower, 220-volt, three-phase, part-winding motor?

21. What is the voltage in the wye winding using a wye-delta motor for reduced starting with the supply voltage rated at 480 volts?

22. Connect the line to the leads of the armature and field windings.

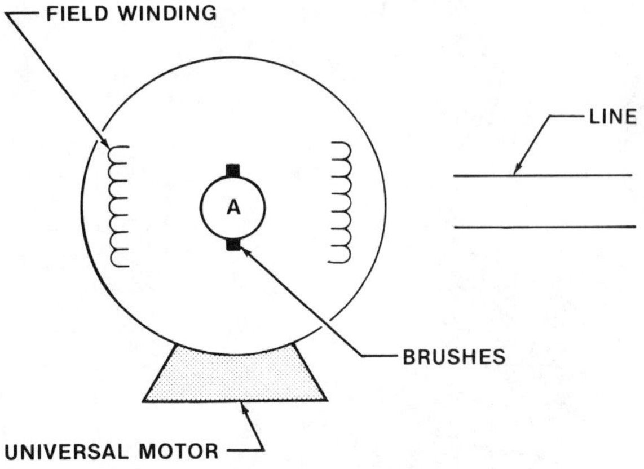

23. Connect the starting and running windings to the line.

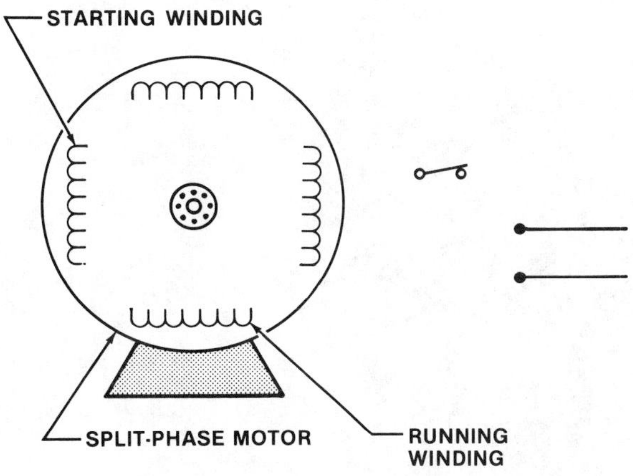

24. Connect the leads of the three-phase motor for wye operation.

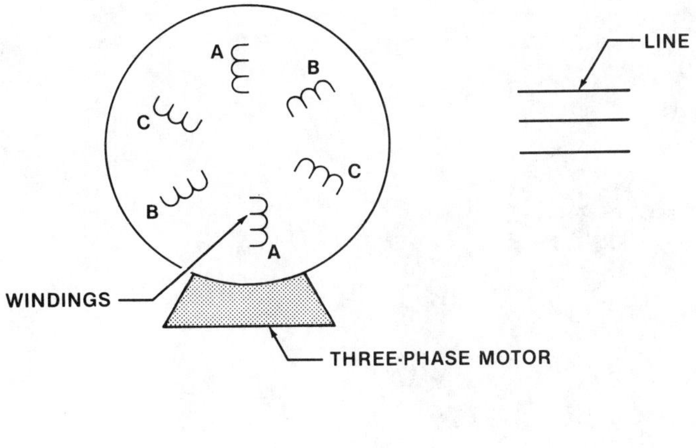

25. What size THWN copper conductors and OCPD is required to supply power to the 1ϕ side of a phase converter that runs a 30 HP, 230 V, 3ϕ induction motor? The phase converter has a nameplate with a 1ϕ input full-load ampacity of 80 A.

TEST 3—CHAPTER 9

Name ______________________________ Date ______________

True-False

T F 1. Running overload protection protects a motor in the event of a short circuit.

T F 2. The service factor of a motor determines the amount of overload the motor can endure before insulation failure.

T F 3. Nonfusible disconnects used to disconnect a motor circuit must be rated at least 110% of the motor's full-load current rating.

T F 4. For a multispeed motor, the conductors between the service equipment and controller are sized on the amperage rating of the largest rpm of the motor.

T F 5. Motors used for varying duty can have their rating reduced by their method of operation.

T F 6. The greater the resistance in the rotor of a wound-rotor motor is, the faster the motor runs.

Completion

__________ 7. Capacitors can be checked to determine if they are defective by using a(n) __________.

__________ 8. The full-load current rating of unlisted motors can be determined by application of a(n) __________.

__________ 9. Capacitors __________ the power factor of motors and lower the current the motor will draw.

__________ 10. Part-winding starting has the disadvantage of utilizing only __________ of the motor's copper during the starting period.

__________ 11. Motors must provide enough __________ to start and drive the loads.

__________ 12. Class __________ motors are the most commonly used motors in industry.

Multiple Choice

__________ 13. When 20-amp, T-rated snap switches are used to control and disconnect motors, they must not be loaded more than ________ amps.

A. 10
B. 15
C. 16
D. 20

__________ 14. Motors with a service factor of 1.10 and a full-load current rating of 22 amps can endure an overload of ________ amps.

A. 22
B. 24.2
C. 28
D. 30

__________ **15.** The starting current of capacitor-start motors is _______% less than for a split-phase motor.
A. 30
B. 40
C. 50
D. 67

__________ **16.** Multiple speeds on shaded-pole motors can be obtained by using taps from a __________.
A. transformer
B. capacitor
C. starting winding
D. none of the above

__________ **17.** The brushes of universal motors are connected in series with the __________.
A. shaded coil
B. field windings
C. both A and B
D. neither A nor B

__________ **18.** A 90-amp time-delay fuse holds 450 amps for approximately __________ seconds.
A. 2
B. 5
C. 10
D. 20

Problems

19. If phases A, B, and C pull 32 amps each and power is lost on phase A, what will phases B and C pull?

20. Connect the windings for high voltage operation.

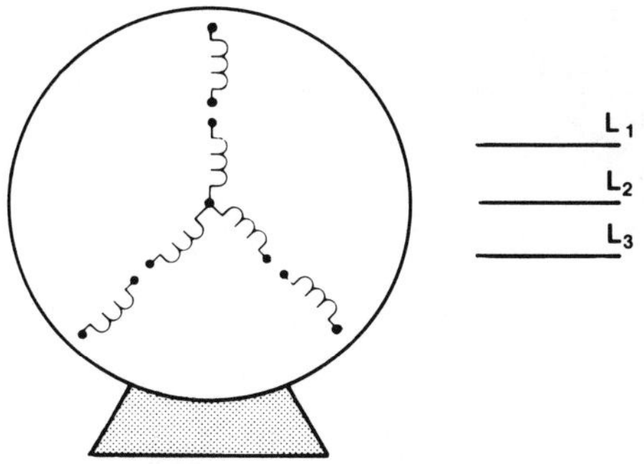

21. What is the reduced starting current for a three-phase induction motor pulling 624 amps of locked-rotor current using resistor starting?

22. What is the reduced starting current for a 30-horsepower, 230-volt, three-phase motor using a 65% tap from an autotransformer?

23. What is the synchronous speed of a four-pole induction motor?

24. What is the actual speed of an induction motor operating with 4% slip and having a synchronous speed of 3600 rpm?

25. What is the maximum size overload protection required for a motor with a temperature rise of 55°C and a nameplate current rating of 48 amps?

TEST 4—CHAPTER 9

Name ____________________ Date ____________________

True-False

T F **1.** Table 430-152 lists four types of reduced starting methods.

T F **2.** Resistor starting uses resistance to reduce the starting current of induction motors.

T F **3.** Reactor starting is used with low-voltage systems and resistor starting is used with high-voltage systems.

T F **4.** The starting winding in a split-phase motor is wound at the bottom of the running winding.

T F **5.** Split-phase motors draw more current when starting than capacitor-start motors.

T F **6.** The direction of rotation of split-phase motors can be reversed by changing the flow of current through the starting winding.

Completion

__________ **7.** Split-phase motors with capacitors have __________ as much starting torque as regular split-phase motors.

__________ **8.** The capacitor in a capacitor-start motor is connected in __________ with the starting switch.

__________ **9.** Permanent capacitor motors can be reversed by using a(n) __________ switch.

__________ **10.** Permanent capacitor motors will not start and drive a(n) __________ load.

__________ **11.** The capacitor is not switched out of the circuit in __________ capacitor motors.

__________ **12.** Shaded-pole motors are designed with __________ coils to provide starting torque.

Mutiple Choice

__________ **13.** The full-load current rating for 120-volt, single-phase motors is determined by multiplying the horsepower by ______.

A. 2.5
B. 3
C. 5
D. 10

__________ **14.** The starting winding of split-phase motors has a resistance that is __________.

A. lower than the running winding
B. higher than the running winding
C. the same as the running winding
D. the same as the rotor winding

__________ **15.** Induction motors are designed to run only on __________ systems.
A. single-phase
B. two-phase
C. three-phase
D. all of the above

__________ **16.** The rotor of a motor usually rotates at about ______% of the synchronous speed.
A. 5
B. 10
C. 12
D. 15

__________ **17.** When motors and other loads are on the same feeder circuit, the largest motor must be multiplied by ______%.
A. 110
B. 115
C. 125
D. 135

__________ **18.** The rating of the multiplier used to determine the full-load current rating of an unlisted 45-horsepower, three-phase, 240-volt motor is ______.
A. 1.2
B. 2.0
C. 2.6
D. 3.2

Problems

19. What size THWN copper conductors and OCPD are required for a 1ϕ feeder circuit supplying power to two phase converters serving a group of 230 V, 3ϕ motors? The motors are rated at 7½ and 10 HP.

20. What size motor is required to drive a load with a 160 foot-pound torque rotating at 1780 rpm?

21. What is the reduced voltage on a three-phase, 480-volt system using an autotransformer with a 50% tap?

22. What size THWN copper conductors are required to supply a 550-volt, three-phase, 125 horse-power synchronous motor operating with an 80% power factor?

23. What is the full-load torque and starting torque of a 15-horsepower, class B motor rotating at 1725 rpm?

24. Connect the windings for 120-volt, single-phase operation.

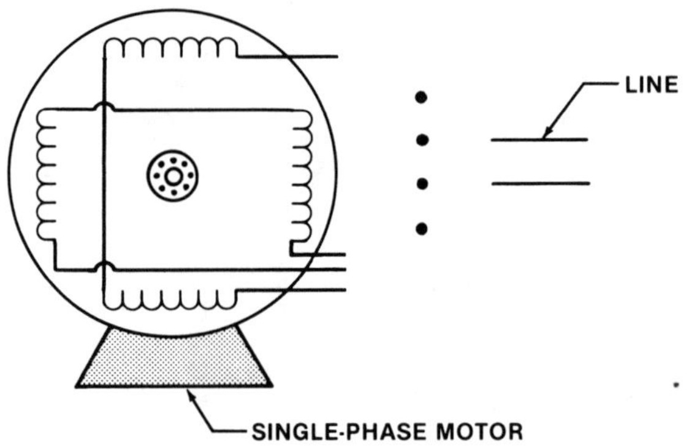

25. What is the actual speed of an induction motor having four poles and operating at 5% slip?

TRANSFORMER OPERATION

Chapter 10

Transformer windings are connected in either series or parallel to obtain the different voltages required for supplying various loads. Basic voltages are 120V, 1ϕ; 120/240V, 1ϕ; 120/240V, 3ϕ; 120/208V, 3ϕ; and 480V, 3ϕ. Higher voltages are available for other specific applications.

Loads or transformer windings must be balanced between phases and from each phase-to-neutral to prevent overloads. Windings are generally connected wye, or open or closed delta. Combination connections may be used on the primary and secondary sides of the transformer to obtain different voltage configurations.

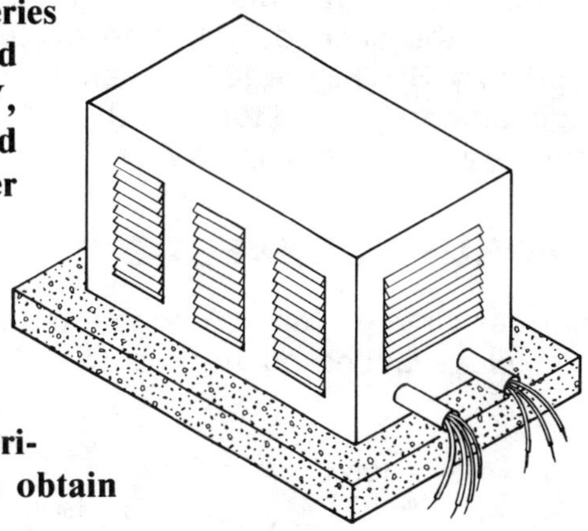

TRANSFORMER PRINCIPLES

The kVA rating of a transformer divided by the primary or secondary voltage determines the current a transformer will deliver. For example, a 30-kVA transformer with a single-phase, 480-volt secondary has a current rating of 62.5 amps.

$$\frac{30\text{kVA} \times 1000}{480\text{V}} = 62.5\text{A}$$

For secondary voltages of single-phase, 240 volts, the current is 125 amps.

$$\frac{30\text{kVA} \times 1000}{240\text{V}} = 125\text{A}$$

The number of turns on the primary or secondary windings of a transformer determines the voltage (V), current (A), and impedance (Z). See Figure 10-1. When the number of turns on the primary and secondary are equal, the input voltage and output voltage are the same, impedance remains constant, and the input current and the output current are the same. When there are fewer turns on the primary than the secondary, the voltage and impedance are stepped up and the current is stepped down. For fewer turns

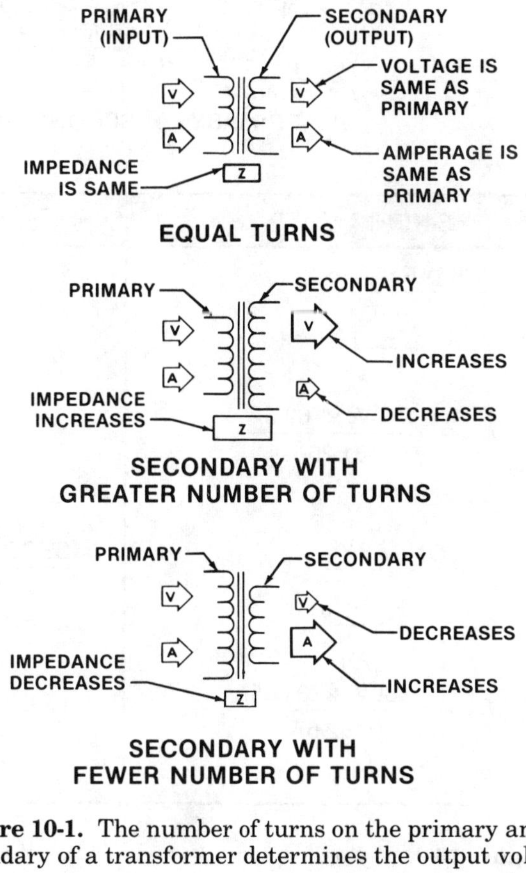

Figure 10-1. The number of turns on the primary and secondary of a transformer determines the output voltage.

on the secondary than the primary, the voltage and impedance are stepped down and the current is stepped up.

The voltage, amperage, and impedance ratios of a transformer are determined by the ratio of the number of turns of the primary windings to the number of turns on the secondary windings. The kVA or volt-amp rating of a transformer is the same value for the primary and secondary. Voltage is stepped down and amperage is increased when the secondary has fewer turns than the primary. See Figure 10-2 for calculations for voltage, amperage, and turns of the primary and secondary with the same kVA rating.

TRANSFORMER CONNECTIONS

A transformer winding may be connected in a number of ways based on the job to be performed. Two or more transformer windings may be connected together in a number of ways based on the voltage desired. Delta and wye connections are commonly used. Voltage configurations for these connections are obtained by connecting the windings in series for high voltages and in parallel for low voltages.

Single-phase Connections

When a transformer's secondary supplies single-phase, 120/240 volts, there will be 120 volts, single-phase between either one of the phase lines and the neutral. At the same time, both phase lines can be tapped to deliver 240 volts, single-phase. The 120-volt lines can supply all lighting, receptacle, and appliance loads. The 240-volt lines can supply all 240-volt, single-phase loads such as water heaters, air condi-

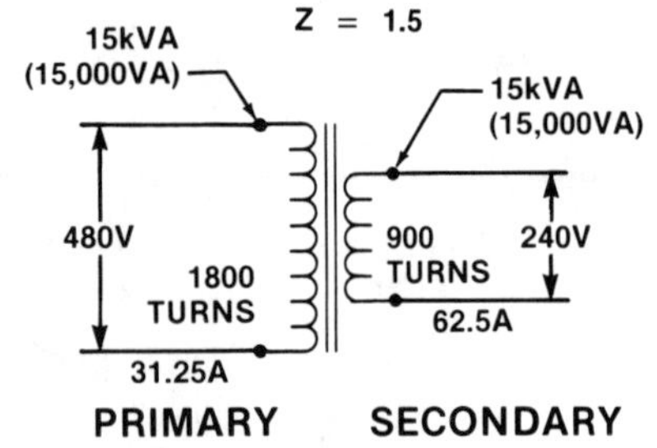

E_P = primary voltage
E_S = secondary voltage
I_P = primary current
I_S = secondary current
T_P = primary turns
T_S = secondary turns

FINDING VOLTAGE	FINDING AMPERAGE	FINDING TURNS
PRIMARY	PRIMARY	PRIMARY
$E_P = \frac{E_S \times I_S}{I_P}$	$I_P = \frac{E_S \times I_S}{E_P}$	$T_P = \frac{E_P \times T_S}{E_S}$
$E_P = \frac{240V \times 62.5A}{31.25A}$	$I_P = \frac{240V \times 62.5A}{480V}$	$T_P = \frac{480V \times 900T}{240V}$
$E_P = 480V$	$I_P = 31.25A$	$T_P = 1800T$
SECONDARY	SECONDARY	SECONDARY
$E_S = \frac{E_P \times I_P}{I_S}$	$I_S = \frac{E_P \times I_P}{E_S}$	$T_S = \frac{E_S \times T_P}{E_P}$
$E_S = \frac{480V \times 31.25A}{62.5A}$	$I_S = \frac{480V \times 31.25A}{240V}$	$T_S = \frac{240V \times 1800T}{480V}$
$E_S = 240V$	$I_S = 62.5A$	$T_S = 900T$

Figure 10-2. Voltage, amperage, and number of turns in transformer windings are determined by applying the proper formula.

tioners, and electrical heating. The transformer may be connected for either 120-volt, single-phase supply or for 120/240-volt, single-phase supply. See Figure 10-3.

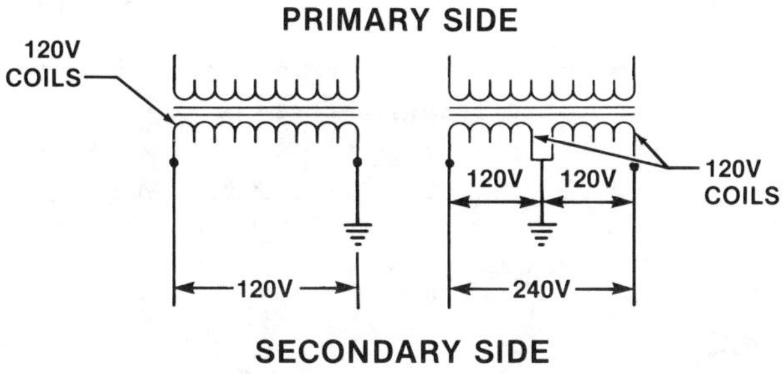

Figure 10-3. Transformers may be connected for 120-volt or 240-volt, single-phase. Tapping the midpoint (center) of a 240-volt transformer produces 120 volts-to-ground.

Neutral Current. The neutral carries the same amount of current as the hot conductor on 120-volt, single-phase systems. If the ungrounded hot conductor carries 25 amps, the grounded neutral conductor will carry 25 amps. The neutral conductor carries the unbalanced current between the two ungrounded conductors of a 120/240-volt, single-phase system.

The 120-volt hot conductors are usually identified as phases A and B. For example, if phase A carries 60 amps and phase B carries 40 amps, the neutral carries 20 amps (60A − 40A = 20A). If phase A carries 60 amps and phase B carries 60 amps, the current in phases A and B would be balanced and the neutral would not pull any current. Neutral current flow can only occur in an unbalanced system. There must be two or more hot conductors with a neutral for an unbalanced condition to occur. See Figure 10-4.

Delta Connections

The delta system is a good short-distance distribution system. It is used for neighborhood and small commercial loads close to the supplying substation. Only one voltage is available between any two wires in a delta system. The coil voltage is the same as the phase-to-phase voltage. For example, if the phase-to-phase voltage is 480 volts, the coil voltage is 480 volts.

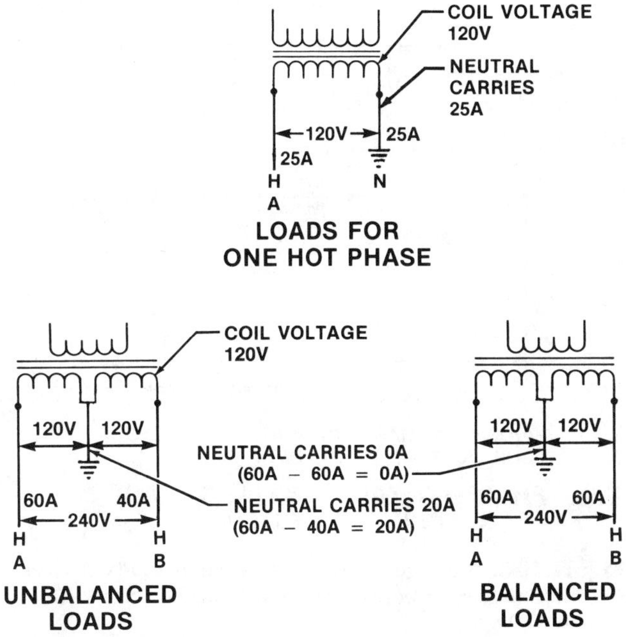

Figure 10-4. There is no current flow in the neutral when phases A and B are balanced. Neutral current flow can only occur in an unbalanced system.

The windings of transformers may be connected for open or closed delta systems. An open delta system consists of only two transformers, one of which is always larger than the other. A closed delta system always consists of three transformers. One transformer may be larger than the other two, depending on the three-phase and single-phase loads served.

The delta system is shown with a triangle. A wire from each point of the triangle represents a three-phase, three-wire delta system. The voltage is the same between any two wires. See Figure 10-5.

Balanced Current. While the phase-to-phase and coil voltage are the same in a delta system, the line current and coil current are not the same. The flow of current in a delta system has two paths to follow at each closed end where the phase conductors terminate. The amount of current in the coils is 58% of the line current measured on each phase. The 58% multiplier is derived from dividing 1 by the square root of 3 ($1/\sqrt{3} = 1/1.732 = 58\%$). For example, if the current in each phase is 100 amps, the coil current would be 58 amps. See Figure 10-6.

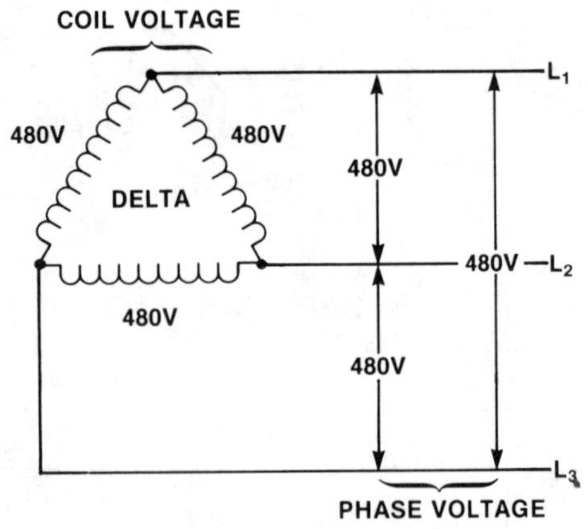

PHASE TO PHASE VOLTAGE = 480 VOLTS
COIL VOLTAGE = 480 VOLTS

PHASE VOLTAGE = COIL VOLTAGE IN DELTA SYSTEMS

Figure 10-5. Voltage is the same between any two wires of a three-phase, three-wire delta system. The coil voltage is also 480 volts.

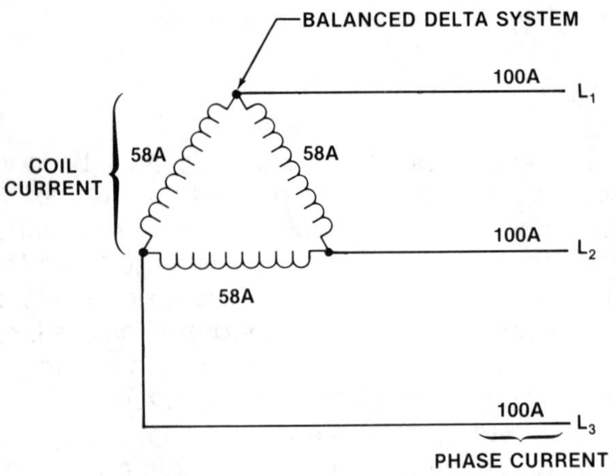

COIL CURRENT = 58% OF PHASE CURRENT

PROBLEM: What is the coil current of a closed delta system with 100 amps on each phase?

Step 1: 100A per ϕ

$$\sqrt{3} = 1.732$$

$$\frac{1}{1.732} = 58\%$$

Step 2: 100A × 58% = 58A

Answer: **58A**

Figure 10-6. The coil current of a balanced delta system is 58% of the phase current.

Unbalanced Current. If the current flow in a delta system is not balanced, the current flow in L_1 is found by applying the appropriate formula as shown in Figure 10-7. *NOTE:* The current of other lines is found by substituting appropriate values of the coil current.

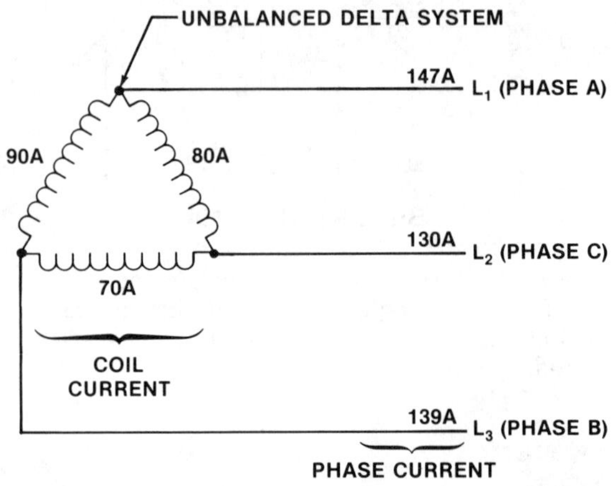

PROBLEM: What is the current in L_1 of an unbalanced delta system when the coil current is 90 amps for Phase B and 80 amps for Phase C?

Step 1: $L_1 = \sqrt{b^2 + c^2 + bc}$

$L_1 = \sqrt{90^2 + 80^2 + (90 \times 80)}$

$L_1 = \sqrt{8100 + 6400 + 7200}$

$L_1 = \sqrt{21{,}700} = 147\text{A}$

Answer: **147A**

Figure 10-7. L_1 current of an unbalanced delta system is found by applying the formula $L_1 = \sqrt{b^2 + c^2 + bc}$. Currents of L_2 and L_3 can be found by substituting appropriate coil current values.

Neutral Current. The neutral in a four-wire, three-phase delta system carries the unbalanced current between phases A and C. Only two ungrounded hot conductors can be derived from a delta system. One of the 240-volt coils is tapped and 120 volts is obtained from the tap to each outside phase leg (A and C). The current flow in phase B must travel through one 240-volt and one 120-volt coil to reach the tap, which is connected to ground. Therefore, the voltage measured to ground is 208 volts [(120V + 240V = 360V). (360V/$\sqrt{3}$ = 208V)]. See Figure 10-8. See 215-8 for marking the high leg, 230-56 for location in service equipment, and 384-3(e)(f) for panelboards.

Wye (Star) Connections

In a wye system, better known as a *star system,* the voltage between any two wires will always be the same on a three-phase, four-wire system. The voltage between any one of the phase conductors and the neutral will always be less than the voltage of the power conductors. For example, if the voltage between the power conductors of any two phases of the four-wire system is 208 volts, the voltage from any phase power conductor to ground will be 120 volts. This voltage is derived from applying the square root of the three-phase power. In a wye system, the voltage between any two power phase conductors will always be 1.732 (the square root of 3) times the voltage between the neutral and any one of the power phase conductors. The phase-to-ground voltage can be found by dividing the phase-to-phase voltage by 1.732. See Figure 10-9.

The difference between single-phase voltage

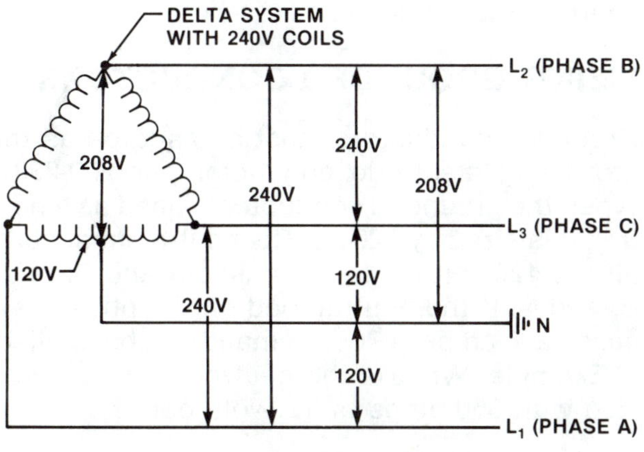

215-8

RELATION OF VOLTAGE		
Phases	High Leg	120 Volts-to-ground
A - B = 240V A - C = 240V B - C = 240V	B - N = 208 volts-to-ground	A - N = 120V C - N = 120V

DERIVING HIGH LEG

$$B = \frac{(B - A) + (A - N)}{\sqrt{3}}$$

$$B = \frac{240V + 120V}{1.732} = 208V$$

B = 208 volts-to-ground

RECOMMENDED COLOR CODING

Phase A — black	Phase C — red or blue
Phase B — orange	Neutral — white or gray

Figure 10-8. A four-wire, three-phase delta system produces only two 120-volt circuits and one 208-volt circuit to ground. Phase voltage equals coil voltage.

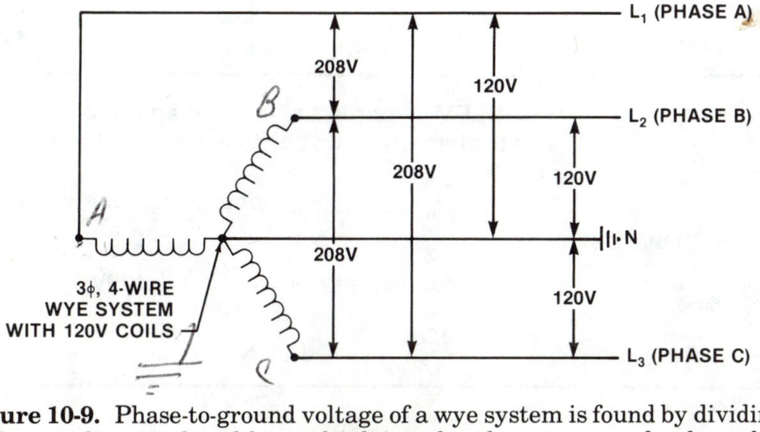

RECOMMENDED COLOR CODING

Low Voltage (120/208V)
- Phase A - black
- Phase B - red
- Phase C - blue
- Neutral - white or gray

High Voltage (277/480V)
- Phase A - brown
- Phase B - orange
- Phase C - yellow
- Neutral - white or gray

Figure 10-9. Phase-to-ground voltage of a wye system is found by dividing the phase-to-phase voltage by $\sqrt{3}$. Phase-to-phase voltage is found by multiplying the phase-to-ground voltage by $\sqrt{3}$.

and three-phase voltage lies in the number of conductors used. When all three power conductors of the system are used, the system is operating on three-phase voltage. If only two power conductors are used, the system would be operating on single-phase voltage. These voltage conductors do not change direction of polarity at the same time, so they are not in phase. If all the three-phase conductors are used, the system would be operating on three-phase voltage.

The phase-to-phase voltage is the same in a wye system, but the coil voltage is equal to 1.732 divided into the phase-to-phase voltage. Each phase leg in a wye system is connected through the coil to a mid-point tap where they all meet. See Figure 10-10.

Balanced Current. The phase-to-phase voltage is not the same as the coil voltage in a wye system. The phase-to-phase voltage is found by multiplying the coil voltage by 1.732, which is the square root of 3, or by dividing the phase-to-phase voltage by 1.732. The flow of current in a wye system is the same in the coils as the phase-to-phase current. See Figure 10-11.

In this type of system, the coil windings will heat because they are pulling the same current as the line. Delta-connected windings do not heat as much as wye-connected windings because they pull only 58% of the line current.

Neutral Current. The flow of current in the neutral conductor of a 120/208-volt or 277/480-volt wye system is determined by using a different procedure than that used for a 120/240-volt, single-phase or 120/240-volt delta system. To find the current flow in the neutral of a 120/208-volt or 277/480-volt wye system, square the current in all lines and add together. Subtract the currents in $L_1 \times L_2$, $L_2 \times L_3$, and $L_3 \times L_1$. The square root of these values is the current flow in the neutral. See Figure 10-12.

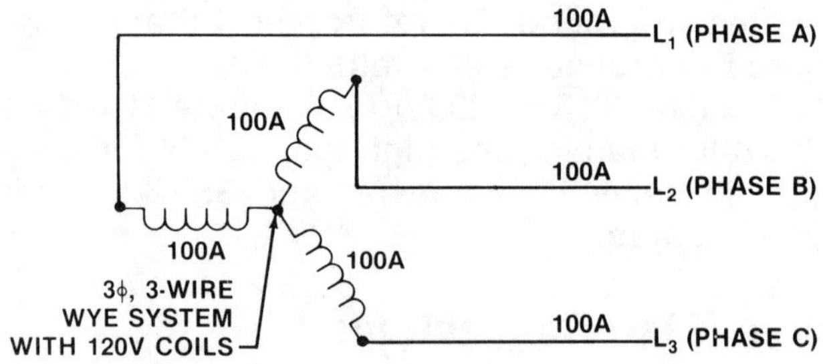

Figure 10-11. The current flow in the coils of a wye system is the same as the phase-to-phase current.

SIZING GROUNDED CONDUCTORS

Grounded conductors must be as large as the grounding electrode conductor per 250-23(b). When the grounded conductor is used as a neutral, it is sized by 220-22. The neutral must carry all the 120-volt loads or phases A and B, or A, B, and C. If the neutral load on the phases exceeds 200 amps, a 70% demand can be applied.

Example: What is the neutral load for a service with 350 amps of 120-volt loads?

Step 1: *220-22.*

First 200A × 100% = 200A
Next 150A × 70% = 105A
Total: 305A

Answer: **305A**

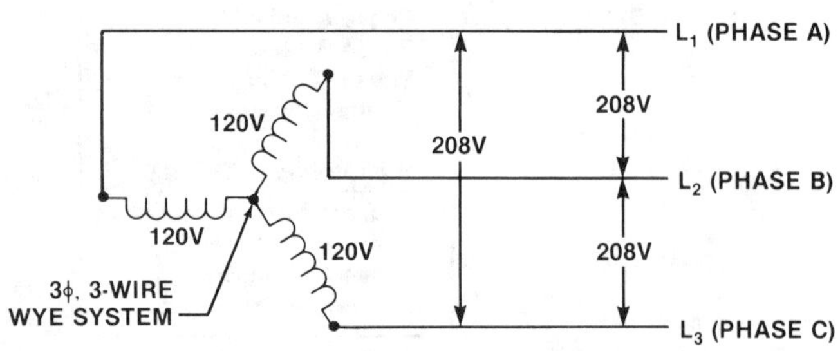

PROBLEM: What is the coil voltage of a 3-phase, wye system when the phase-to-phase voltage is 208 volts?

Step 1: $\textit{coil voltage} = \dfrac{\textit{phase-to-phase voltage}}{\sqrt{3}}$

$\textit{coil voltage} = \dfrac{208V}{1.732} = 120V$

Answer: **120V**

Figure 10-10. The coil voltage in a wye system is found by dividing the phase-to-phase voltage by $\sqrt{3}$.

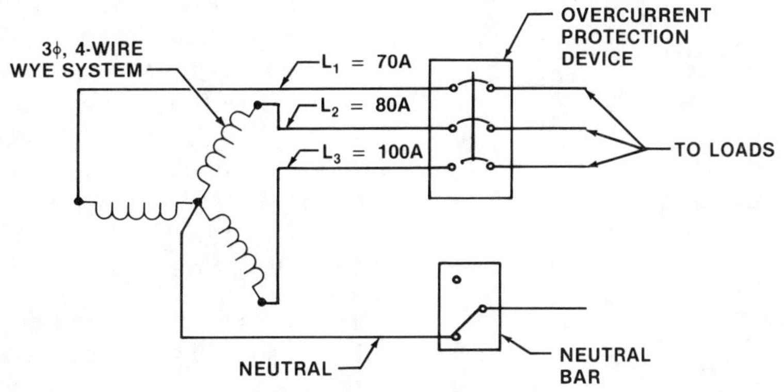

PROBLEM: What is the current flow in the neutral with L_1 pulling 70 amps, L_2 pulling 80 amps, and L_3 pulling 100 amps?

Step 1: $amps = \sqrt{I^2_{L1} + I^2_{L2} + I^2_{L3} - (I_{L1} \times I_{L2}) - (I_{L2} \times I_{L3}) - (I_{L3} \times I_{L1})}$

$amps = \sqrt{70^2 + 80^2 + 100^2 - (70 \times 80) - (80 \times 100) - (100 \times 70)}$

$amps = \sqrt{4{,}900A + 6{,}400A + 10{,}000A - 5{,}600A - 8{,}000A - 7{,}000A}$

$amps = \sqrt{21{,}300 - 20{,}600}$

$amps = \sqrt{700} = 26.5A$

Answer: **26.5A**

Figure 10-12. The neutral load on a three-phase, four-wire wye system is found by taking the square roots of the phases.

The neutral cannot be reduced for 120-volt loads consisting of electric-discharge lighting. Note 10 to Tables 310-16 through 310-19 shall be applied when the neutral supplies loads such as electric-discharge lighting, data processing, and similar equipment.

Example: What is the neutral load for a service supplying 280 amps of fluorescent lighting fixtures in an office building?

Step 1: *220-22; Note 10 of Notes to Ampacity Tables of 0 to 2000 Volts.*
280 A × 100% = 280 A

Answer: **280 A**

BALANCING LOADS ON A SINGLE-PHASE TRANSFORMER

Single-phase transformers have a secondary voltage of 120/240 volts and utilize a three-wire system. If the load is not distributed as evenly as possible on each 120-volt winding, one of the windings can be overloaded. Where a 20-kVA transformer with a 240-volt winding is center-tapped and carried to ground, there will be two 120-volt windings. The 20 kVA must be divided by 2 (20 kVA/2 = 10 kVA). Each 120-volt winding can be loaded to 10 kVA or less.

For example, a 20-kVA transformer with a 120/240-volt secondary is to serve a 16-kVA load at 240 volts and two 2-kVA loads at 120 volts. Each 120-volt load must be divided as evenly as possible on each 120-volt winding to balance the load. The same procedure must be applied for the 240-volt loads. Proper balancing of loads prevents the windings from overheating. See Figure 10-13. If the two 120-volt, 2-kVA loads are connected to one winding, the load is unbalanced. Unbalanced loading overloads the winding and causes insulation failure, which creates a short circuit.

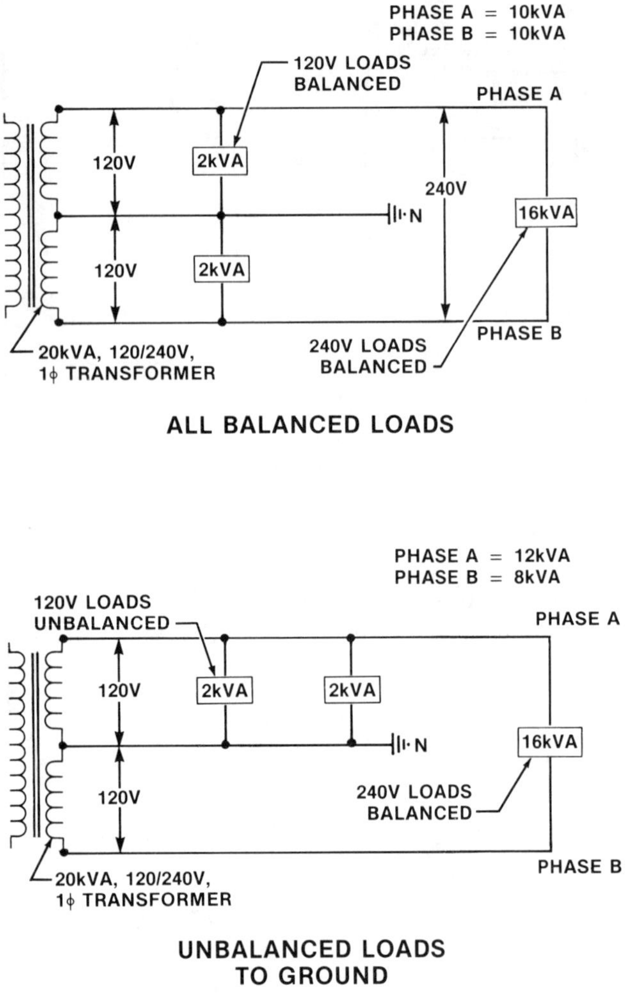

Figure 10-13. Single-phase loads must be balanced to ground and phase-to-phase loads must be balanced to prevent overheating of windings.

BALANCING LOADS ON A THREE-PHASE TRANSFORMER

Three-phase transformers have a secondary that operates on a four-wire system. The voltage is usually 120/208-volt or 277/480-volt. Each phase of a three-phase transformer must be considered as a single-phase transformer when balancing the load. For example, a 50-kVA transformer with a 120/208-volt secondary is to serve five loads at 120-volt, single-phase. The loads are 15-kVA, 4-kVA, 10-kVA, 8-kVA, and 5-kVA. Each 120-volt phase can be loaded to 16.7 kVA (50 kVA/3 = 16.7 kVA). See Figure 10-14.

Trade Tip: When trimming out a panelboard, loads should be divided evenly between phases to assure that transformer windings are not overloaded. Branch-circuit loads should be connected to CBs or fuses to evenly distribute loads on phases A, B, and C.

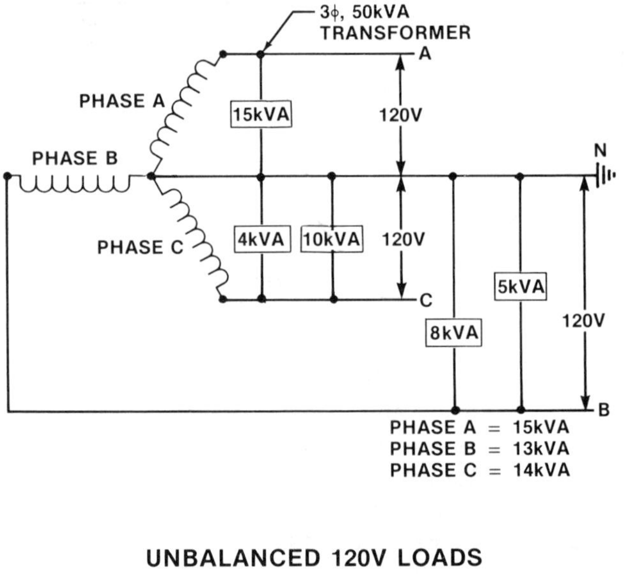

Figure 10-14. Any one phase of a 50-kVA transformer must not exceed 16.7kVA [(50kVA/3) = 16.7kVA]. An unbalanced load of less than 16.7kVA is permitted.

REVIEW—CHAPTER 10

Name Date

True-False

T F 1. Amperage for the secondary side of a transformer can be determined by dividing the kVA by the primary voltage.

T F 2. Coil current in a delta-connected system is 58% of the line current.

T F 3. The secondary of transformers can be stepped up or down to provide various voltages.

T F 4. Transformers with higher voltage deliver less amps for the load to be served.

T F 5. Secondaries with more turns than the primary step down voltages.

T F 6. Transformer windings must be connected in series to obtain higher voltages.

T F 7. The number of turns on the primary and secondary determine voltage output.

T F 8. Three transformers are required for an open delta system.

T F 9. Output terminals are located on the primary side of a transformer.

T F 10. The core of transformers is magnetized by the current flow through the primary winding.

T F 11. The core of transformers limits the voltage applied to the primary coil to be transmitted by induction to the secondary coil.

T F 12. The voltage of the secondary of transformers supplying single-phase, 120/240 voltage measures 240 volts phase-to-phase.

T F 13. Where one leg of the secondary of a 120-volt coil is grounded on a transformer supplying single-phase current, 120 volts is measured between the conductors.

T F 14. With unbalanced current on a neutral conductor, there is no current flow on the grounded leg.

T F 15. In a delta-connected system, the phase-to-phase voltage is opposite the coil voltage.

Completion

__________ 1. Transformers connected for delta operation have the same coil voltage and __________________ voltage.

__________ 2. The amount of current in the coils is _______% of the line current measured in each phase of a balanced system.

__________ 3. The flow of current in delta systems takes __________________ paths at each corner to the coils.

_____ **4.** The high leg voltage-to-ground, single-phase on delta-connected transformers is _____ than on the other two systems.

_____ **5.** The phase-to-ground voltage of wye-connected transformers is determined by _____ the phase-to-phase voltage by $\sqrt{3}$.

_____ **6.** The coil voltage of wye-connected transformers is found by dividing the phase-to-_____ voltage by $\sqrt{3}$.

_____ **7.** Wye-connected transformers with the wye connected to ground have _____ supply wires.

_____ **8.** In a three-wire, single-phase system, the neutral conductor carries the _____ current.

_____ **9.** The neutral conductor must be the same size as the phase conductors where electric-discharge lighting is supplied by a(n) _____-wire wye system.

_____ **10.** Transformers connected to operate on closed or open delta systems have only _____ 120-volt conductors to ground.

_____ **11.** A grounded neutral conductor must be identified with a white or _____ color wire.

_____ **12.** The voltage on the secondary can be stepped up or stepped down by adding or subtracting the number of _____.

_____ **13.** The impedance of transformers determines the amount of _____ that will be delivered.

_____ **14.** If the primary has the same number of _____ as the secondary, the supply source is the same.

_____ **15.** Wye systems are often utilized in locations having numerous _____-volt loads.

Multiple Choice

_____ **1.** A three-wire, single-phase system has two conductors that are _____ volts-to-ground.

A. 120
B. 194
C. 208
D. 240

_____ **2.** The 120-volt conductors of three-wire, single-phase systems are identified with the letters _____.

A. A and B
B. A and C
C. C and D
D. X and Y

_______ **3.** The phase-to-phase voltage on a three-wire, three-phase system is usually _______ volts for higher voltage motors.
A. 120
B. 208
C. 240
D. 480

_______ **4.** The multiplier used for 1/1.732 is _______%.
A. 42
B. 58
C. 60
D. 75

_______ **5.** If the phase-to-phase voltage in a delta-connected bank of transformers is 480 volts, the voltage of the coils is _______ volts.
A. 120
B. 208
C. 240
D. 480

_______ **6.** The voltage of the high leg in a delta-connected bank of transformers is _______ volts-to-ground.
A. 120
B. 184
C. 208
D. 240

_______ **7.** The phase-to-phase voltage in a wye-connected bank of transformers used to supply 120 volts-to-ground is _______ volts.
A. 200
B. 208
C. 240
D. 480

_______ **8.** Neutral current exceeding 200 amps on loads not containing electric-discharge lighting can be reduced _______%.
A. 50
B. 60
C. 70
D. 75

_______ **9.** The coil voltage of wye-connected transformers having phase-to-phase voltage of 480 volts is _______ volts.
A. 120
B. 208
C. 277
D. 480

_______ **10.** The recommended color coding for phases A, B, and C in wye-connected, low-voltage systems is _______ for the lower voltage.
A. black, red, blue
B. brown, orange, yellow
C. blue, brown, pink
D. black, pink, blue

Problems

1. Determine the following for the transformer.
   A. secondary amperage = __________ amps
   B. secondary coil = __________ turns

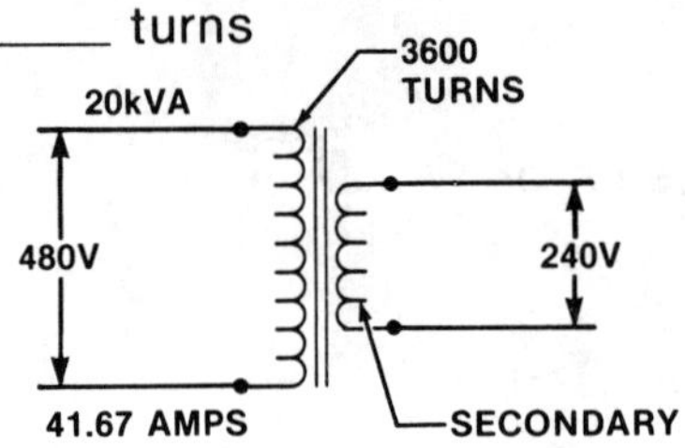

2. What is the neutral current if phase A pulls 72 amps and phase B pulls 96 amps?

3. What is the primary full-load current rating of a 30kVA transformer with a single-phase voltage of 480 volts?

4. What is the coil current in a bank of three-phase, delta-connected transformers having a phase-to-phase current of 87 amps?

5. What is the phase-to-phase voltage of a wye-connected transformer that has a coil voltage of 277 volts?

6. What is the phase current in a three-phase wye-connected transformer if the coil current is 92 amps?

7. What is the neutral current in a wye-connected system where the current in L_1 is 110 amps, L_2 is 140 amps, and L_3 is 160 amps?

8. What size transformer is required to supply the following loads? _______ kVA

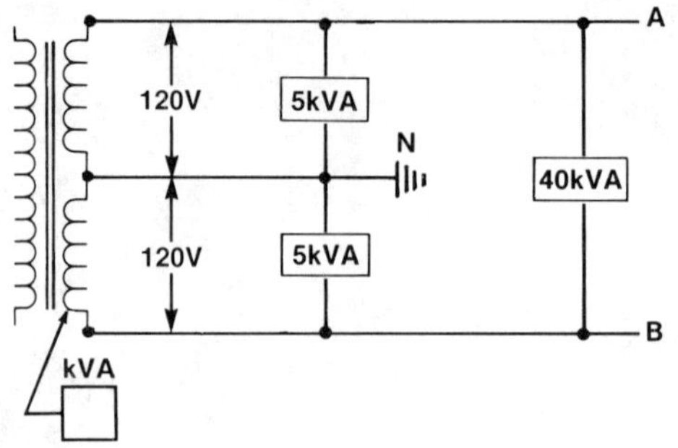

9. What size transformer is required to supply the following loads? _______ kVA

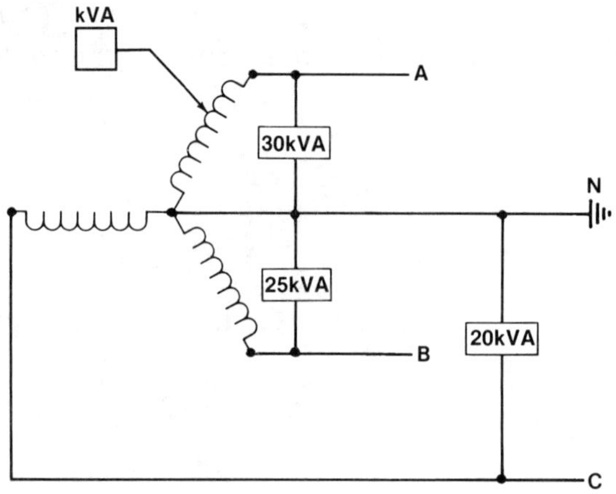

10. Show the method for finding the value of the high leg using coil voltages of a 120/240-volt, 4-wire delta system.

TRANSFORMER INSTALLATION

Chapter 11

Transformers and transformer vaults are covered by Article 450. Transformers must be installed to provide accessibility to qualified personnel and safety to non-qualified personnel and adjacent buildings and property.

Transformer vaults must be ventilated and walls, doors, and roofs must meet minimum fire resistance standards. Three hours is specified in 450-42. Exceptions may be applied. The ventilation requirements of vaults are determined by transformer capacity with the minimum net ventilation area of a vault being one square foot.

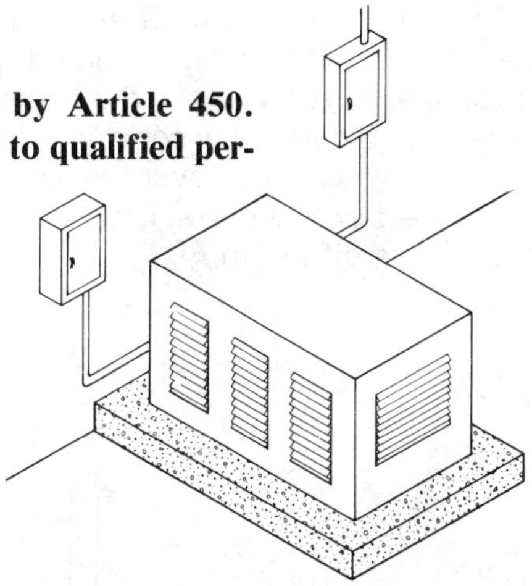

INSTALLING TRANSFORMERS—450

Article 450 permits transformer installation and protection. Transformers may be installed inside or outside a building. They may be installed on a pole, at grade level, or in other accessible locations outside a building. Inside locations commonly used for placing transformers include mechanical rooms, walls, and ceiling areas. Article 450 specifies location, ventilation, guarding, grounding, accessibility, and other factors pertaining to transformers.

Setting or Mounting Dry-type Transformers—450-13

Transformers must be located to provide easy access to qualified personnel for maintenance and inspection. If a ladder must be used to reach the transformer, the transformer is not considered readily accessible.

450-13, Ex. 1. Transformers mounted on the wall or hung from ceilings are not considered readily accessible by the Code because a ladder is needed to reach the transformer. However, since the transformer is in the open and can be seen, Exception 1 to 450-13 will allow such an installation. See Figure 11-1.

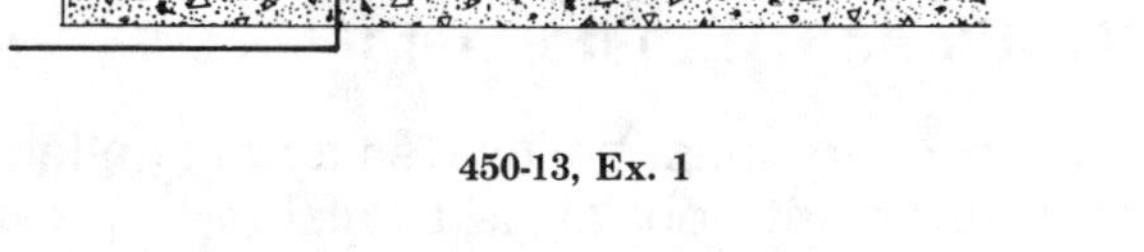

450-13, Ex. 1

Figure 11-1. Transformers may be hung on the wall or ceiling if mounted in the open where visible.

Transformers can be mounted on the floor in mechanical rooms or switchgear rooms. However, the clearances per 110-16 and 110-34 must be applied. Condition 1 of 110-16 or 110-34 must be applied if there is a wall in front of the transformer. Condition 2 must be applied where the wall in front of the transformer is grounded to earth. Walls grounded to earth include concrete, brick, and tilt-slab. If a switch or panelboard with live parts is opposite the transformer, Condition 3 must be applied. Note that panelboards must not be mounted directly above the transformer. See Figure 11-2. Electricians servicing the panelboard would have to reach over the transformer to work on dead or live parts, creating a potentially dangerous situation.

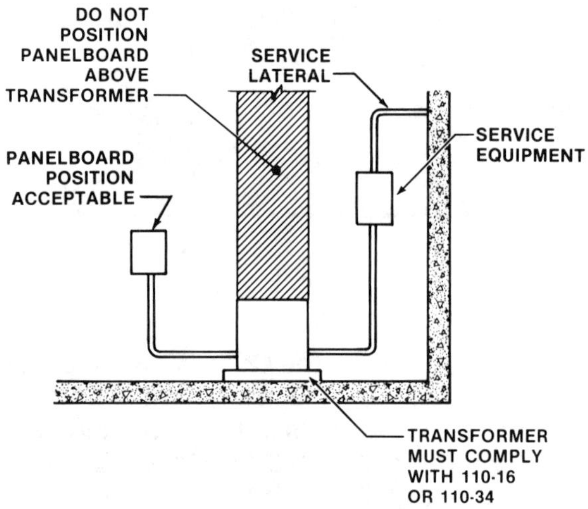

Figure 11-2. Panelboards must not be mounted directly above transformers.

450-13, Ex. 2. Transformers of 600 volts or less and 50 kVA or less may be mounted above suspended ceilings or fire-resistant hollow spaces of buildings not permanently closed in by the structure. These transformers are not required to be readily accessible.

Ventilating Transformers—450-9

Transformers must be located and installed so that sufficient cool air is available to prevent overheating. Either natural or conditioned air is suitable to prevent overheating. Transformers should never be located in rooms or areas subjected to exceedingly high temperatures.

Guarding Transformers—450-8

To prevent accidental contact with live parts that could cause damage or injury, transformers should either be isolated in a room or be located in an area accessible only to qualified personnel. Transformers may also be elevated to safeguard live parts from possible damage. Sections 110-17 and 110-34(e) specify the following as acceptable means of safeguarding:

1. Transformers located in a room or place accessible only to qualified personnel.
2. Permanent partitions or screens may be used to isolate transformers.
3. Transformers may be elevated at least 8′ above the floor to prevent unauthorized contact.

Grounding Transformers—450-10

Transformer cases must be grounded per the provisions of 250-42. Fences or guards around transformers are also required to be grounded. See Figure 11-3.

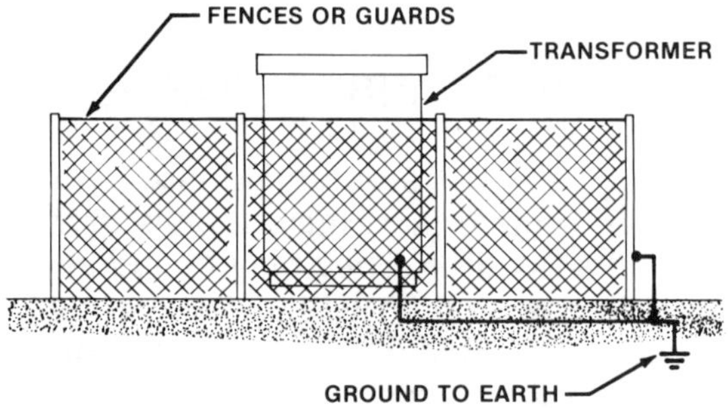

Figure 11-3. Transformer case, fences, and guards must be grounded.

Dry-type Transformers Installed Indoors—450-21

Dry-type transformers installed indoors must meet the following conditions. See Figure 11-4.

1. Dry-type transformers rated at 112.5 kVA or less must have a fire-resistant, heat-insulating barrier between transformers and combustible material. If no barrier is used and the voltage is

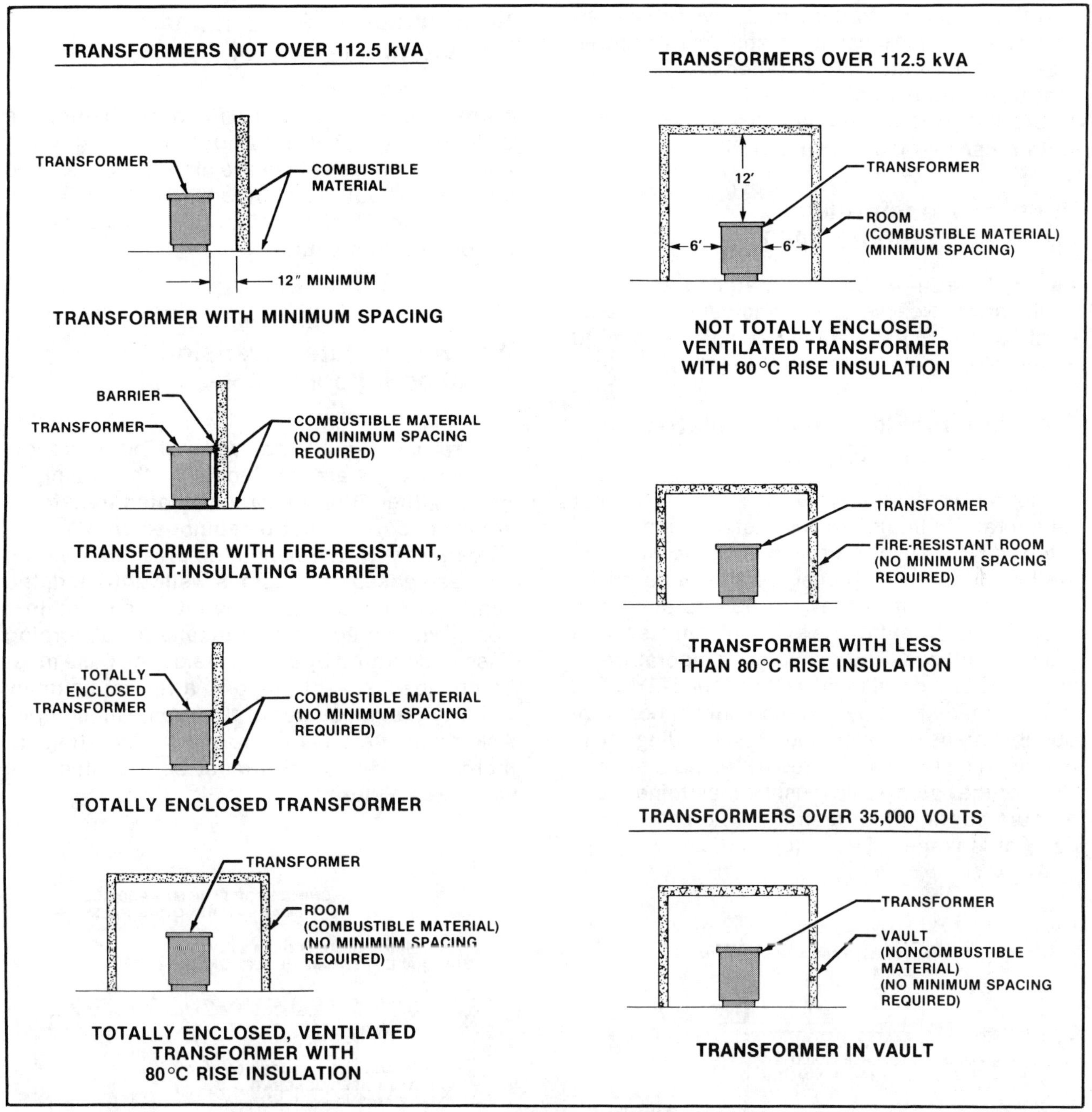

Figure 11-4. Dry-type transformers are installed indoors per 450-21.

over 600 volts, a 12″ minimum separation from combustible material is required.

2. Dry-type transformers are not required to have a 12″ minimum separation or barrier if they are completely enclosed except for ventilation openings, and if they are rated 112.5 kVA or less and operate at 600 volts or less.

3. Dry-type transformers that have Class B or Class H insulation and are rated more than 112.5 kVA must have a fire-resistant, heat-insulating barrier between transformers and combustible material. Without a barrier a 6′ horizontal and 12′ vertical separation from combustible material is required.

4. Dry-type transformers that have other than Class B or Class H insulation and are rated more than 112.5 kVA must be installed in a fire-resistant transformer room.

5. Dry-type transformers rated over 35,000 volts must be installed in a vault.

Dry-type Transformers Installed Outdoors—450-22

All dry-type transformers installed outdoors must have weatherproof enclosures. Transformers over 112.5 kVA must not be located within 12″ of any combustible material.

Less-flammable, Liquid-insulated Transformers—450-23

Transformers insulated with a high fire point (temperature) liquid may be installed indoors if rated at 35,000 volts or less. Transformers installed indoors with higher voltages must be installed in a vault. If they are not installed in a vault, an automatic fire-extinguishing system must be utilized. The minimum temperature at which the liquid will ignite is 300 °C or 572 °F. The transformers must be installed only in noncombustible areas of noncombustible buildings. The minimum clearance required by the heat release rates of the listed liquid must be maintained, and a liquid confinement area must be provided in case of leakages. See Figure 11-5.

Nonflammable Fluid-insulated Transformers—450-24

Transformers insulated with a nonflammable dielectric fluid that does not have a flash point and is not flammable in the air may be installed indoors or outdoors in any location. Transformers over 35,000 volts installed indoors must be located in a vault.

Askarel-insulated Transformers Installed Indoors—450-25

Askarel is a liquid that will not burn. Nonexplosive gases are produced by the arcing in Askarel; therefore, Askarel-insulated transformers over 25 kVA must be equipped with a pressure-relief vent to release the pressure built up by these gases. If a room is well-ventilated, the vent may discharge directly into the room. In a poorly ventilated room, a means for absorbing gases generated by arcing inside the case must be provided. A vent piped to a flue or chimney to carry the gases out of the room will comply. Askarel-insulated transformers operating at more than 35,000 volts must be installed in a vault. See Figure 11-6.

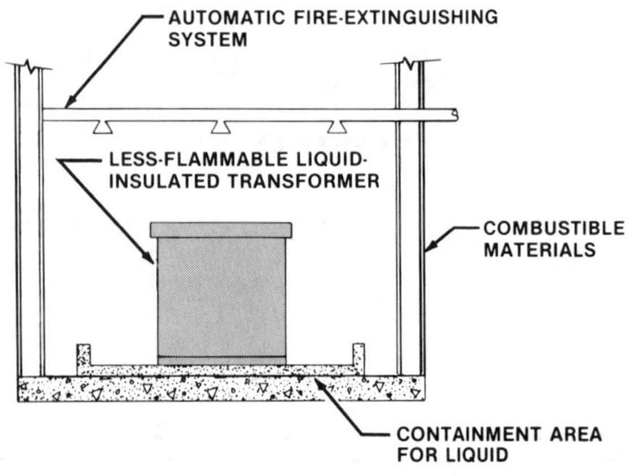

Figure 11-5. Less-flammable liquid-insulated transformers of 35,000 volts or less may be installed indoors. For higher voltages, either a vault location or an automatic fire-extinguishing system is required.

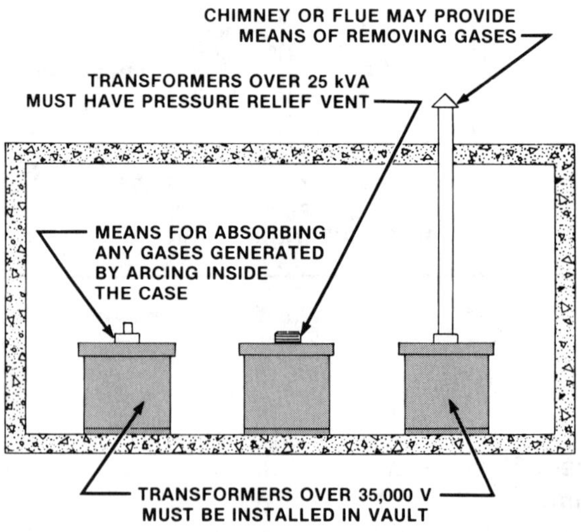

Figure 11-6. Askarel-insulated transformers of more than 35,000 volts must be installed in a vault.

Oil-insulated Transformers Installed Indoors—450-26

Generally, oil-insulated transformers must be in a vault when installed indoors according to the provisions of Article 450, part C. See Figure 11-7. The exceptions to the rule are:

1. If the total capacity does not exceed 112.5 kVA, one or more units may be installed in a vault constructed of reinforced concrete not less than 4″ thick.
2. Oil-insulated transformers installed in detached buildings used only for providing electric service and accessible only to qualified persons do not require a vault. *NOTE:* Precautions must be taken to prevent a fire hazard from being created by this type of installation.
3. Electric furnace transformers rated 75 kVA or less may be located in a fire-resistant room if provisions are made to prevent an oil fire from spreading to other combustible material.
4. When provisions are made to prevent an oil fire from spreading to other combustible materials, oil-insulated transformers of 600 volts or less may be installed without a Code-constructed unit. The total number of kVA of the transformers allowed in a room or section of a building is limited to 10 kVA for nonfire-resistant buildings and 75 kVA for fire-resistant buildings.

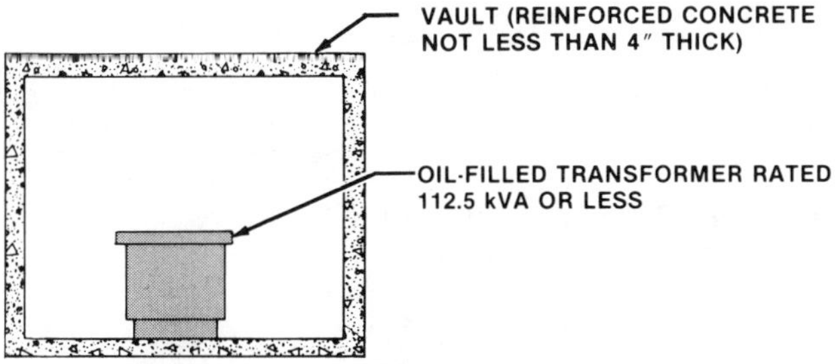

Figure 11-7. Oil-insulated transformers must be installed in a vault or a detached building accessible only to qualified personnel.

Oil-insulated Transformers Installed Outdoors—450-27

Oil-insulated transformers installed on or adjacent to buildings with combustible material must be provided with a means of protecting the building from fire hazards that could be caused by leaking oil. Fire-resistant barriers, water spray systems, and approved enclosures are considered approved safeguards for oil-insulated transformers. See Figure 11-8.

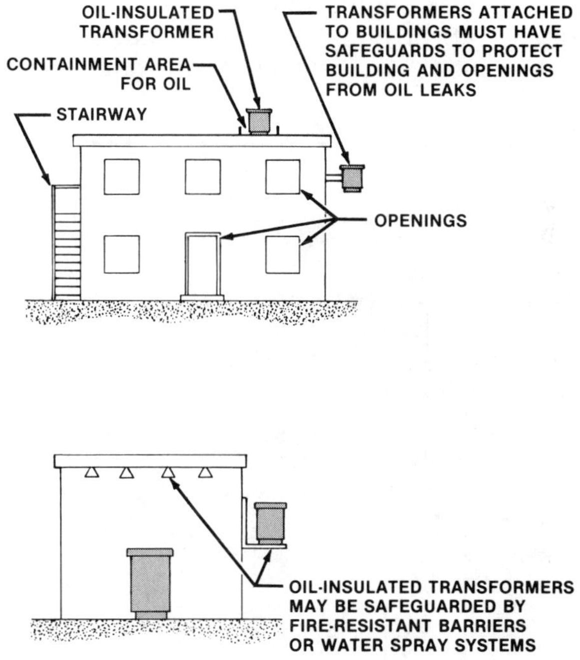

Figure 11-8. Oil-insulated transformers on or adjacent to buildings must be installed in a manner that provides protection for the building.

TRANSFORMER VAULTS

To permit ventilation directly to the outside without the use of ducts or flues, transformer vaults must be located near an outside wall of the building per 450-41. The walls, roofs, and floors of transformer vaults must be of fire-resistant material (such as concrete, brick, or tile blocks) that will withstand and contain a fire for at least three hours. A minimum of 6″ of concrete is required for the walls and roof. A 4″ thickness is required for a slab floor per 450-42.

The National Fire Protection Association requires that the door to a transformer vault be built to certain specifications. A 4″ high door sill must be provided to prevent oil from spreading to other areas. The door must have a minimum fire rating of three hours. It must be locked at all times to prevent access of unqualified persons per 450-43.

Ventilation openings directly to the outside without ducts or flues must have an area of at least 3 square inches for each kVA of transformer capacity. The ventilation openings must be at least 1 square foot in area. A screen or grating (with an automatic closing damper rated for at least 1½ hours) must be provided for the ventilation opening. One or more openings are permitted; however, if a single opening is used, it must not be installed near the floor. A single ventilation opening must be installed near the roof of the vault.

TRANSFORMER AND FIRE PUMP INSTALLATION—450-3(d)

A secondary OCPD is not required for a dedicated transformer supplying power to a fire pump. The primary OCPD shall be provided per 450-3(a)(b). The primary shall be designed and installed to carry indefinitely the sum of the secondary LRCs of the fire pump motors and associated equipment. See Figure 11-9.

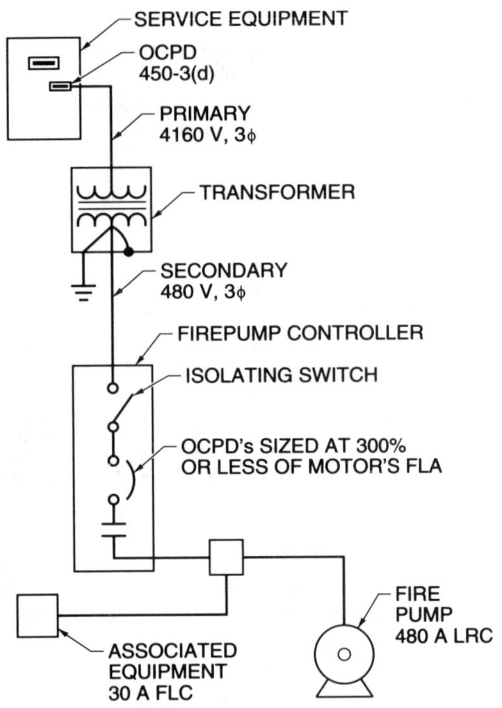

PROBLEM: What size OCPD is required in the primary side of the transformer, which is capable of holding the secondary amps indefinitely?

Step 1: *450-3(d).* Finding FLA.

$$480\text{ A} + 30\text{ A} = 510\text{ A}$$

Step 2: *450-3(d).* Finding ratio.

$$\text{Ratio} = \frac{4160\text{ V}}{480\text{ V}} = 8.67$$

Step 3: Sizing primary OCPD.

$$\text{OCPD} = \frac{510\text{ A}}{8.67}$$

$$\text{OCPD} = 58.8\text{ A}$$

Answer: Primary OCPD = **60 A**

Figure 11-9. The primary OCPD of a transformer shall be sized to hold the secondary full load amps (FLA) indefinitely.

REVIEW—CHAPTER 11

Name ______________________________ Date ____________

True-False

T F 1. Transformers must be located where they are accessible for maintenance and inspection.

T F 2. Transformers can be hung from ceilings if they are visible and can be reached with a ladder for servicing.

T F 3. Transformers mounted on the floor in mechanical rooms and other working spaces do not have to comply with the clearances of Section 110-16 for systems of 600 volts or less.

T F 4. Panelboards must not be mounted directly over transformers mounted on the floor.

T F 5. Transformers of 600 volts or less and 60kVA or less can be hung above suspended ceilings.

T F 6. Transformer cases are not required to be grounded.

T F 7. Transformers and components are safeguarded from damage if they are properly elevated.

T F 8. Transformers must be located and installed so that sufficient cool air is available to prevent overheating of windings.

T F 9. Fences around transformers are not required to be bonded and grounded.

T F 10. Dry-type transformers rated over 35,000 volts are not required to be installed in a vault.

Completion

__________ 1. Completely enclosed transformers are not required to have a(n) __________″ minimum separation from combustible material if rated 112.5kVA or less and operating at less than 600 volts.

__________ 2. All dry-type transformers installed outdoors must have __________ enclosures.

__________ 3. Askarel-insulated transformers over 25kVA must be furnished with a(n) __________ vent.

__________ 4. The walls and roof of transformer vaults are required to have a minimum of __________″ concrete.

__________ 5. The floor of a transformer vault is required to have at least a(n) __________″ thick slab of concrete.

__________ 6. The door of a transformer vault must have a minimum fire rating of __________ hours.

__________ 7. Transformer vault doors must be __________ at all times to prevent access to unqualified personnel.

__________ **8.** A(n) __________″ high door sill must be provided to prevent oil from spreading to other areas.

__________ **9.** Condition __________ to Table 110-16(a) is applied where a concrete wall is in front of transformers.

__________ **10.** Transformers containing liquid must have a liquid __________ area in case leaks occur.

Multiple Choice

__________ **1.** Transformers with insulated walls directly in front of and opposite the transformers and equipment must have a clearance of 3′ at __________ volts-to-ground.
A. 0–150
B. 151–600
C. 2400–4160
D. 12,000–13,500

__________ **2.** Transformers installed indoors and rated at 112.5 kVA or less must have a separation of at least __________″ from combustible material.
A. 8
B. 10
C. 12
D. 18

__________ **3.** Transformers that are rated over __________ volts must be installed in vaults.
A. 600
B. 4160
C. 13,500
D. 35,000

__________ **4.** Dampers must have a standard fire rating of not less than __________ hour(s).
A. $\frac{1}{2}$
B. 1
C. $1\frac{1}{2}$
D. 2

__________ **5.** Door sills in transformer vaults are required to be at least __________″ high.
A. 2
B. 4
C. 6
D. 8

__________ **6.** Reinforced concrete at least 6″ thick has a(n) __________ hour construction.
A. 1
B. 2
C. 3
D. 4

__________ **7.** Doors for transformer vaults must have at least a(n) __________ hour fire rating.
A. 1
B. 2
C. $2\frac{1}{2}$
D. 3

______ **8.** Entrance doors to transformer vaults must be kept ______.
A. open
B. closed
C. locked
D. closed but not locked

______ **9.** Drains are required in transformer vaults where the kVA rating is more than ______.
A. 50
B. 60
C. 75
D. 100

______ **10.** Completely enclosed and ventilated dry-type transformers can be installed in a room of nonfire-resistant construction if they are constructed with insulation of ______°C rise.
A. 40
B. 50
C. 70
D. 80

Problems

1. What is the minimum spacing between the combustible wall and a 100kVA transformer rated at 480 volts?

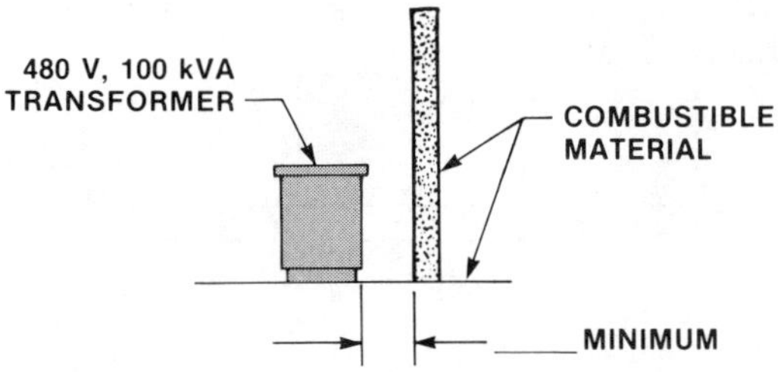

2. What are the minimum clearances for a 150kVA transformer rated at 4160 volts? The transformer is not totally enclosed, not ventilated, and has a 80°C rise.

A. ______

B. ______

C. ______

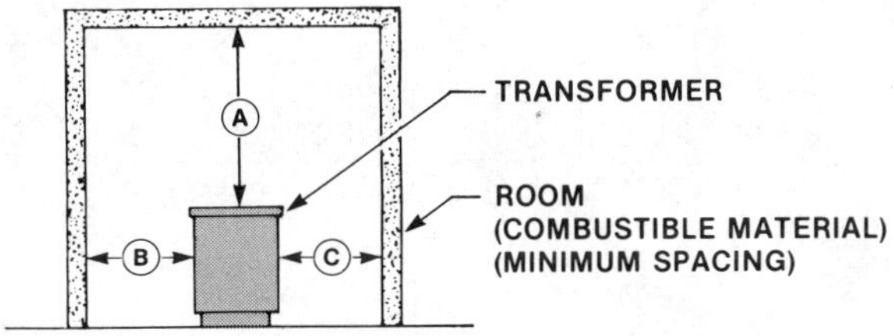

3. What is the maximum voltage and kVA ratings of a transformer that can be installed above a suspended ceiling?

A. __________ volts

B. __________ kVA

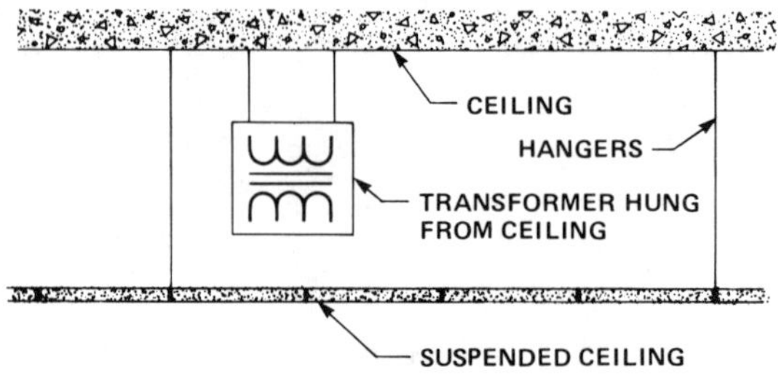

4. Panelboards may be located in positions __________, __________, and __________.

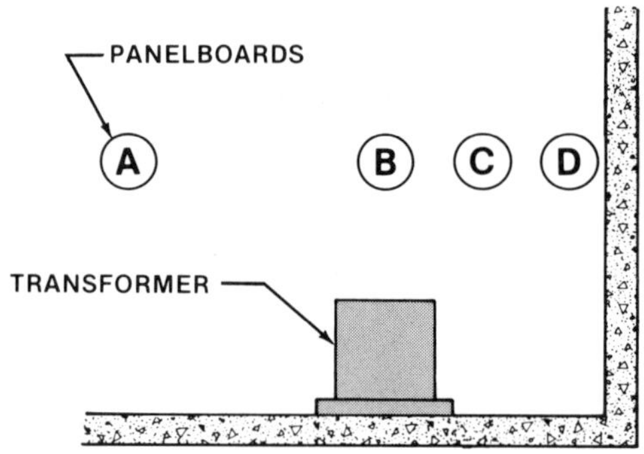

5. Ground all necessary elements of this transformer installation.

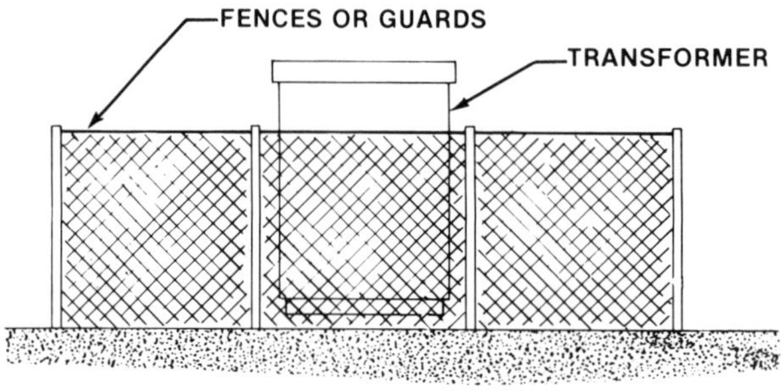

TRANSFORMER SIZING AND PROTECTION

Chapter 12

Transformers are sized and selected based upon the types of loads to be served. The overcurrent protection device for the primary of a transformer is selected based upon transformers rated above 600 volts or at 600 volts or less.

For transformers rated above 600 volts, individual protection may be placed on the primary side or combination protection may be provided on both the primary and secondary sides per 450-3(a). For transformers rated at 600 volts or less, individual protection may be placed on the primary side based on the full-load current. Combination protection may be provided on the primary and secondary sides with 125% to 250% of the full-load current placed on the primary side per 450-3(b).

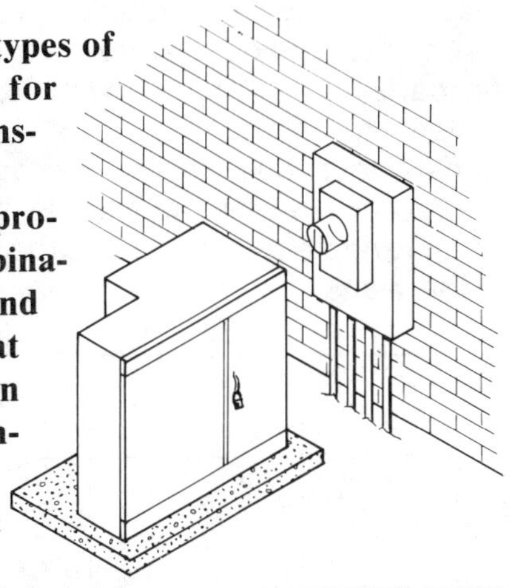

SIZING TRANSFORMERS

Transformers are sized based on the total volt-amps that a building requires to supply all loads used. A building may be supplied by single-phase or three-phase current, depending on the load requirements. The load of a building is determined by adding all of the single-phase and three-phase loads together and dividing by the supply voltage.

Example: A building has a total connected load of 16,000 volt-amps, single-phase, and 28,000 volt-amps, three-phase. The supply voltage is 120/208 volts. What is the load in amps?

Step 1: 16,000VA + 28,000VA = 44,000VA

Step 2: $\dfrac{44{,}000\text{VA}}{208 \times \sqrt{3}} = 122\text{A}$

Answer: **122A**

Sizing Transformers for Single-phase Connected Secondary

The size transformers needed to supply a single-phase connected secondary system can be determined by adding all the single-phase 120- and 240-volt loads together. The amperage of the load is found by dividing the total load by the voltage. See Figure 12-1.

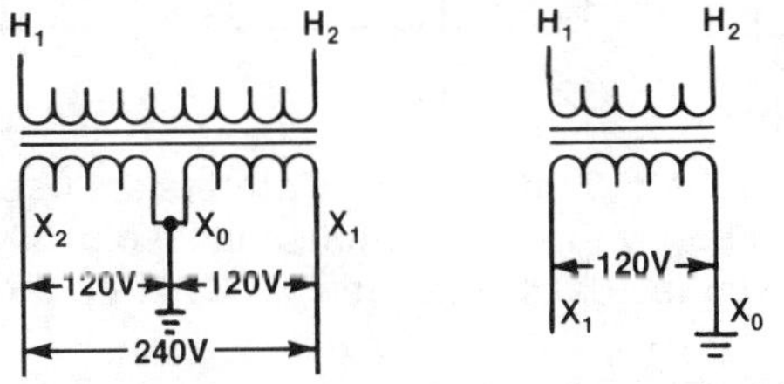

PROBLEM: What size transformer is needed for a building with a total connected load of 8 kVA for the 120-volt load and 15 kVA of 240-volt loads?

Step 1: Size load for transformer:

Total 120-volt load =	8 kVA
Total 240-volt load =	15 kVA
	23 kVA

Step 2: The lighting and power transformer are to be the same size.

Answer: **25 kVA**

Figure 12-1. Transformers for a single-phase connected secondary system are sized by adding all 120- and 240-volt loads together. (To find the amperage of the load, divide total load by voltage.)

Sizing Transformers for Wye-connected Secondary

The size transformers required to supply a wye-connected secondary load can be found by (1) adding all the single-phase and three-phase loads together; and (2) multiplying this total load by ⅓ (or multiplying by .33). The size of the transformers can also be found by (1) multiplying the single-phase load by ⅓ (and the three-phase load by ⅓); and (2) adding the two computations. Again, the loads may be multiplied by .33 instead of by ⅓. See Figure 12-2.

One transformer may be sized by adding all single-phase and three-phase loads together. The volt-amp rating obtained equals the volt-amp rating of the transformer.

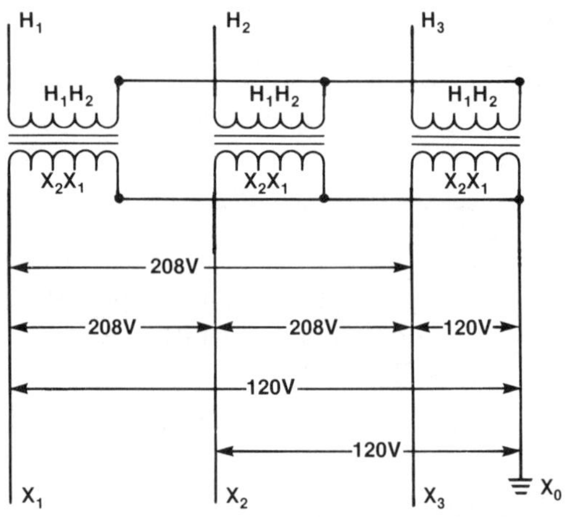

PROBLEM: What size transformers are needed for a building with a total connected load of 80 kVA for the single-phase load and 60 kVA for the three-phase load?

Step 1: Size load for each transformer:

Total single-phase kVA:	
80 kVA × .33 =	26.4 kVA
Total three-phase kVA:	
60 kVA × .33 =	19.8 kVA
	46.2 kVA

Step 2: Lighting and power transformers are to be the same size.

Answer: **Three 50 kVA transformers or one 150 kVA transformer**

Figure 12-2. Transformers for a wye-connected secondary load are sized by adding all loads and multiplying by ⅓ (or multiply by .33). One transformer is sized by adding all single-phase and three-phase loads.

Sizing Transformers for Closed Delta-connected Secondary

The size transformers needed to supply a closed delta-connected secondary system can be determined by (1) adding all single-phase loads and multiplying by ⅔ (.67); and (2) adding all three-phase loads and multiplying by ⅓ (.33). These two totals size the mid-tap lighting transformer. The size of the two power transformers can be found by adding the total single-phase and three-phase loads and multiplying by ⅓. See Figure 12-3.

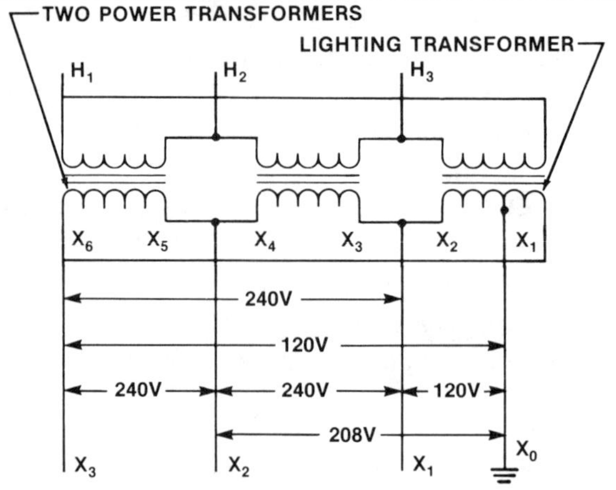

PROBLEM: A building has a total connected load of 28 kVA for the single-phase load and 68 kVA for the three-phase load. What size transformers is required to supply this load, using a closed delta system on the secondary?

Step 1: Lighting transformer:

Total single-phase kVA:	
28 kVA × .67 =	18.76 kVA
Total three-phase kVA:	
68 kVA × .33 =	22.44 kVA
	41.2 kVA

Answer: **45 kVA lighting transformer**

Step 2: Power transformers:

Total single-phase kVA:	
28 kVA × .33 =	9.24 kVA
Total three-phase kVA:	
68 kVA × .33 =	22.44 kVA
	31.68 kVA

Answer: **Two 37½ kVA power transformers**

Figure 12-3. The mid-tap transformer is found by multiplying the single-phase load by ⅔ and the three-phase load by ⅓. The two power transformers are found by multiplying the single- and three-phase loads by ⅓.

If one of the single-phase transformers in a closed delta hookup should become defective, the transformer bank may be connected open delta. The two single transformers would supply 58% of the normal capacity of the three transformers connected three-phase, closed delta.

For example, if three 125-kVA, single-phase transformers were connected three-phase closed delta and one were lost, what capacity could an open delta hookup supply? Fifty-eight percent of the normal three-phase closed delta capacity [375kVA (3 × 125kVA)] is 217.5kVA. The unessential loads could be shut down and the system could continue to operate until another transformer is installed. See Figure 12-4.

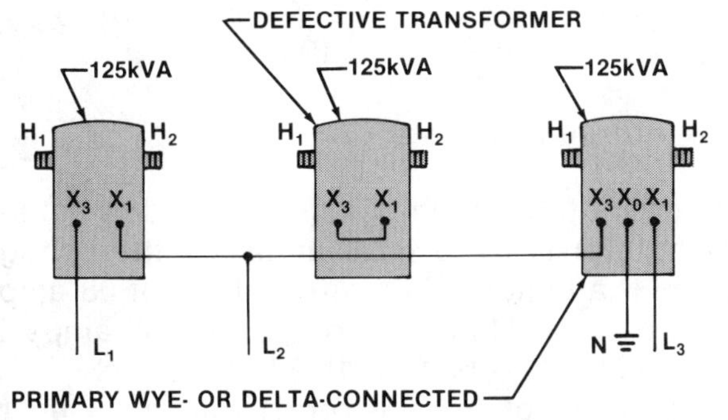

PROBLEM: If three 125kVA, single-phase transformers are connected closed delta and one is lost, what capacity would an open delta hookup supply until the defective transformer is replaced?

Step 1: 125 kVA × 3 = 375 kVA

Step 2: 375 kVA × 58% = 217.5 kVA

Answer: **217.5 kVA**

Figure 12-4. To size the capacity of an open delta system losing a defective transformer in a closed delta system, multiply the kVA of the three transformers by 58%.

ADVANTAGE OF DELTA OR WYE SYSTEM

The closed delta system is used to supply large three-phase loads with a small, 120-volt, neutral load. Greater capacity is supplied to serve the loads on a closed delta system because three transformers are utilized. When a large amount of three-phase current is required, the closed delta system should be utilized.

The open delta system is used to supply large 120-volt neutral loads and single-phase, 240-volt loads. Only two transformers are required for an open delta system. When a small amount of three-phase current is required, the open delta system should be utilized.

The advantage of the wye system is that 120 volts can be obtained from three-phase conductors for the system. The wye system is often used for commercial buildings because three hots can be used with one neutral.

Open Delta-connected Secondary System

The size transformers needed to supply an open delta-connected system can be determined by taking 100% of the single-phase load and 58% of the three-phase load. The size of the mid-tap transformer can be found by adding these two loads together. The size of the power transformer can be found by taking 58% of the three-phase load. [The reciprocal of 1.732 ($\sqrt{3}$) is 58%.] See Figure 12-5.

The lighting transformer carries 100% of the 120/240-volt, single-phase loads plus 58% of the three-phase load due to the transformers being 60° out-of-phase with each other. The power transformer carries 58% of the total three-phase load. Note that the 240-volt, single-phase load can be connected to the power transformer, but care must be taken not to supply items such as 120-volt clocks and timers with the high leg (208 volts-to-ground).

SIZING PRIMARY AND SECONDARY CURRENT OF TRANSFORMER

The primary and secondary of transformers are used to step the voltage up or down. When the voltage is stepped up, the current is reduced. When the voltage is stepped down, the current is increased. Currents that are reduced due to the higher voltage permit smaller conductors and conduit to be used. Currents that are increased provide more amps to supply loads.

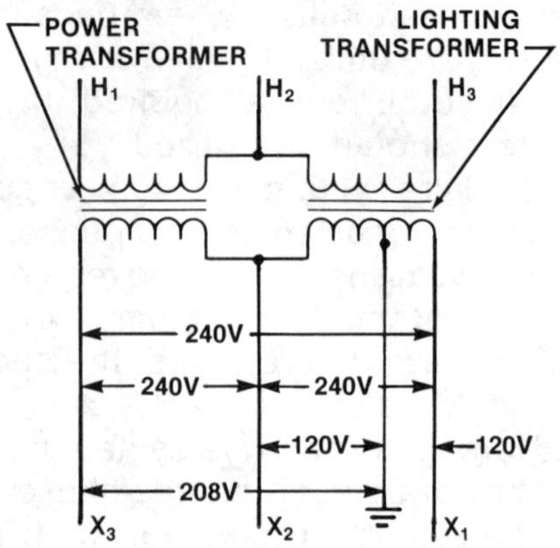

PROBLEM: A building has a total connected load of 21 kVA for the single-phase load and 35 kVA for the three-phase load. What size lighting transformer is required? What size power transformer is required?

Step 1: Lighting transformer:

Total single-phase kVA:	21 kVA
Total three-phase kVA: 35 kVA × 58% =	20.3 kVA
	41.3 kVA

Answer: **45 kVA lighting transformer**

Step 2: Power transformer:

Total three-phase kVA: 35 kVA × 58% =	20.3 kVA

Answer: **25 kVA power transformer**

Figure 12-5. The lighting transformer of an open delta-connected secondary system is sized by multiplying the three-phase loads by 58% and adding to the single-phase loads. The power transformer is sized by multiplying the three-phase loads by 58%.

Finding Amperage of Transformers

For single-phase systems, the amperage of a transformer can be found for the primary or secondary by using the following formulas:

$$kVA = \frac{volts \times amps}{1000}$$

$$amps = \frac{kVA \times 1000}{volts}$$

The ratio of a transformer's windings is determined by dividing the primary voltage by the secondary voltage as follows:

$$\frac{primary}{secondary} = \frac{480V}{240V} = 2{:}1 \text{ ratio}$$

The kVA value is found by dividing the total volt-amps by 1000. By dividing 1000 into the volt-amps, the kVA *rating* can be obtained. For example, a building has a total load of 280 amps, which includes both single-phase 120- and 240-volt loads. The building is served by a 120/240-volt, single-phase supply. The kVA can be determined as follows:

$$\text{Step 1: } kVA = \frac{volt \times amps}{1000}$$

$$\text{Step 2: } kVA = \frac{240V \times 280A}{1000} = 67.2 \text{ kVA}$$

Answer: **kVA = 67.2**

Note that the volt-amps of a load is determined by multiplying the total amps by the voltage used. If a building has a total load of 68 amps served by a 240-volt supply, the volt-amps is 16,320 (240V × 68A = 16,320VA).

The ratio of a transformer can be used to quickly check the amount of amperage that the primary or secondary will deliver. For example, a 480/240-volt, 24-kVA transformer with a 50-amp primary and 2 to 1 ratio will deliver 100 amps on the secondary side [(480V/240V) = 2 × 50A = 100A].

The amperage can be determined by dividing the volt-amps by the voltage. If 16,320 volt-amps is divided by 120-volt, single-phase, the amperage will be 136 amps [(16,320VA/120V) = 136A] on one hot leg only. If the 16,320 volt-amps is divided by 240-volt, single-phase, the amperage will be 68 amps [(16,320VA/240V) = 68A] on two hot legs. A 120-volt, single-phase system has only one hot conductor to the grounded neutral while the 240-volt, single-phase system has two hot conductors to the grounded neutral. Figure 12-6 shows the method used to determine a transformer's primary and secondary amperage ratings for a single-phase system.

For three-phase systems, the amperage can be found for the primary or secondary by using the same formula used for single-phase systems;

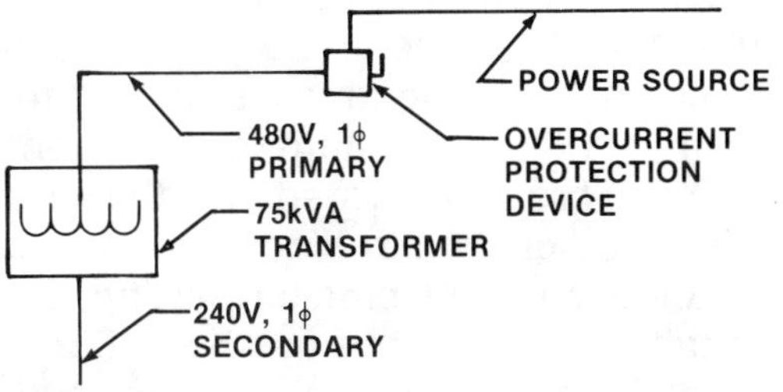

PROBLEM: What are the primary and secondary amperage ratings for a 480/240-volt, single-phase, 75 kVA transformer?

Step 1: Primary amperage:

$$A = \frac{kVA \times 1000}{V}$$

$$A = \frac{75 \text{ kVA} \times 1000}{480 \text{ V}} = 156 \text{ A}$$

Answer: **156 A primary rating**

Step 2: Secondary amperage:

$$A = \frac{kVA \times 1000}{V}$$

$$A = \frac{75 \text{ kVA} \times 1000}{240 \text{ V}} = 312.5 \text{ A}$$

Answer: **312.5 A secondary rating**

Figure 12-6. A transformer's primary and secondary amperage ratings for single-phase systems are found by dividing the total kVA by the voltage.

however, the square root of 3 (1.732) must be included in the formula as follows:

$$kVA = \frac{volts \times 1.732 \times amps}{1000}$$

$$amps = \frac{kVA \times 1000}{volts \times 1.732}$$

The ratio of a transformer's windings is determined by dividing the primary voltage by the secondary voltage as follows:

$$\frac{primary}{secondary} = \frac{480V}{240V} = 2{:}1 \text{ ratio}$$

The square root of 3 is used to evenly divide the amperage per phase (each hot leg) where a three-phase system is used.

Example: A building is supplied by a 120/208-volt system with a connected load of 72,000 volt-amps. What is the amperage on phases A, B, and C?

Step 1: $A = \dfrac{VA}{V \times \sqrt{3}}$

Step 2: $A = \dfrac{72{,}000VA}{208V \times 1.732} = 200A$

Answer: **200A per phases A, B, and C**

Figure 12-7 illustrates the method used to determine a transformer's primary and secondary amperage rating for a three-phase system.

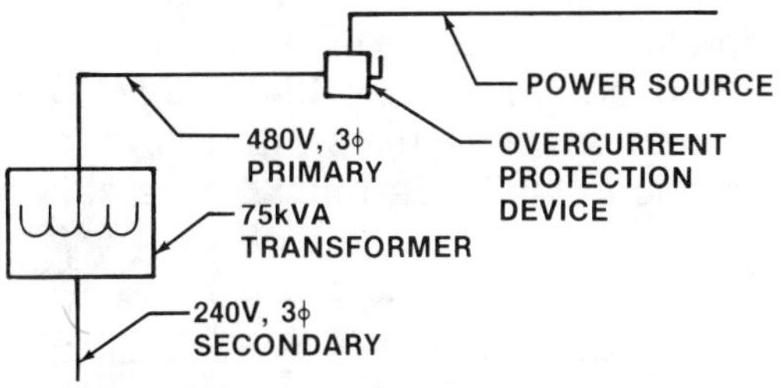

PROBLEM: What are the primary and secondary amperage ratings for a 480/240-volt, three-phase, 75 kVA transformer?

Step 1: Primary amperage:

$$A = \frac{kVA \times 1000}{V \times \sqrt{3}}$$

$$A = \frac{75 \text{ kVA} \times 1000}{480 \text{ V} \times 1.732} = 90 \text{ A}$$

Answer: **90 A primary rating**

Step 2: Secondary amperage:

$$A = \frac{kVA \times 1000}{V \times \sqrt{3}}$$

$$A = \frac{75 \text{ kVA} \times 1000}{240 \text{ V} \times 1.732} = 180 \text{ A}$$

Answer: **180 A secondary rating**

Figure 12-7. A transformer's primary and secondary amperage ratings for three-phase systems are found by dividing the total kVA by the voltage $\times \sqrt{3}$.

Designing Overcurrent Protection for Primary of Transformer—450-3

Two sets of rules are used when sizing and selecting the overcurrent protection device for the primary of a transformer. One set of rules applies to transformers rated above 600 volts, and the other applies to transformers rated at 600 volts and less.

Transformers Rated over 600 Volts—450-3(a)(1). Protection may be provided by a combination of overcurrent protection devices on the primary and secondary of the transformer. If the fuse is rated by the percentages listed in Table 450-3(a)(1) and does not correspond to a standard fuse rating, the next higher standard rating is permitted by Exception 1. If the circuit breaker is rated by the percentages listed in Table 450-3(a)(1) and does not correspond to a standard circuit breaker setting, the next higher trip setting must be used. See Figure 12-8. *NOTE:* Overcurrent protection for systems with 600 volts or less on the secondary side must be calculated at 125%.

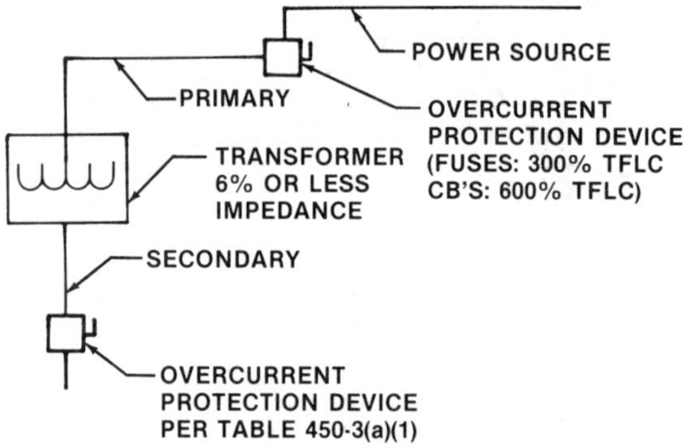

Figure 12-8. Combination overcurrent protection is provided on the primary and secondary for transformers located in unsupervised locations per Table 450-3(a)(1).

Transformers Rated over 600 Volts—450-3(a)(2)b. Protection may be provided either by an overcurrent protection device on the transformer's primary or by a combination of overcurrent protection devices on the primary and secondary of the transformer. For primary protection only, a fuse rated at 250% or less of the transformer's primary full-load current must be used, or a circuit breaker rated at 300% of the transformer's primary full-load current must be used. If the fuse is rated at 250% of the primary full-load current and does not correspond to a standard fuse rating, the next higher standard rating is permitted. If the circuit breaker is rated at 300% of the primary full-load current and does not correspond to a standard circuit breaker setting, the next higher trip setting must be used. See Figure 12-9.

If a combination of overcurrent protection devices of specified ratings are used on the primary and secondary to provide transformer protection, they must be selected and sized according to the percentage ratings of Table 450-3(a)(2)b. The size

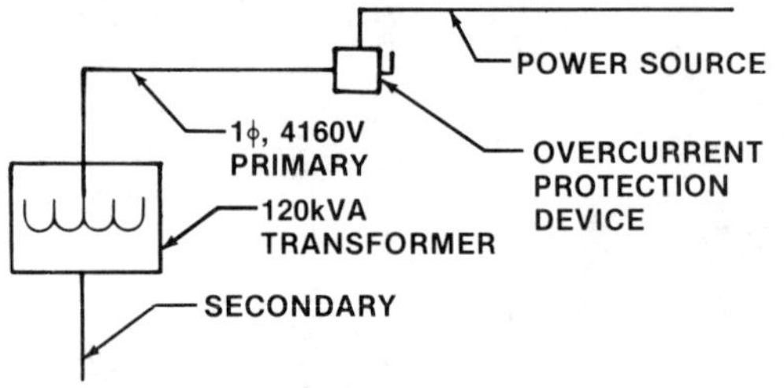

PROBLEM: What size fuses are required to protect the transformer?

Step 1: $A = \dfrac{kVA \times 1000}{V}$

$A = \dfrac{120{,}000 \text{ VA}}{4160 \text{ V}} = 29 \text{ A}$

Step 2: *450-3(a)(2)a.*

29 A × 250% = 72 A

Answer: *240-6.*
80 A fuses

PROBLEM: What size circuit breaker is required to protect the transformer?

Step 1: $A = \dfrac{kVA \times 1000}{V}$

$A = \dfrac{120{,}000 \text{ VA}}{4160 \text{ V}} = 29 \text{ A}$

Step 2: *450-3(a)(2)a.*

29 A × 300% = 87 A

Answer: *240-6.*
90 A cb

Figure 12-9. To size fuses for the primary side of a transformer in a supervised location, multiply total amps by 250%. For circuit breakers, multiply total amps by 300%.

of the overcurrent protection device depends on the impedance of the transformer, whether a circuit breaker or fuse is selected. From the rated impedance of the transformer, for the primary side, select a circuit breaker or fuse based on the percentage ratings from Column 1 or 2 of Table 450-3(a)(2)b. For the secondary protection, select a circuit breaker or fuse based on the percentage ratings from Column 3, 4, or 5 of Table 450-3(a)(2)b.

An individual overcurrent protection device on the primary side is not required if protection according to Table 450-3(a)(2)b is placed in the feeder circuit, or if the transformer has coordinated thermal overload protection furnished by the manufacturer. See Figure 12-10.

Combination protection in a transformer's primary and secondary must be sized from Table 450-3(a)(2)b for primary voltage over 600 volts. The secondary must be equipped with coordinated thermal overload protection or have an overcurrent protection device sized from 450-3(a)(2)b and Table 450-3(a)(2)b on the secondary.

Section 230-207 and 240-3(i) are the only parts of the NEC® that consider the primary overcurrent protection devices to protect the secondary conductors with no limitation of length and with no protection applied on the secondary side. Exception No. 2 to 240-3 permits the primary to protect the secondary conductors if the primary overcurrent protection device is set or rated at not more than 125% of the transformer.

Section 230-207 permits the overcurrent protection device to be omitted for vaults having the primary load-interrupter switch manually operable from outside the vault. There can be only one set of secondary conductors connected to a common bus. If there are two or more sets of secondary conductors, the overcurrent protection device must be placed in the secondary side of the transformer.

Transformers Rated at 600 Volts or Less—450-3(b)(1). Section 240-3(i) allows the primary protection device to protect the secondary of the transformer only when the transformer is two-wire–to–two-wire. For this rule to be applied, the primary overcurrent device must be set at no more than 125% of the full-load current of the transformer's primary. Overcurrent protection may be provided on the primary side only according to 450-3(b)(1).

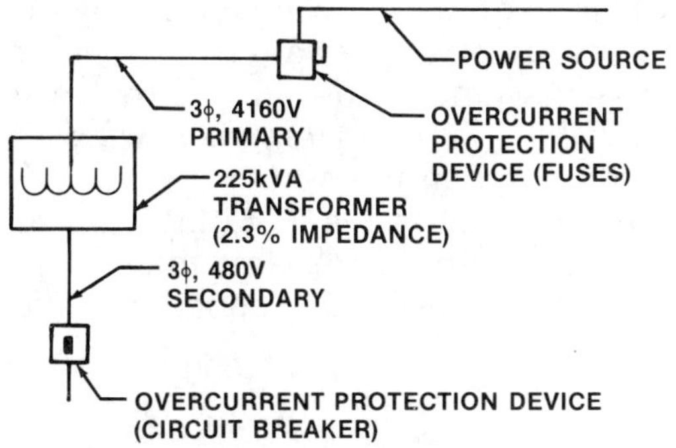

SEE TABLE 450-3(a)(2)b FOR SUPERVISED LOCATIONS.

PROBLEM: What size fuses are required to protect the transformer's primary? What size circuit breaker is required to protect the transformer's secondary?

Step 1: Primary:

$$A = \frac{kVA \times 1000}{V \times \sqrt{3}}$$

$$A = \frac{225{,}000\ VA}{4160\ V \times 1.732} = 31\ A$$

Step 2: *Table 450-3(a)(2)b, Column 2.*

31 A × 300% = 93 A

Answer: *240-6.*
90 A fuses

Step 3: Secondary:

$$A = \frac{kVA \times 1000}{V \times \sqrt{3}}$$

$$A = \frac{225{,}000\ VA}{480\ V \times 1.732} = 271\ A$$

Step 4: *Table 450-3(a)(2)b, Column 5.*

271 A × 250% = 677 A

Answer: *240-6.*
600 A cb

Figure 12-10. To size overcurrent protection devices for primary and secondary, use combination protection per Table 450-3(a)(2)b.

IMPEDANCE

Impedance is the current-limiting characteristic of a transformer at its terminals. The impedance of a transformer (always expressed as a percentage) is used to determine the *interrupting capacity* (IC) *rating* of fuses and circuit breakers used to protect the primary of a transformer. The IC rating is the amount of fault current a transformer will deliver at its terminals when a short circuit develops in the system.

Example: What is the IC rating of a 20-kVA transformer with 1.5% impedance and a 120/240-volt, single-phase secondary?

Step 1: $flc = \dfrac{kVA \times 1000}{V}$

$flc = \dfrac{20 \times 1000}{240V} = 83A$

Step 2: $IC = \dfrac{flc}{impedance}$

$IC = \dfrac{83A}{.015} = 5533A$

Answer: **IC = 5533A**

Note that the greater the impedance rating is, the lower the interrupting capacity will be, in amps. If the transformer previously discussed has an impedance of 4%, the interrupting capacity would be 2075 amps [(83 amps/.04) = 2075 amps)].

If the overcurrent protection device is placed on the primary side and sized at no more than 125% of the transformer's full-load current, the feeder conductors may be run any distance to be connected to the transformer terminals. See Figure 12-11.

Exception 1 to 450-3(b)(1) permits the next higher standard size overcurrent protection device where the 125% times the full-load rating of the transformer does not correspond to a standard size device.

Combination protection on both the primary and secondary sides may be provided according to 450-3(b)(2). If, for any reason, 125% of the primary full-load current is not sufficient, a value of not more than 250% of the rated primary current may be used, provided overcurrent protection is installed on the secondary at 125% of the secondary full-load current. See Figure 12-12.

Additional taps on motor loads can be taken from the secondary terminals of a transformer per 240-21(b)(j). These taps can trip open the primary overcurrent protection device sized and selected at 125%. Section 450-3(b)(2) permits the 125% to be increased to 250% if 125% protection is installed on the secondary side of the transformer. The 125% must be provided *either* on the primary or secondary side.

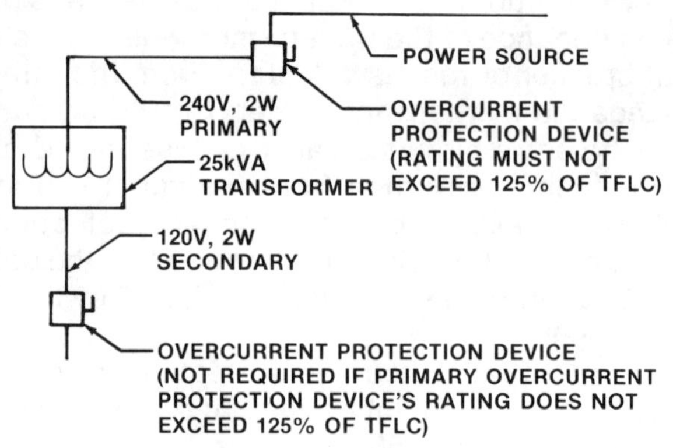

450-3(b)(1)

PROBLEM: What size overcurrent protection device is required in the primary?

Step 1: $A = \dfrac{kVA}{V}$

$A = \dfrac{25{,}000 \text{ VA}}{240 \text{ V}} = 104 \text{ A}$

Step 2: *450-3(b)(1), Ex. 1.*

104 A × 125% = 130 A

Answer: *240-6.*
150 A cb

Figure 12-11. An overcurrent protection device placed in the primary of a transformer rated at 600 volts or less is calculated at 125% of the TFLC.

Section 450-3(b)(2) requires the next lower size overcurrent protection device for the primary side where the 250% does not correspond to a standard overcurrent protection device. The exception to 450-3(b)(2) permits the next higher size overcurrent protection device on the secondary side where the 125% does not correspond to a standard device.

The secondary OCPD shall consist of not more than six CBs or six sets of fuses grouped together in any one location. If multiple OCPDs are used, the total rating of all devices shall not exceed the total rating of one OCPD. If CBs and fuses are used as the OCPD, the total rating of all devices shall not exceed the total rating of one fuse. These rules apply where the primary protection exceeds 125% × primary FLC, but is 250% or less of primary FLC.

When the primary current of a transformer is 9 amps or more, and 125% of this current does not correspond to a standard rated fuse or circuit breaker, the next higher size should be used. If the amperage rating is less than 9 amps, but

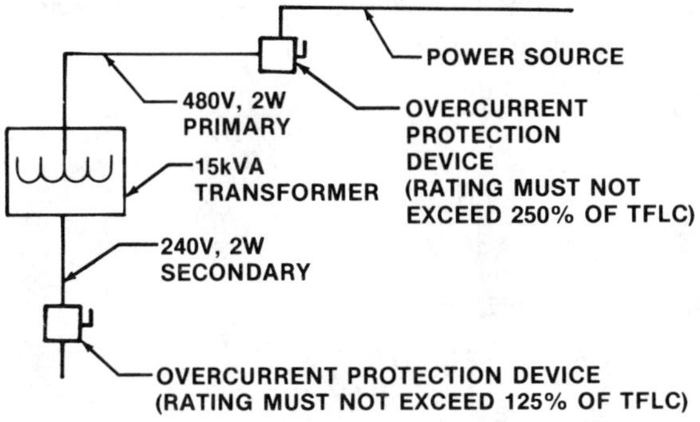

450-3(b)(2)

PROBLEM: What size primary and secondary overcurrent protection devices are required for the transformer?

Step 1: Primary:

$$A = \frac{kVA}{V}$$

$$A = \frac{15{,}000 \text{ VA}}{480 \text{ V}} = 31 \text{ A}$$

Step 2: *450-3(b)(2).*

$$31 \text{ A} \times 250\% = 77.5 \text{ A}$$

Answer: *240-6.*
70 A

Step 3: Secondary:

$$A = \frac{kVA}{V}$$

$$A = \frac{15{,}000 \text{ VA}}{240 \text{ V}} = 62.5 \text{ A}$$

Step 4: *450-3(b)(2), Ex.*

$$62.5 \text{ A} \times 125\% = 78 \text{ A}$$

Answer: *240-6.*
80 A

Figure 12-12. Overcurrent protection devices placed in the primary and secondary of a transformer rated at 600 volts or less are figured at 250% for the primary and 125% for the secondary.

more than 2 amps, 167% of the primary current rating may be used. When the amperage rating is less than 2 amps, 300% of the primary current rating may be used. In each case, respectively, 167% or 300% of the rated primary full-load current shall not be exceeded. See 430-72(c), Ex. 2 for a control circuit.

For two-wire-to-three-wire or three-wire-to-four-wire transformers, the primary overcurrent protection device must not protect the secondary. The overcurrent protection device must always be provided on the secondary side of transformers as required by 240-3(i).

For transformers with a primary full-load current rating of 9 amps or more, the overcurrent protection device must not be rated or set at more than 125% of the primary full-load current rating. See Figure 12-13. The next higher standard rated overcurrent protection device may be used when 125% of the primary current rating does not correspond to a standard rated device from 240-6.

For transformers with a primary full-load current rating of 2 amps, but not more than 9 amps, the overcurrent protection device may be set or rated at 167% of the primary full-load current rating. The next lower standard rated overcurrent protection device must be used when 167% times the primary current rating does not correspond to a standard rated device from 240-6. Apply the exception for fuses smaller than 15 amps and use the method shown in Figure 12-14.

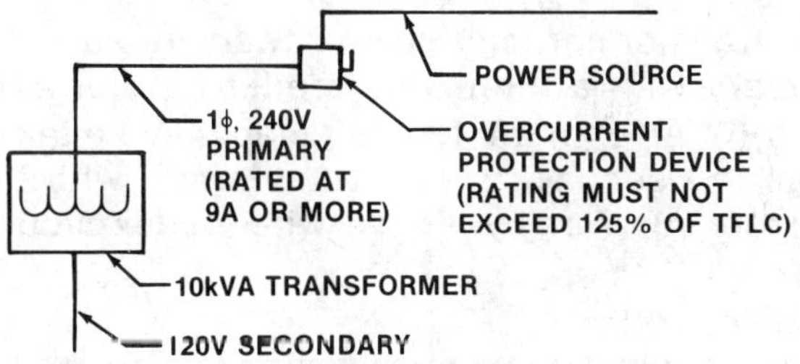

450-3(b)(1), Ex. 1

PROBLEM: What size overcurrent protection device is required in the primary?

Step 1: $A = \frac{kVA}{V}$

$$A = \frac{10{,}000 \text{ VA}}{240 \text{ V}} = 42 \text{ A}$$

Step 2: *450-3(b)(1), Ex. 1.*

$$42 \text{ A} \times 125\% = 52.5 \text{ A}$$

Answer: *240-6.*
60 A

Figure 12-13. When protecting the primary side of a transformer rated at 9 amps or more, the next higher size device may be used.

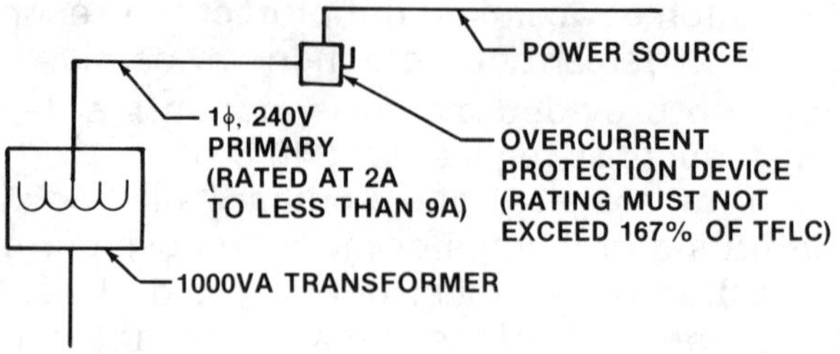

450-3(b)(1), Ex. 1

PROBLEM: What size overcurrent protection device is required in the primary?

Step 1: $A = \frac{kVA}{V}$

$$A = \frac{1000\ \text{VA}}{240\ \text{V}} = 4\ \text{A}$$

Step 2: *450-3(b)(1), Ex. 1.*

$$4\ \text{A} \times 167\% = 6.7\ \text{A}$$

Answer: *240-6, Ex.*
6 A fuse

Figure 12-14. When protecting the primary side of a transformer rated at 2 amps to less than 9 amps, the next lower size device must be used.

For transformers with a primary full-load current rating of less than 2 amps, the overcurrent protection device may be set or rated at 300% of the primary full-load current rating. The next lower standard rated overcurrent protection device must be used when 300% times the primary current rating does not correspond to any standard rated device from 240-6. Apply the exception for fuses smaller than 15 amps and use the method shown in Figure 12-15.

A transformer may have a two-wire supplied primary with a three-wire feeder tap per 240-21(b)(j). A 10′ or 25′ four-wire tap may be taken from the secondary of a transformer with the primary supplied by a three-wire feeder circuit.

SIZING SECONDARY FEEDER TAPS—240-21(b)(d)(j)

For sizing transformer circuits, 450-3 shall be used along with 240-21(b). Special rules are provided in 240-21(b) for tap conductors used in conjunction with transformers.

240-21(b)

A 10′ tap may be made from the secondary of a transformer in the same way a feeder is tapped. The tap conductors must not extend more than 10′ from the point of tap. Their ampacity must not be less than the amperage rating of the switchboard, panelboard, motor control center, or other equipment that is being supplied.

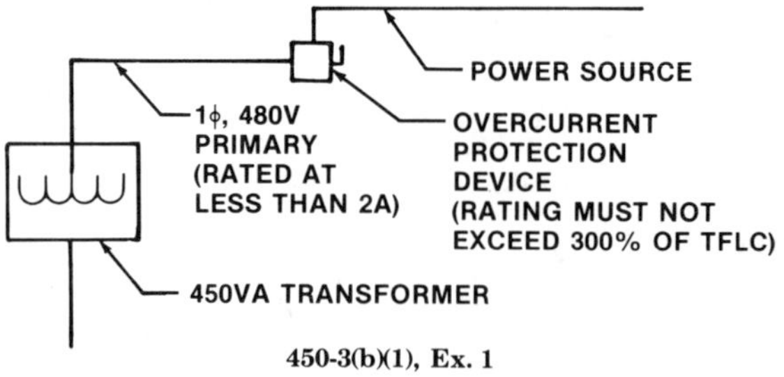

450-3(b)(1), Ex. 1

PROBLEM: What size overcurrent protection device is required in the primary?

Step 1: $A = \frac{kVA}{V}$

$$A = \frac{450\ \text{VA}}{480\ \text{V}} = .94\ \text{A}$$

Step 2: *450-3(b)(1), Ex. 1.*

$$.94\ \text{A} \times 300\% = 2.82\ \text{A}$$

Answer: *240-6, Ex.*
1 A fuse

NOTE: If transformer supplies control circuit, 500% of FLC may be used per 430-72(c), Ex. 2.

Figure 12-15. When protecting the primary side of a transformer rated at less than 2 amps, the next lower size device must be used.

Tap conductors may terminate in a main set of single fuses or circuit breaker rated at not more than the ampacity of the tap conductors. The rules for a 10′ tap are

(1) Smaller conductors must have a current rating not less than the combined, computed

loads of the circuits supplied by the tap conductors. Their ampacity must not be less than the rating of the overcurrent protection device that may be installed at the termination of the tap conductors.

(2) The tap must not extend beyond the switchboard, panelboard, or control device that it supplies.

(3) Tap conductors must be enclosed in conduit, EMT, metal gutter, or other approved raceway when not a part of the switchboard or panelboard.

When the 10′ tap rule is used to connect the secondary of a transformer to a power panel, the tapped conductors may terminate at the lugs of the bus or be connected to a main overcurrent protection device in the panel. If the panel is classified as a power panel, the main overcurrent protection device is not required. If the panel is classified as a lighting and appliance panel, the main overcurrent protection device is always required, regardless of tap length.

For example, a panelboard with a bus and lugs rated at 225 amps requires a #4/0 THWN copper conductor (230 amps per Table 310-16); however, if the 225-amp rated panelboard had a 150-amp main circuit breaker, the secondary conductors can be #1/0 THWN copper (150 amps per Table 310-16). See Figure 12-16. The tap conductors can be any size when applying the 10′ tap rule per 240-21(b).

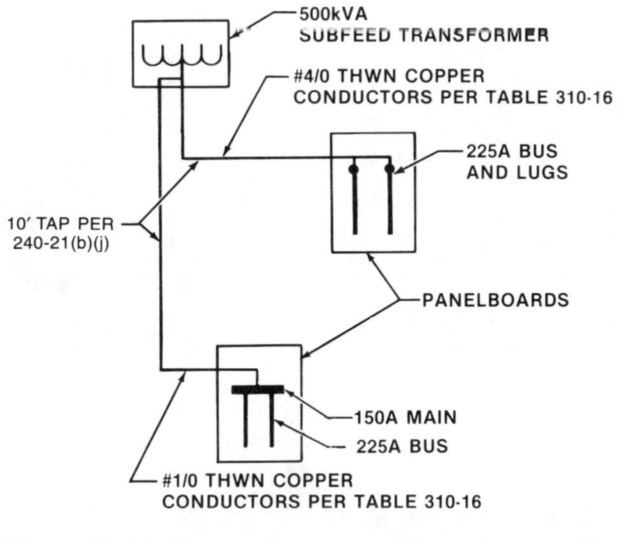

Figure 12-16. Tap conductors must be as large as the lugs or terminals where they terminate. Tapped conductors can be any size where the 10′ tap rule is applied.

Sections 384-14 and 384-15 are used to determine panel classification. If more than 10% of the overcurrent protection devices are rated at 30 amps or less, and the panel has a solid grounded neutral connection, the panel is classified as a lighting and appliance branch-circuit panelboard and could have any combination of 30-amp, 20-amp, and 15-amp devices. See Figure 12-17.

For the purpose of classifying a power or lighting and appliance panelboard, a single cir-

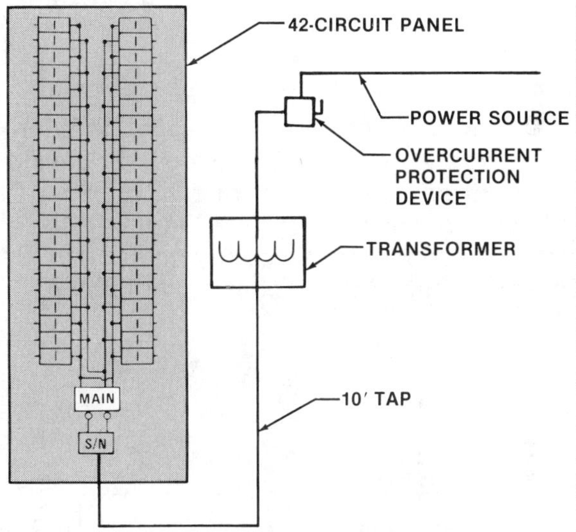

PROBLEM: Is a main overcurrent protection device required in this panel? The panel has six single-pole circuit breakers rated 30 amps or less and neutral-connected loads.

Step 1: *384-14.*

42 cb's × 10% = 4.2 poles

(Four circuit breakers with neutral-connected loads are permitted.)

Step 2: *384-14; 384-15.*

The panel is classified as a lighting and appliance panel.

Answer: **Yes**

Figure 12-17. For classification as a lighting and appliance panelboard, not more than 10% of the total number of overcurrent protection devices can exceed 30 amps.

cuit breaker counts as one, a two-pole circuit breaker counts as two, and a three-pole circuit breaker counts as three circuit breakers. The main does not count as one of the breakers when classifying lighting and appliance panelboards. Note that no mains are installed ahead of the system in a power panel.

Care must be exercised not to overfill the panelboard with overcurrent protection devices that will exceed the 10% limitation. For example, a 32-circuit panelboard can house no more than three devices rated at 30 amps or less (32 × 10% = 3.2) with neutral load. The panelboard cannot house two 30-amp, one 20-amp, and one 15-amp devices with neutral load. The 10% fill is exceeded by one overcurrent protection device. Only three overcurrent protection devices are permitted without a main in this example.

240-21(d)

A smaller conductor may be tapped to a larger conductor. An overcurrent protection device is not required at the point of tap, provided the following rules are met. See Figure 12-18.

(1) Tap conductor ampacity must be at least one-third that of the feeder being tapped.

(2) Secondary conductor ampacity must be at least one-third that of the feeder, based on the primary-to-secondary voltage ratio.

(3) Total length of the tap (primary plus secondary) must not be over 25′.

(4) All conductors must be protected from physical damage.

(5) Secondary conductors must terminate in a main (single set of fuses or circuit breaker) sized to protect the secondary.

240-21(j)

In industrial locations, taps up to 25′ in length can be made from the secondary side of transformers which are classified as *separately derived systems.* See Figure 12-19. The following conditions must be met:

(1) Secondary taps must not exceed 25′ in length.

(2) Ampacity of tapped conductors must be equivalent to the transformer's output, and the overcurrent protection devices must not exceed output.

(3) Tapped conductors must be protected from physical damage.

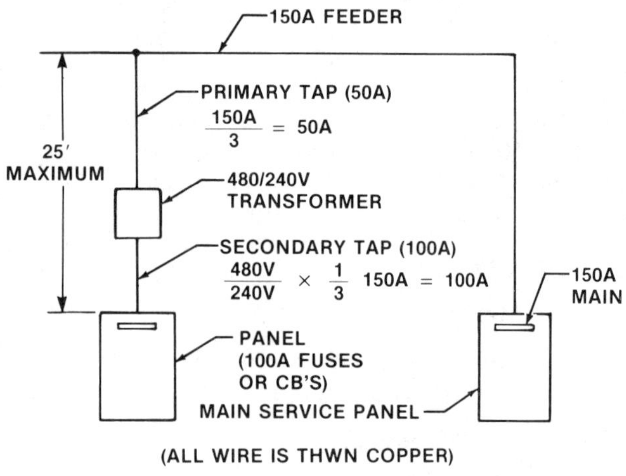

240-21(d)

Figure 12-18. Primary taps to the feeder circuit are sized from the ampacity of the feeder circuit conductors or overcurrent protection device. Secondary taps are sized from the ratio of secondary to primary voltage.

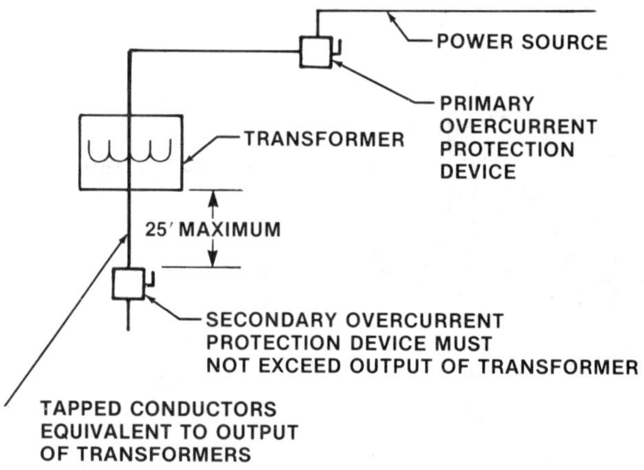

240-21(j)

Figure 12-19. When applying the 25′ tap rule to the secondary of transformers, the overcurrent protection device(s) must not exceed the secondary output of the transformer.

Trade Tip. The tapped conductors per 240-21(d) can be one-third the rating of the overcurrent protection device or the feeder circuit conductors. If the feeder circuit conductors are increased to correct voltage drop, and the tapped conductors are sized at one-third of the increased rating, the tapped conductors must be larger in size.

For example, #1/0 THWN copper conductors are required to supply a feeder circuit. However, the conductors are increased to #4/0 to correct for voltage drop. What is the minimum size of the tapped conductors? The size of the tapped conductors is 50 amps (150A × ⅓ = 50A) if they are based on the #1/0 THWN per Table 310-16. The tapped conductors are selected at 76.7 amps (230A × ⅓ = 76.7A) based on the #4/0 THWN copper conductors.

Size the tap from the smallest rating obtained from the feeder circuit conductors or the overcurrent protection device protecting the feeder circuit. In the preceding example, the 150A overcurrent protection device requires a 50A tap.

If a panelboard or switch must be mounted over 10′, but not more than 25′, the tap rule per 240-21(j) must be applied. For example, a 200-amp panelboard cannot be mounted within the required 25′ or less limitation. In order for the panelboard to be installed, a 200-amp fused disconnect switch must be installed and located in the 25′ boundary. See Figure 12-20.

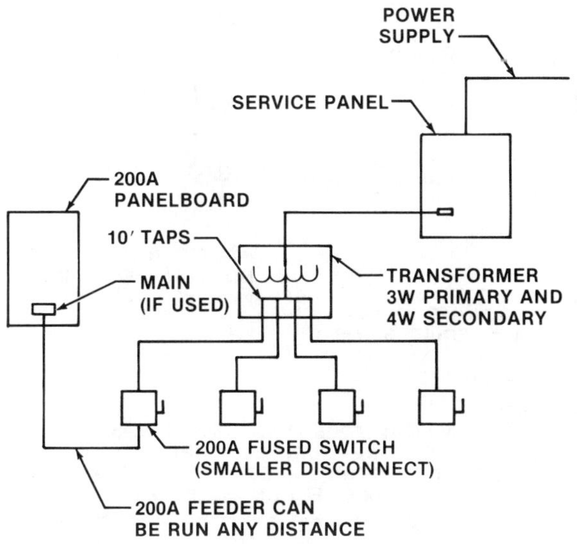

Figure 12-20. Where wall space is not available to mount a large panelboard, a smaller fused disconnect switch can be installed.

TAPPING SECONDARY SIDE OF TRANSFORMER—240-21(b)(d)(j)

The secondary of a two-wire–to–two-wire transformer can be run any distance without placing overcurrent protection within 25′ of the secondary side. However, the primary side must be protected at 125% of the transformer's full-load current rating per 450-3(b)(1), Exception 1 and 240-3(i). The secondary side of a two-wire–to–three-wire or three-wire–to–four wire transformer must have overcurrent protection placed in the secondary no further than 25′ from the secondary terminals of the transformer per 240-21(b)(d)(j). The overcurrent protection should be placed as close as possible to the secondary terminal of the transformer.

Trade Tip. If a panelboard is to be installed on the floor above the switchgear room and supplied by the secondary of a step-down transformer, overcurrent protection must be provided within 25′ of the transformer's secondary terminals. The feeder circuit cannot be routed from the secondary of the transformer to the panelboard on the second floor without providing overcurrent protection within the 25′ limitation. See Figure 12-21. *NOTE:* The 25′ tap rule is limited to industrial locations.

DIAGRAMS OF PRIMARY OVERCURRENT PROTECTION FOR TRANSFORMERS

Overcurrent protection devices may be installed in the primary or secondary of a transformer. See 240-3(i), 240-21(b)(d)(j), and 384-16(a). For rapid and easy installation of overcurrent protection devices, follow the diagrams in Figure 12-22. Note that ratings are given for the overcurrent protection devices. Also, each panel is classified as a power panel or lighting and appliance panel.

For example, a power panelboard feeder circuit is tapped from the secondary of a transformer. The feeder circuit conductors are routed less than 10′ from the transformer to the panelboard. Illustration A in Figure 12-22 is an example of a 10′ tap supplying a power panel. Illustration D is an example of a 25′ tap supplying a lighting panel. Illustration H is applied for protecting primary and secondary using a two-wire primary to two-wire secondary.

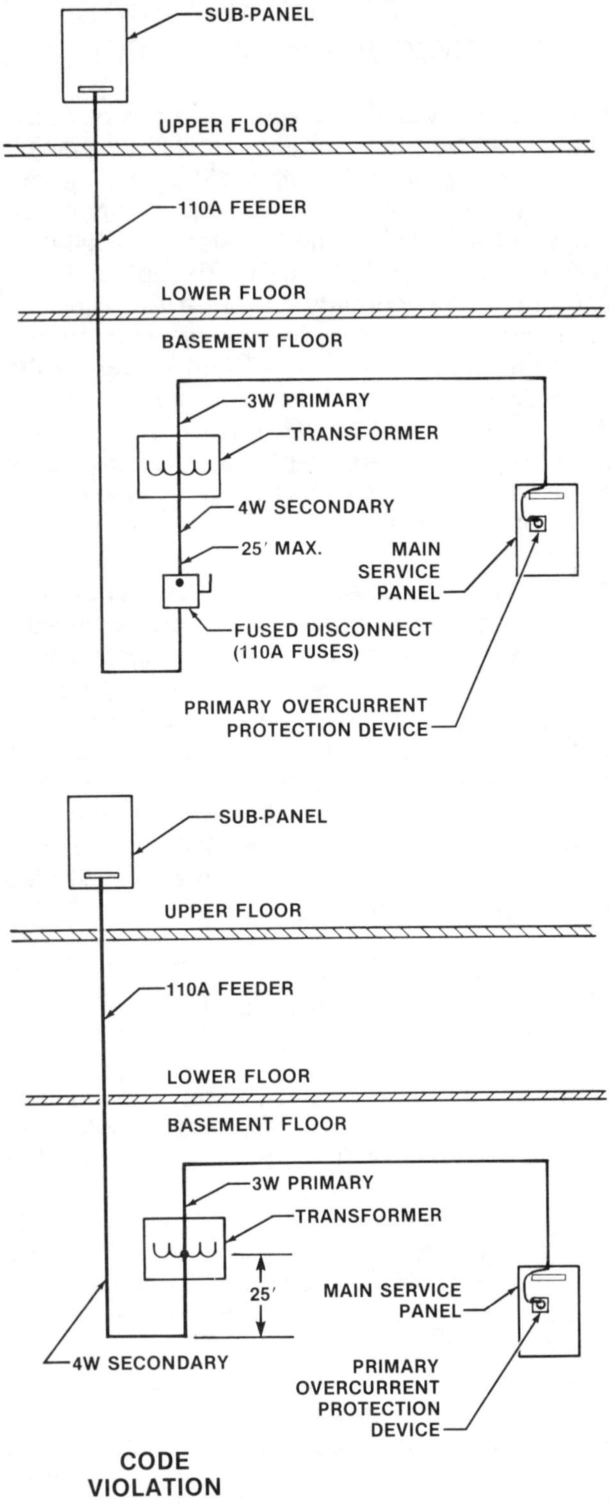

Figure 12-21. Three- or 4-wire secondary tapped conductors must not exceed 25′ without overcurrent protection on the secondary side. Two-wire secondaries with primary protection do not require overcurrent protection on the secondary side.

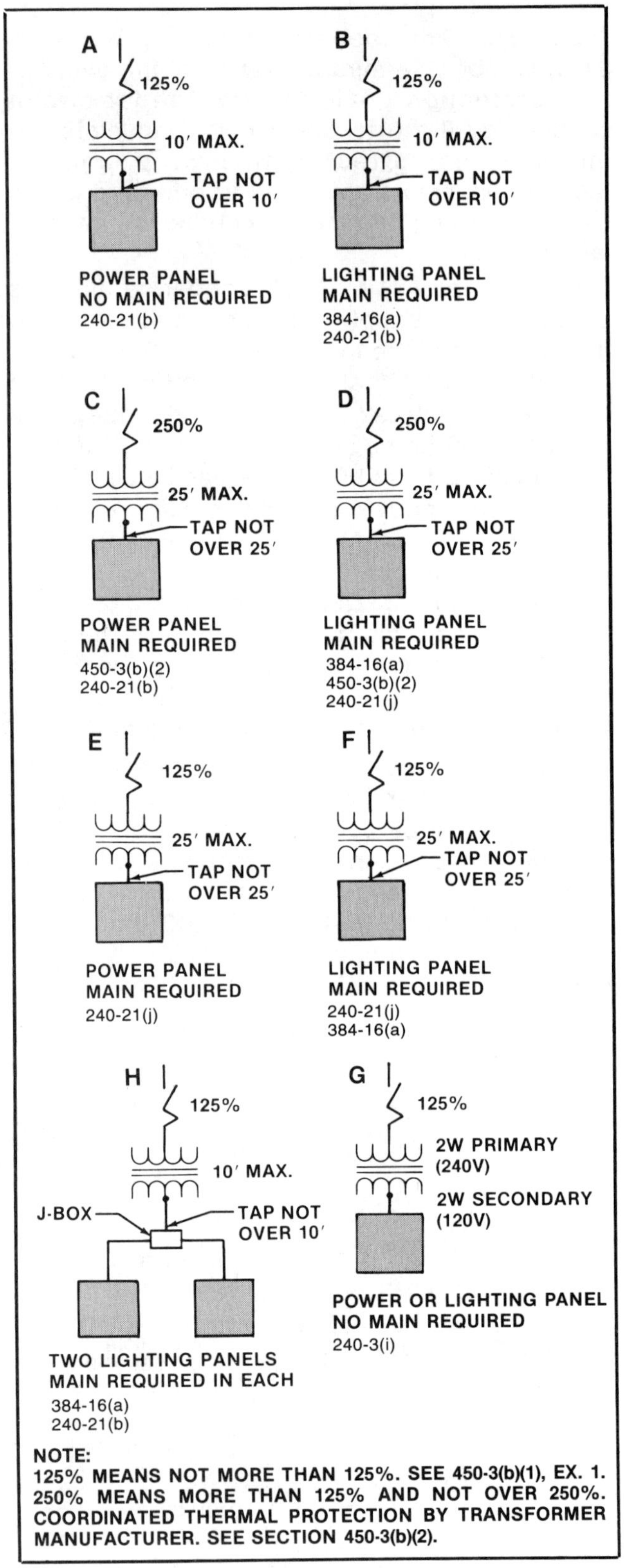

Figure 12-22. Overcurrent protection may be installed in the primary or secondary of a transformer.

REVIEW—CHAPTER 12

Name **Date**

True-False

T F 1. A building can be served by single-phase or three-phase systems, depending on the types of loads.

T F 2. Single-phase systems provide three 120-volt conductors to ground.

T F 3. The load of a building is determined by adding all the single-phase loads to the three-phase loads where a wye system is used.

T F 4. Voltage determines the amount of amperage that transformers deliver to the loads served.

T F 5. Three-phase transformers connected for wye operation have three 120-volt conductors to ground.

T F 6. The kVA of wye-connected transformers is derived by adding the single-phase and three-phase VA loads together and dividing by 100.

T F 7. When computing the single-phase load to size the lighting transformer for a closed delta system, the single-phase load must be multiplied by 67%.

T F 8. When computing the three-phase load to size the power transformer for an open delta system, the three-phase load must be multiplied by 67%.

T F 9. A primary overcurrent protection device can be used to protect the secondary conductors of a two-wire-to-two-wire system.

T F 10. A primary 480 V overcurrent protection device is sized at not more than 300% of the primary full-load current rating to avoid tripping open due to inrush current.

T F 11. Individual overcurrent protection devices such as fuses are selected at 250% or less of the primary full-load current rating for systems over 600 volts located in supervised areas.

T F 12. Overcurrent protection can be provided in the primary and secondary of transformers rated over 600 volts.

T F 13. The 10′ tap rule can be applied when tapping the secondary of transformers.

T F 14. The main or lugs of panelboards can be used to terminate tap conductors when the 25′ tap rule is applied where the overcurrent protection device does not exceed the output.

T F 15. When applying the 10′ tap rule, the conductors can be longer than 10′ if the conduit does not exceed 10′ in length.

Completion

__________ 1. Ampacity of tapped conductors, when applying the 25′ plus tap rule, must be at least ______________ of either the feeder circuit conductors or the overcurrent protection device (25′ tap).

_______ 2. The maximum setting of the overcurrent protection device for a transformer with a primary full-load current rating over 9 amps is ________%.

_______ 3. Secondary conductors of transformers are sized from the _______ served.

_______ 4. The conductors from a 10′ tap must terminate in a(n) ________ when supplying a lighting and appliance panelboard.

_______ 5. Tapped primary and secondary conductors and the transformer can extend no more than _______′ from the feeder circuit.

_______ 6. The secondary conductors of transformers cannot extend more than _______′ without installing a main or approved overcurrent protection.

_______ 7. Transformers having a three-wire secondary must be provided with ________ protection.

_______ 8. Lighting and appliance panelboards can have neutral loads served by ________ amps or less of overcurrent protection devices.

_______ 9. A closed delta system cannot have more than ________ lighting and power transformer(s).

_______ 10. The lighting and power transformer of an open delta system is larger than the ________ transformer.

_______ 11. Phase-to-phase voltage of single-phase, three-wire systems with 120 volts-to-ground is _______ volts.

_______ 12. One advantage of closed delta systems is that if one transformer becomes defective, the remaining two can be connected for _______ delta operation.

_______ 13. _______-wire, wye-connected transformers are required when used for 120-volt lighting loads.

_______ 14. Wye-connected transformer systems are usually used where there are many _______-volt loads.

_______ 15. Closed delta systems are used where ________-phase loads predominate.

Multiple Choice

_______ 1. When determining the number of circuit breakers in panelboards, three-phase, three-pole breakers count as ________ circuit breaker(s).

A. one
B. two
C. three
D. three two-pole

_______ 2. Where overcurrent protection devices protect primary current ratings of less than 2 amps, the setting of the devices is _______% of the full-load current rating.

A. 125
B. 150
C. 200
D. 300

________ **3.** Where the rated primary full-load current is 2 amps or more, but less than 9 amps, the setting of the overcurrent protection device must not exceed ______%.
A. 125
B. 150
C. 167
D. 175

________ **4.** A 15kVA, three-phase, 120/208-volt transformer delivers _______ amps.
A. 35
B. 42
C. 55
D. 60

________ **5.** The ratio of a transformer having a 480-volt primary and 240-volt secondary is ____________.
A. 1 to 1
B. 2 to 2
C. 2 to 1
D. 2½ to 1

________ **6.** Any size tap can be made using the _______′ tap rule.
A. 10
B. 25
C. 75
D. 100

________ **7.** For a 25′ tap from a feeder circuit, the tapped conductors must be _______ the ampacity of the conductors from which they are tapped.
A. one-third
B. one-half
C. two-thirds
D. three-fourths

________ **8.** Where circuit breakers are used to protect the primary of transformers rated over 600 volts, the setting must not exceed _______% of the full-load current rating.
A. 125
B. 175
C. 250
D. 300

________ **9.** Where overcurrent protection is provided in the primary and secondary of transformers rated over 600 volts with 2% impedance, the setting of the protection must not exceed ____________ in unsupervised locations.
A. 250% for fuses
B. 300% for fuses
C. 325% for circuit breakers
D. 350% for circuit breakers

________ **10.** The size of the power transformer in open delta systems can be determined by multiplying the three-phase load by _______%.
A. 42
B. 50
C. 58
D. 70

Problems

1. How many amps does a 125-kVA, 480-volt, three-phase transformer deliver?

2. What is the kVA rating of a three-phase, 208-volt transformer with an output amperage of 312 amps?

3. What is the primary full-load current rating for a 75-kVA, single-phase, 240-volt transformer?

4. What size 120/208-volt transformer is needed to supply a building with a single-phase load of 27,000 volt-amps and a three-phase load of 40,000 volt-amps?

5. What size three-phase, 277/480-volt transformers are needed to supply a building with a single-phase load of 198 amps and a three-phase load of 70 amps?

6. What size lighting and power transformers connected for closed delta operation is required to supply a building with a single-phase load of 36 kVA and a three-phase load of 78 kVA?

7. What size fuses are required to provide protection on the primary side of a 1000-kVA transformer supplied by a three-phase, 4160-volt system? The impedance of the transformer is less than 6%. The transformer is located in an unsupervised area.

8. What size overcurrent protection device is required for a transformer with a primary full-load current rating of 83 amps (600 V or less)?

9. What size THWN copper conductors are required to supply a 225-amp panelboard with a 175-amp main? Use the 10′ tap rule.

10. Fill in the blanks for the proper operating voltage of the secondaries for the following transformers.

A. ____________

B. ____________

C. ____________

D. ____________

E. ____________

F. ____________

G. ____________

H. ____________

I. ____________

J. ____________

K. ____________

L. ____________

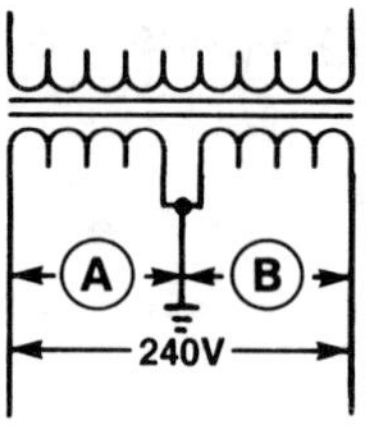

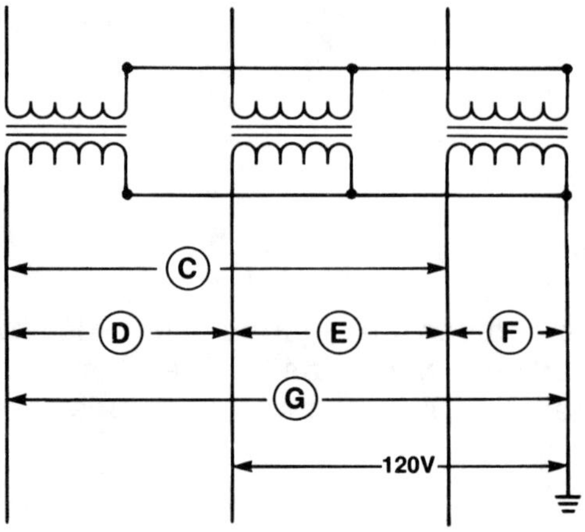

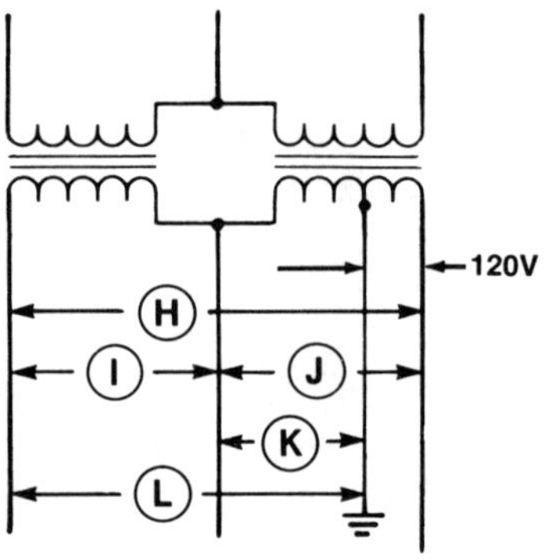

TRANSFORMER CONNECTIONS AND TESTING

Chapter 13

Transformer windings are connected for either additive or subtractive polarity to deliver desired voltage. The windings can be connected in a delta or wye configuration to supply single-phase or three-phase voltage to service equipment or other electrical equipment. Care must be exercised to see that the windings are connected for the proper polarity so the current will flow through the windings in the proper direction.

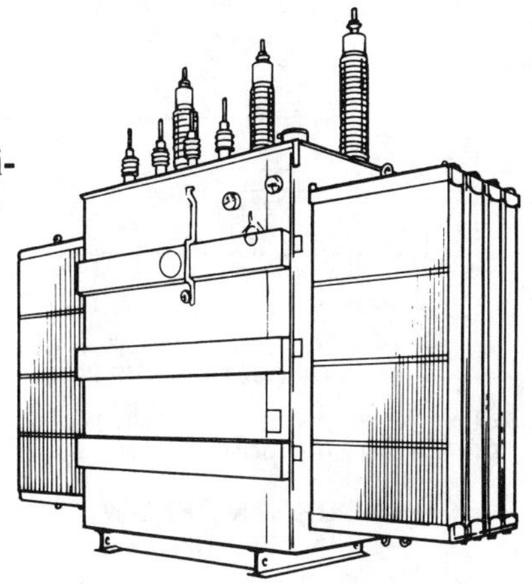

CONNECTIONS FOR ADDITIVE OR SUBTRACTIVE POLARITY

When transformer windings are connected in additive polarity, the induced voltage in the primary and secondary windings would be in the opposite directions. The windings are wound in the same direction for additive-connected transformers. A subtractive transformer can be connected and used as additive by reversing the flow of current through the windings. If one of the windings is erroneously connected in subtractive polarity, either the transformer will trip the overcurrent protection device or the transformer will operate improperly. See Figure 13-1.

When the transformer windings are all connected in subtractive polarity, the induced voltage in the primary and secondary windings would be in the same direction. Likewise, a subtractive transformer can be connected and used as an additive transformer by reversing the flow of current through the windings. If one of the windings is erroneously connected in additive polarity, either the transformer will trip the overcurrent protection device or the transformer will operate improperly. See Figure 13-2.

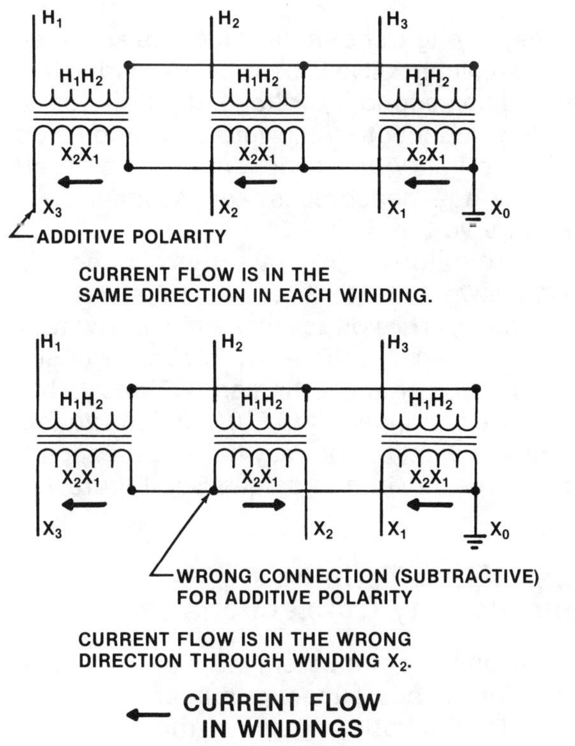

Figure 13-1. Connecting transformers for additive polarity.

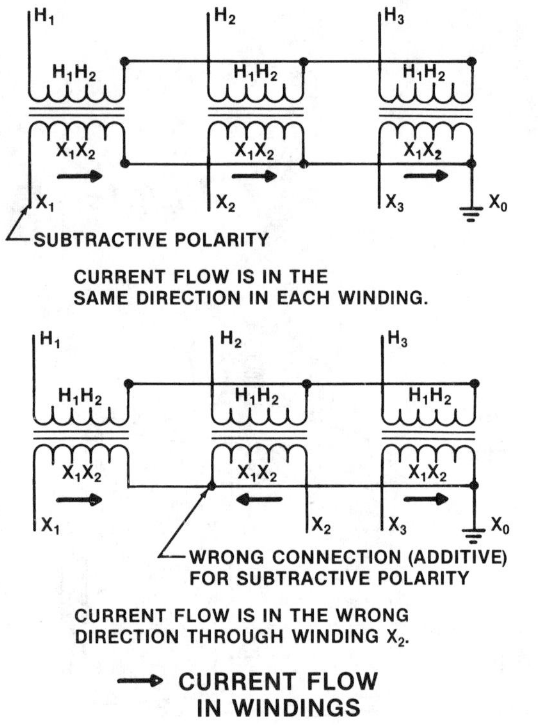

Figure 13-2. Connecting transformers for subtractive polarity.

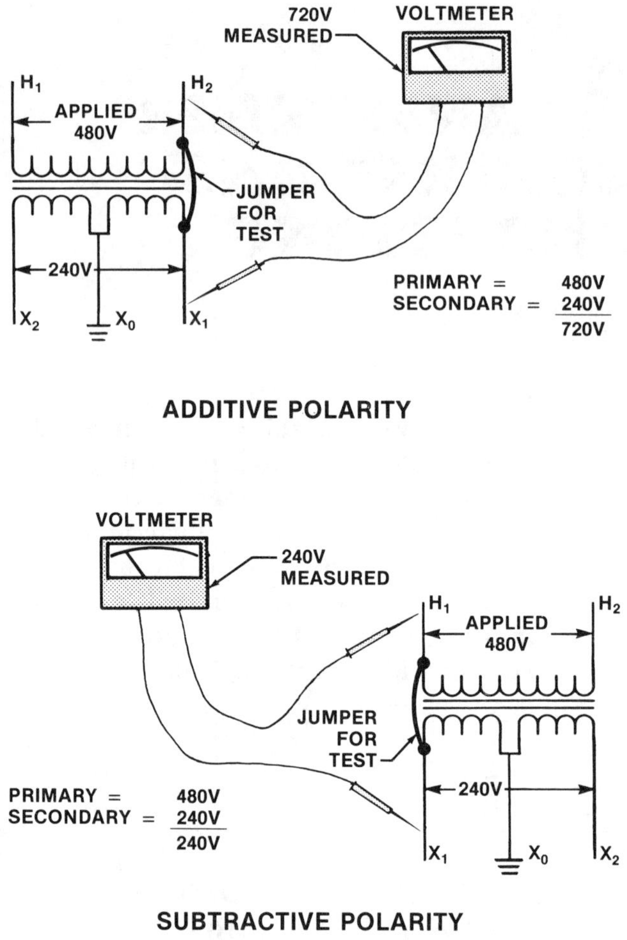

Figure 13-3. Measuring the voltage for additive or subtractive polarity.

Identifying transformer windings as either additive or subtractive is done by measuring the primary and secondary voltage. If a transformer is additive-connected, the voltage would be equal to the primary and secondary added together. The voltage measured is always higher than the primary voltage.

If a transformer is subtractive-connected, the voltage would be equal to the primary minus the secondary. The voltage measured is always less than the primary voltage. *NOTE:* This is under the assumption that the primary voltage is higher than the secondary. *WARNING:* If the voltage is high, a lower voltage may have to be applied to measure the voltage safely. See Figure 13-3.

Identifying Terminals and Polarity Connections

High-voltage or input terminals are identified by the letter *H* and subscript numerals located at the left of the primary side of the transformer's windings. Low-voltage or output terminals are identified by the letter *X* and subscript numerals. The primary or secondary side of a transformer can be used as input or output terminals.

On transformers with additive polarity, the *X* terminal is located at the left of the secondary side of the transformer. If the transformer is subtractive, H_1 on the primary side will align with X_1 on the secondary side of the transformer. The current in the primary of additive-connected transformers flows in the opposite direction. The current in subtractive-connected transformers flows in the same direction. See Figure 13-4.

Connecting Secondary Terminals of Transformer for 120/240-volt, Single-phase

To connect a transformer for a single-phase system, connect X_3 of the first transformer winding

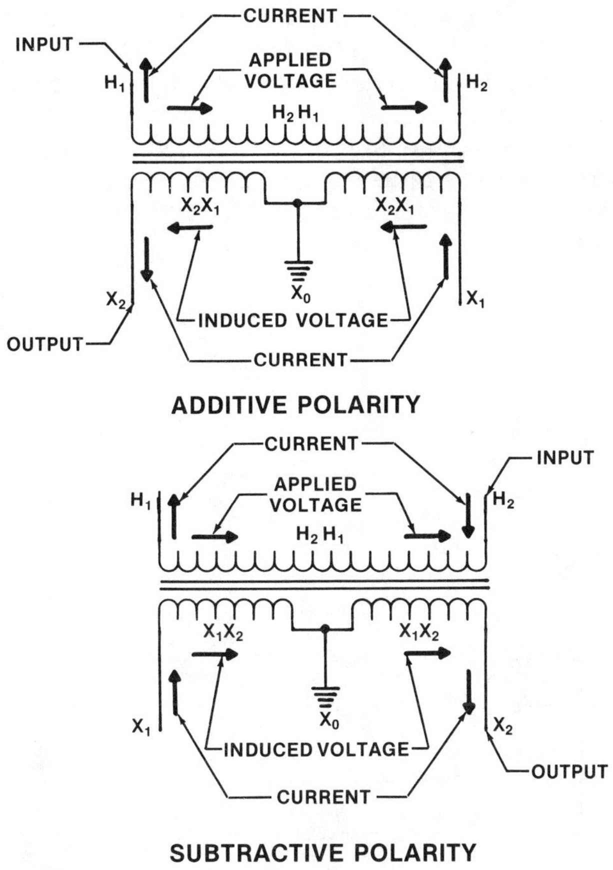

Figure 13-4. If the windings of the two coils are in the opposite direction, the transformer is in additive polarity. If the windings of the two coils are in the same direction, the transformer is in subtractive polarity.

to X_2 of the second winding, making the *neutral connection* X_0. The connection between X_3 to X_2 will series the two 120-volt windings to derive 240 volts from L_1 to L_2. The jumper from X_3 to X_2 is tapped and connected to ground per 250-26. From L_1 to N, and L_2 to N 120 volts is obtained. See Figure 13-5.

Connecting Secondary Terminals of Transformer for Three-phase, Closed Delta System

To connect a transformer for a three-phase closed delta system, connect X_6 of the first transformer winding in series with a jumper to X_1 of the third transformer; X_5 of the first transformer

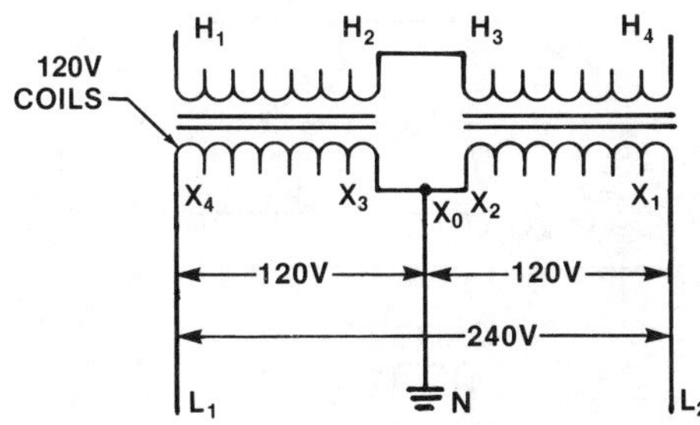

Figure 13-5. Connecting the terminals of a transformer to drive a single-phase, 120/240-volt system.

winding in series with a jumper to X_4 of the second transformer; and X_3 of the second transformer winding in series with a jumper to X_2 of the third transformer.

The center of any 240-volt winding can be tapped for a neutral connection. From the outside lines of the tap, which is connected to ground, 120 volts can be measured. The high leg will measure 208 volts-to-ground because the current must travel through one full winding and one-half of the other winding to ground. The transformer windings of a closed delta system are rated at 240 volts each, for 120/240-volt, four-wire hookup. See Figure 13-6.

Connecting Secondary Terminals of Transformer for Three-phase, Open Delta System

To connect a transformer for a three-phase, open delta system, connect X_3 of the first transformer winding in series with a jumper to X_2 of the second transformer. Either one of the 240-volt windings can be tapped and connected to ground to obtain 120 volts-to-ground. The phase-to-phase voltage is 240 volts from L_1 to L_2, L_1 to L_3, and L_2 to L_3. The high leg voltage is 208 volts-to-ground. See Figure 13-7.

Connecting Secondary Terminals of Transformer for Three-phase, Wye System

To connect a transformer for a three-phase wye system, connect X_5 of the first transformer winding in series with a jumper to X_3 of the second transformer and to X_1 of the third transformer,

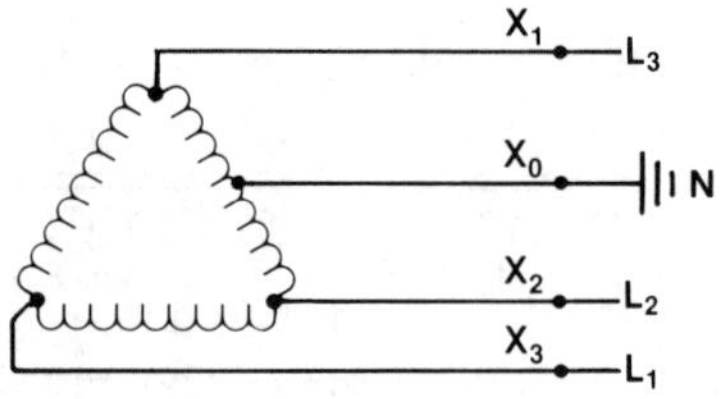

CLOSED DELTA

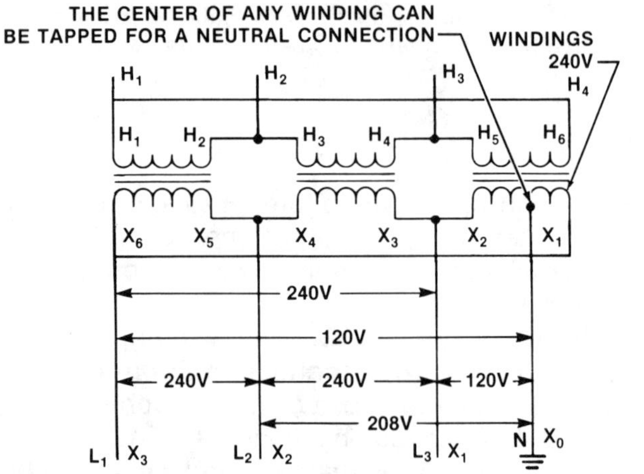

Figure 13-6. Connecting the terminals of a transformer to derive a three-phase, closed delta system.

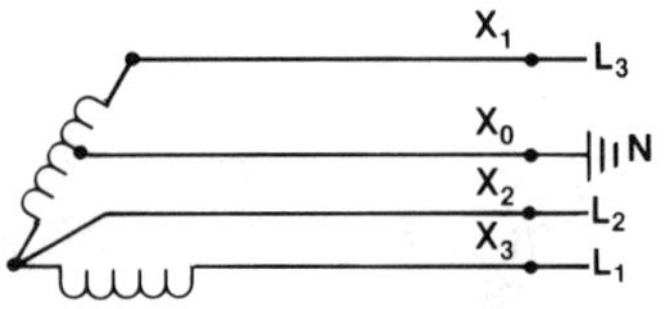

OPEN DELTA

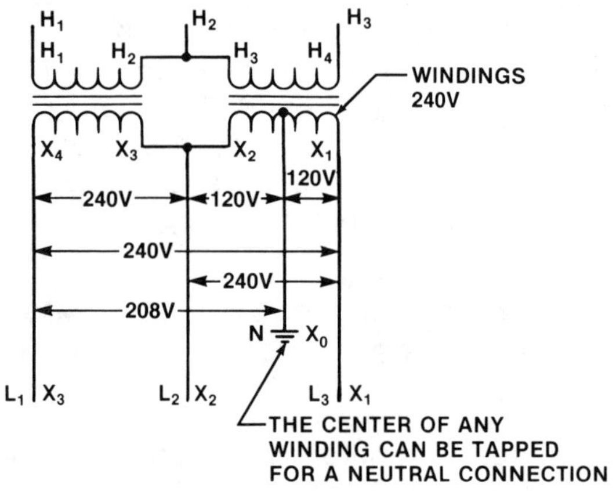

Figure 13-7. Connecting the terminals of a transformer to derive a three-phase, open delta system.

making the neutral connection. In this system, X_6, X_4, and X_2 will be Phases 1, 2, and 3. X_0 is the grounded neutral conductor, which is connected to ground.

There is a measurement of 120 volts to X_0 (N) from phases L_1 to L_2, L_1 to L_3, and L_2 to L_3. The voltage of each winding is 120 volts. See Figure 13-8.

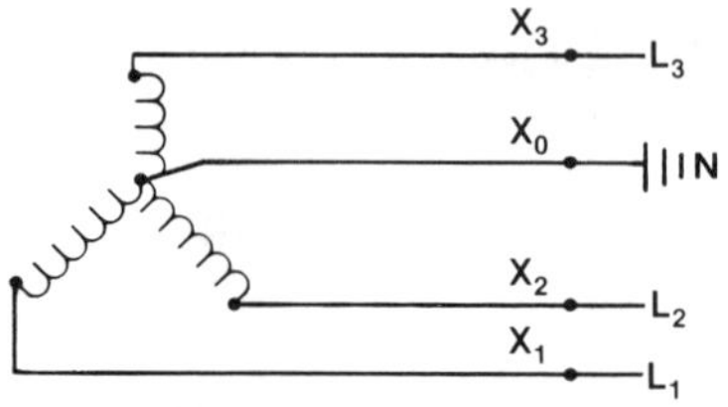

WYE

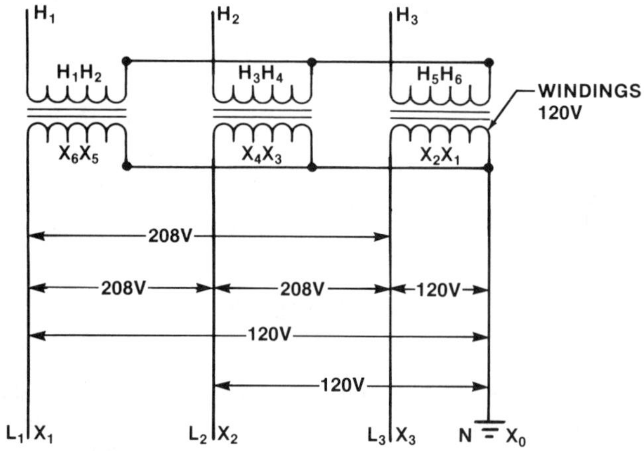

Figure 13-8. Connecting the terminals of a transformer to derive a three-phase, wye system. The voltage on a wye-connected transformer can be 277/480 volts. Each winding is rated at 277 volts.

Connecting Secondary Terminals of Transformer for Three-phase, Corner-grounded Delta System

The grounded conductor of a corner-grounded delta system is derived by tapping any one of the phase legs. If the voltage of the windings is 480 volts, the phase-to-phase voltage would be 480 volts. The voltage to ground is 0 volts from the grounded phase leg.

The grounded phase leg must be white or gray in color per 200-7. It must never be fused per 230-90(b) except motor circuits, per 430-36. The grounded phase conductor must be connected to ground in the panelboard or switchgear to which it enters. See Figure 13-9.

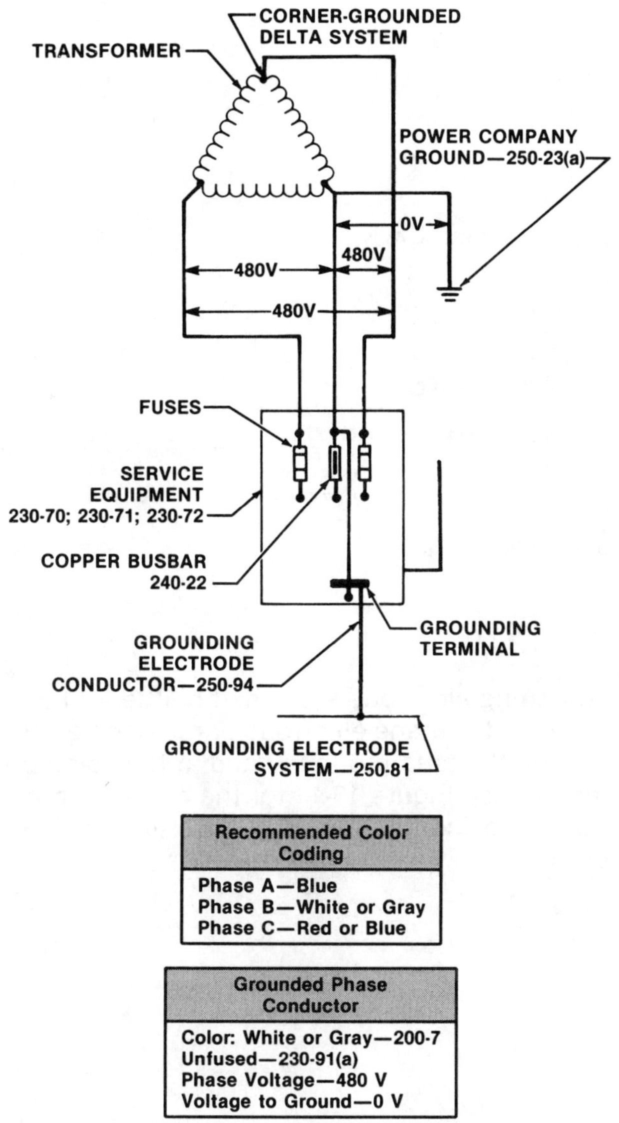

Figure 13-9. Voltage relationships and connections of a three-phase, corner-grounded delta system.

T- or Scott Connection

The T- or Scott connection is used mostly where the supply voltage from the utility company is a three-phase system. It is not economical for the utility company to install a two-phase system to supply a special two-phase motor or piece of equipment. The existing three-phase system, however, can be used to derive a two-phase system using two transformers.

The primary windings must be tapped with one winding center-tapped and the other winding tapped at 86.6% of the winding voltage. The center-tapped transformer is the *main transformer* and the transformer with the winding tapped at 86.6% is the *teaser transformer.*

For example, the main transformer winding is 480 volts, and the center tap provides 240 volts from each side of the tapped winding. The teaser transformer winding is rated at 480 volts, and a tap at 86.6% derives 416 volts (480V × 86.6% = 416V). The T-connection has a power limitation because it has a 15½% higher rating than the load served. If, for example, the load served is 20 kVA, the transformer must be rated at 23.1 kVA. Figure 13-10 shows a Scott or T-connection with values.

SEPARATELY DERIVED SYSTEM

A separately derived system is a system with windings having no direct electrical connections per 250-5(d). The neutral is a solidly connected grounded conductor. Separately derived systems are used when transformers are needed to step up or step down voltages inside a building. Taps from the secondary sides of separately derived systems must comply with 240-21(b)(d)(j).

Bonding Jumper—250-26(a)

The bonding jumper for a separately derived system is the same size as the grounding electrode up to #3/0 copper or 250 kcmil aluminum per 250-79(d) and 250-94. If any one of the phase conductors exceeds 1100 kcmil, the bonding jumper must be equivalent to 12½% of the largest phase conductor. The bonding jumper transfers fault current from metal raceway systems and enclosures to the grounded terminal bar and grounded conductor. From this point the current flows to the overcurrent protection device and opens the shorted circuit to ground.

Grounding Electrode Conductor—250-26(b)

The grounding electrode conductor is sized from the phase conductors supplying the panelboard or equipment per 250-94 and Table 250-94. The grounding electrode conductor grounds the grounded conductor, metal raceways, and enclosures to the earth. It is used to provide a common path to ground for the grounded conductor, metal raceways, and enclosures.

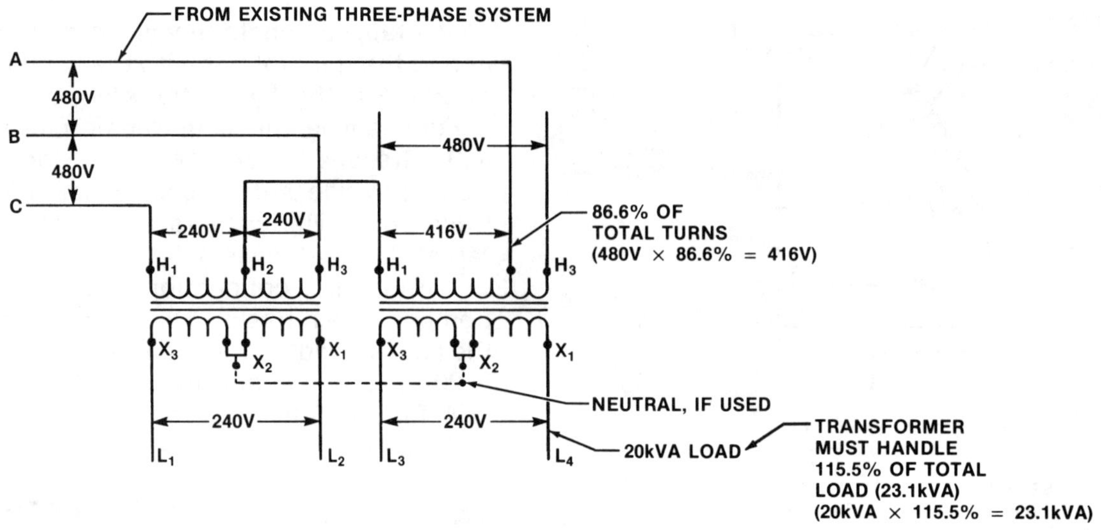

Figure 13-10. Determining the connections and values for a T- or Scott system.

Grounding Electrode—250-26(c)

The grounding electrode must be located as close as possible to the transformer (separately derived system). Building steel is the first choice and metal water pipe is the second choice as the grounding electrode system. If neither of these is present, a made electrode or rod can be used to keep the area grounded and at the same potential. See Figure 13-11 for the proper connections and installation of a separately derived system.

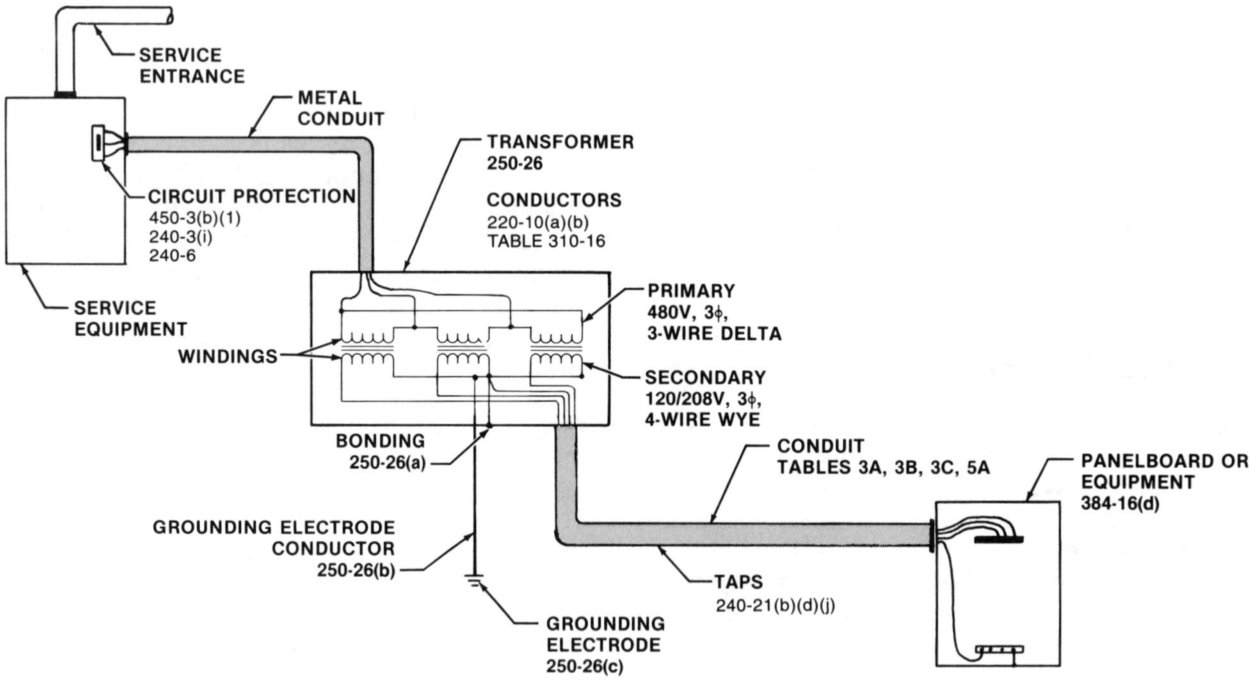

Figure 13-11. Installation requirements for installing a separately derived system.

REVERSE-CONNECTING TRANSFORMER WINDINGS

Reverse-connecting is the connecting of secondary output windings with the power supply and using the primary input windings to supply the load. Dry-type transformers can be reverse-connected and still supply the same kVA. Single-phase transformers rated at 1 kVA and larger and three-phase transformers rated at 15 kVA and larger can be reverse-connected without losing any kVA capacity. The kVA rating is limited to this value because the *turns ratio* is the same as the *voltage ratio*.

Single-phase transformers rated below 1 kVA have a turns ratio that compensates for the low voltage winding. The compensation becomes greater as the kVA rating becomes smaller. When the transformer is reverse-connected, the voltage is less at full load than at no load.

Three-phase transformers rated below 15 kVA are T-connected and are not actual delta-delta or delta-wye connected windings. Three-phase, 120/208-volt transformers rated at 3, 6, and 9 kVA can be reverse-connected if the secondary is T-connected and the neutral is not connected to the supply neutral. Three-phase, 240-volt transformers with the secondary connected T can be reverse-connected because no neutral conductor is used. See the T-connected transformers in Figure 13-12 for reverse-connecting the windings of a transformer.

TRANSFORMER DIAGRAMS

The diagrams in Figure 13-13 show various connections of transformer windings to derive operating voltages. The drawings of the transformer and its windings illustrate a transformer that an electrician will connect for various voltages. For simplicity, the leads for the primary are at the top and the secondary leads are at the bottom of the windings. Most transformers will have the primary and secondary terminals at the bottom for easy connecting. The diagrams in Figure 13-14 show connections for various voltages used by utility companies.

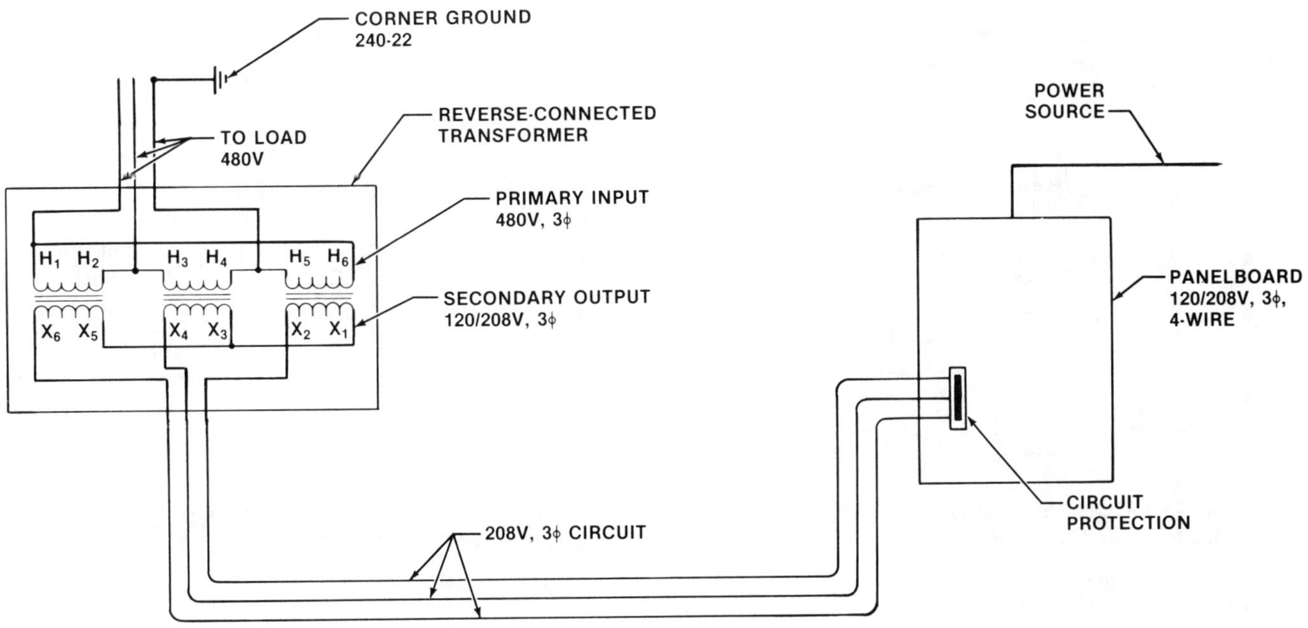

Figure 13-12. Reverse-connecting the windings of a transformer to step up voltage from 208 volts to 480 volts.

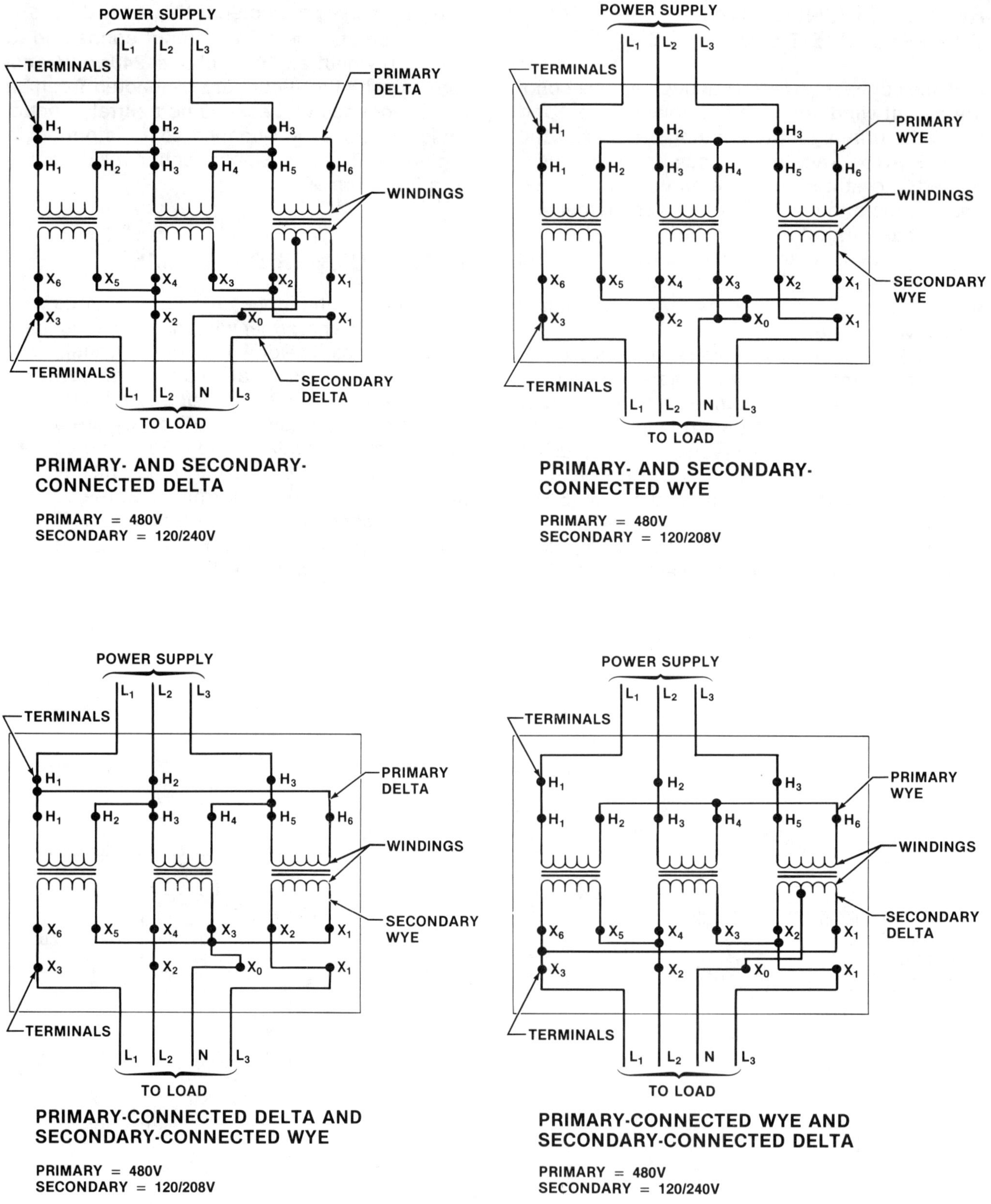

Figure 13-13. Transformers are connected to obtain various voltages required to supply loads.

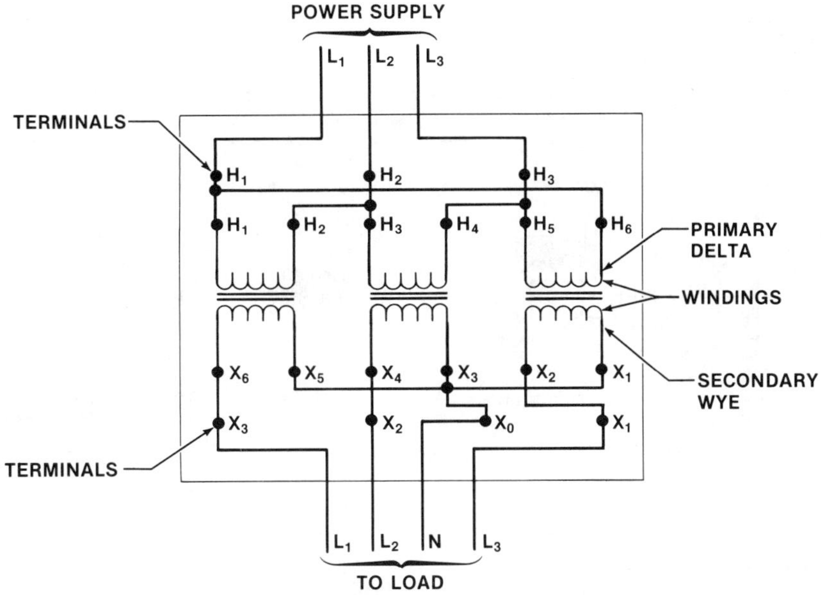

PRIMARY-CONNECTED DELTA AND SECONDARY-CONNECTED WYE

PRIMARY = 4160V
SECONDARY = 480/277V

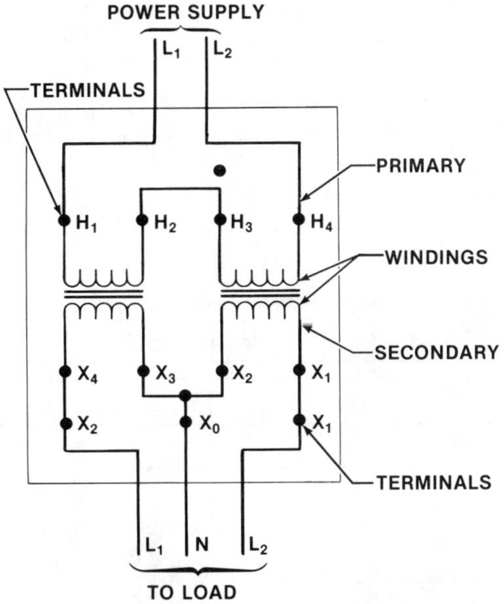

PRIMARY-CONNECTED, 2-WIRE, 1ϕ AND SECONDARY-CONNECTED 3-WIRE, 1ϕ PHASE

PRIMARY = 480V
SECONDARY = 120/240V

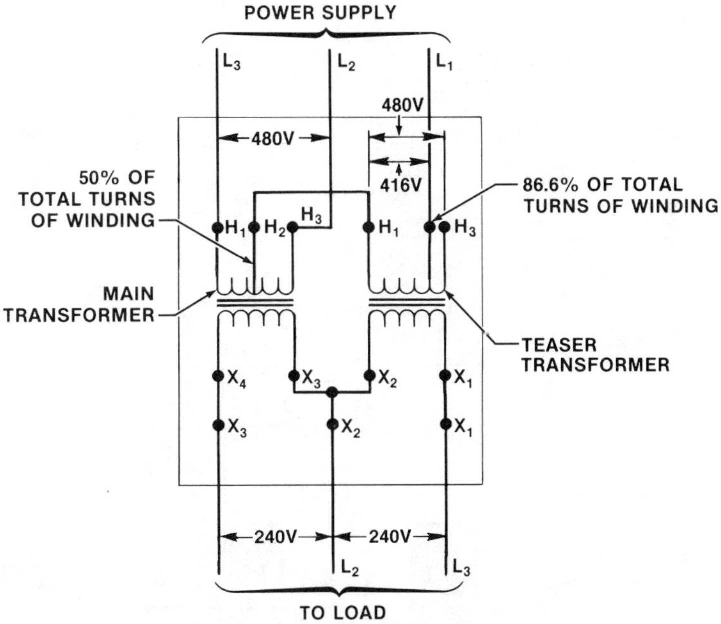

SCOTT OR T-CONNECTION WITH 2ϕ FROM A 3ϕ SYSTEM

PRIMARY = 480V
SECONDARY = 240V

Figure 13-13 (continued).

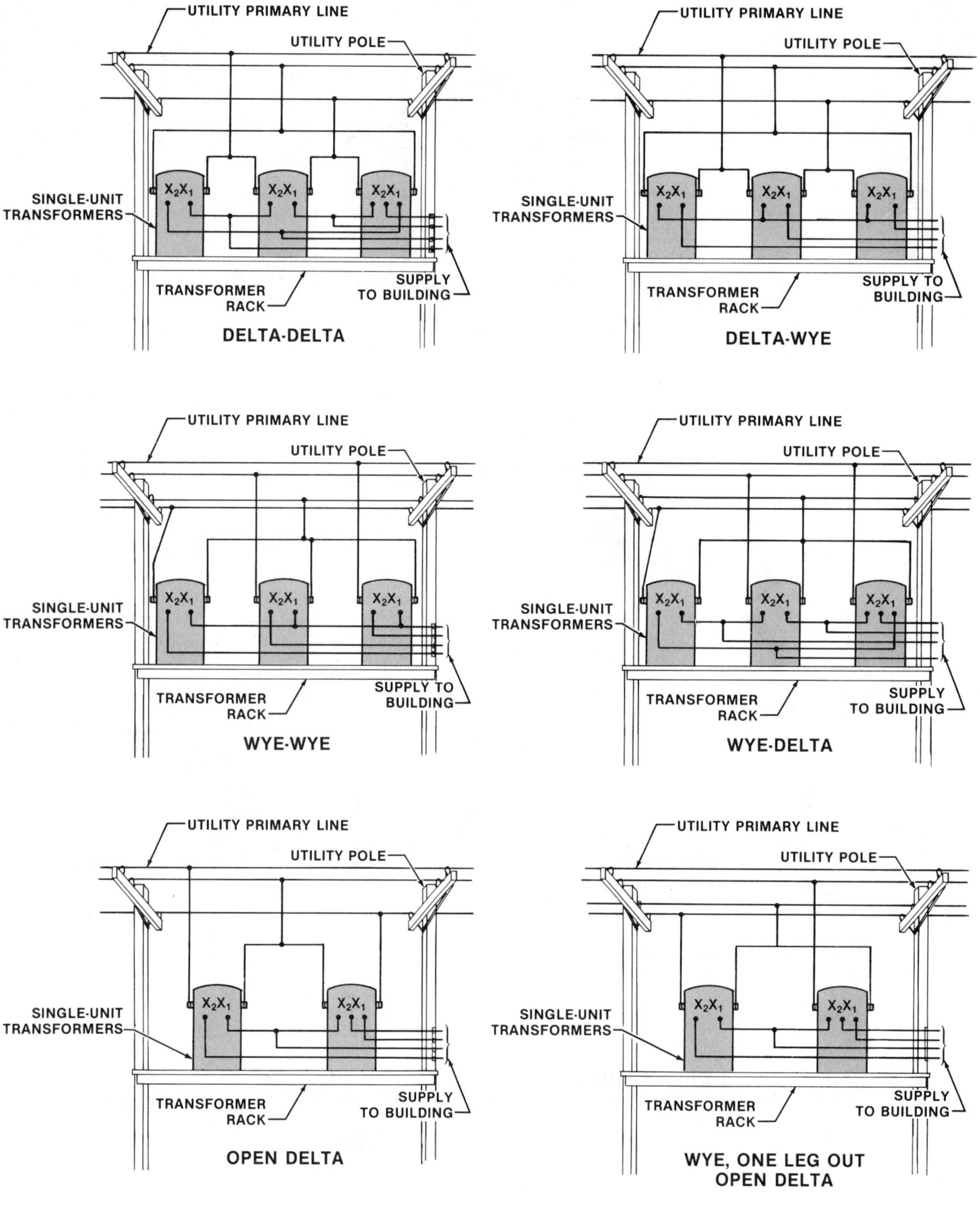

Figure 13-14. Transformers are connected for various voltages by utility companies.

REVIEW—CHAPTER 13

Name ______________________________ **Date** ______________

True-False

T F **1.** Transformer windings are connected for either additive or subtractive polarity.

T F **2.** If the windings of transformers are connected for additive polarity, the voltage will be in the same direction in relation to the primary and secondary.

T F **3.** Transformer windings that are wound additive must be connected additive for proper voltage.

T F **4.** Terminal H_1 of the primary will align with terminal X_1 of the secondary when windings are connected for additive polarity.

T F **5.** If windings are connected for subtractive polarity, the current will flow in the same direction in each winding in relation to the primary and secondary.

T F **6.** Terminal H_3 will align with terminal X_3 if the windings are connected for subtractive polarity.

T F **7.** If the secondary is connected for additive polarity, the secondary voltage will be less than the primary voltage.

T F **8.** Input terminals of the primary are identified by the letter *H* plus a subscript number.

T F **9.** Current in primaries of additive-connected windings flows in the opposite direction in relation to the secondary.

T F **10.** Either one of the 240-volt windings of a closed delta system can be tapped to provide 120 volts-to-ground.

T F **11.** The volts-to-ground on a 480-volt, wye system is 208 volts.

T F **12.** The color of the grounded conductor for a corner-grounded delta system is red.

T F **13.** The grounded conductor of a corner-grounded system must not be protected by a fuse (general rule).

T F **14.** The enclosure for the windings of a separately derived system must be bonded to the grounded conductor and equipment grounding conductors of the system.

T F **15.** Where certain factors are adhered to, dry-type transformers can be reverse-connected and can supply the same kVA.

Completion

__________ **1.** Reverse-connecting of transformer windings is the connecting of secondary output windings to the ______________ and the primary input to the load.

__________ **2.** Teaser transformer windings rated at 480 volts deliver __________ volts with an 86.6% tap.

__________ **3.** T-connected transformers are required to have __________% added to their rating to supply the load.

______ **4.** Scott connections can be used to derive ______-phase systems from an existing three-phase system.

______ **5.** X_0 terminals of transformer windings are connected to ______.

______ **6.** If transformer windings are connected delta, terminal X_5 is connected to ______, and line L_2 is tapped from the series jumper.

______ **7.** In delta-connected transformers, terminal H_1 is connected to H_6 and L_1 is connected to the ______ to the jumper.

______ **8.** Wye-connected windings have X_5, X_3, and X_1 connected to ______ to form the wye configuration.

______ **9.** If transformer windings are connected for additive polarity and one is connected for subtractive, the ______ will not measure correctly.

______ **10.** When checking for the polarity of transformer windings and the voltage is high, a(n) ______ voltage should be used for safety.

______ **11.** For dual windings used to derive 120/240-volt, single-phase systems, terminal X_3 parallels with X_2 and ______.

______ **12.** Transformer windings that are connected in the ______ direction produce additive or subtractive polarity.

______ **13.** Any winding with ______ can be tapped for different voltages.

______ **14.** The corner-grounded conductor can be fused for circuits supplying ______.

______ **15.** A corner-grounded conductor measures ______ volts-to-ground.

Multiple Choice

______ **1.** If the phase conductors of corner-grounded systems are 480 volts between conductors, the grounded corner conductor will measure ______ volts to another phase conductor.

A. 208
B. 240
C. 360
D. 480

______ **2.** Where Phase B is used for the grounded conductor of a corner-grounded system, the color code for the grounded conductor is ______.

A. black or red
B. blue or orange
C. white or gray
D. green or yellow

______ **3.** Current flow through three windings of additive-connected windings is in ______.

A. the opposite direction
B. the same direction
C. either direction
D. there is no current flow through three windings when additive-connected

__________ **4.** If the primary voltage is 220 volts and the secondary voltage is 24 volts, the total voltage for the additive-connected windings is __________ volts.
A. 24
B. 150
C. 196
D. 244

__________ **5.** When the primary voltage is 240 volts and the secondary is 120 volts, the total voltage for subtractive-connected windings is __________ volts.
A. 120
B. 240
C. 300
D. 480

__________ **6.** The low voltage or output terminals of three transformers are identified by the markings __________.
A. H_1, H_2, H_3, and H_4
B. X_0, X_1, X_2, and X_3
C. L_0, X_1, X_2, X_3, and X_4
D. X_0, H_1, X_2, H_2, and X_3

__________ **7.** The mid-point tap for the grounded conductor of a 120/240-volt system is connected to the jumper between __________.
A. X_4 and X_1
B. X_2 and X_4
C. X_1 and X_3
D. X_3 and X_2

__________ **8.** The high leg of delta-connected systems measures __________.
A. 120 volts-to-ground
B. 208 volts from phase-to-phase
C. 208 volts-to-ground
D. 240 volts-to-ground

__________ **9.** The copper bonding jumper for separately derived systems is the same size as the grounding electrode conductor for sizes up to __________.
A. #3/0
B. #4/0
C. 250 kcmil
D. 300 kcmil

__________ **10.** The grounded neutral conductor of a 480-volt, four-wire wye system is __________ volts.
A. 180
B. 200
C. 277
D. 300

Problems

1. Connect the following windings for additive polarity and add the subscript numerals of the secondary terminals using wye connections.

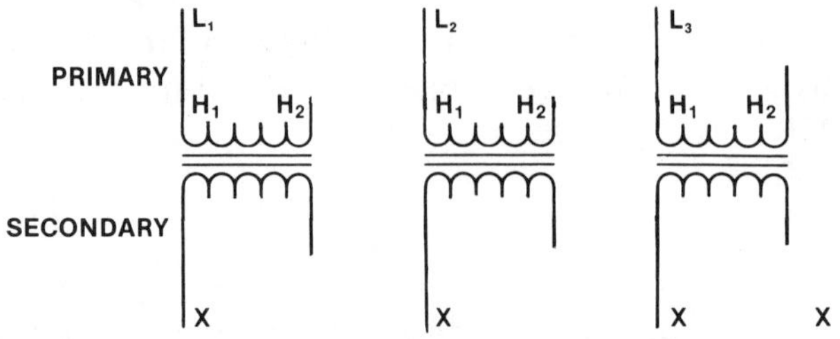

2. Reconnect the following windings for the proper operation using subtractive-connected windings.

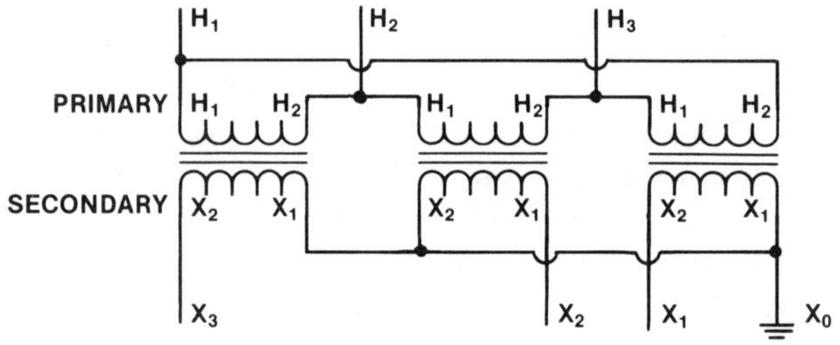

3. Determine the voltage from the primary to the secondary for additive polarity.

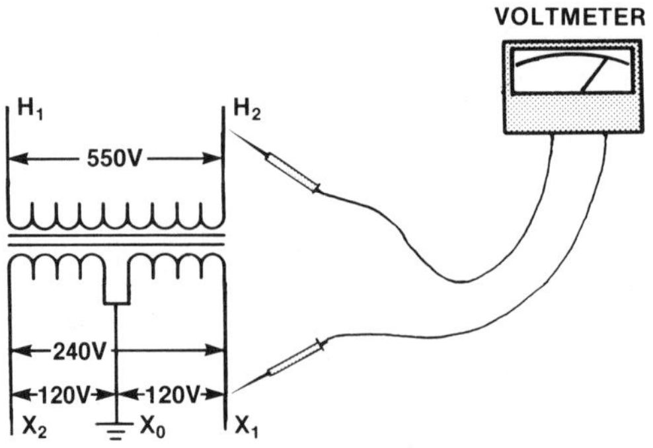

4. Determine the voltage from the primary to the secondary for subtractive polarity.

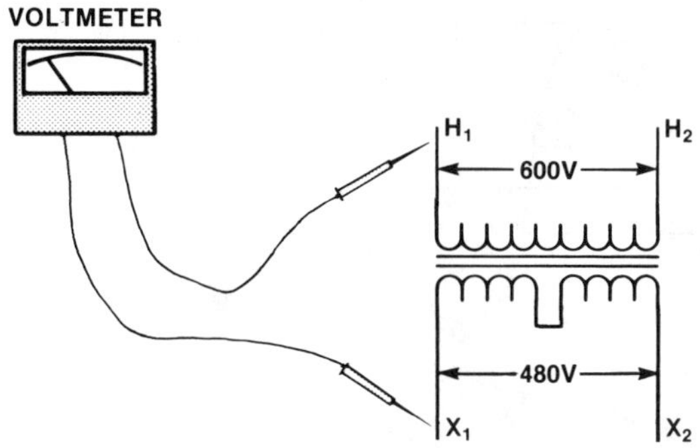

5. Fill in the blanks to show the flow of current through the additive-connected windings.

A. ____________

B. ____________

C. ____________

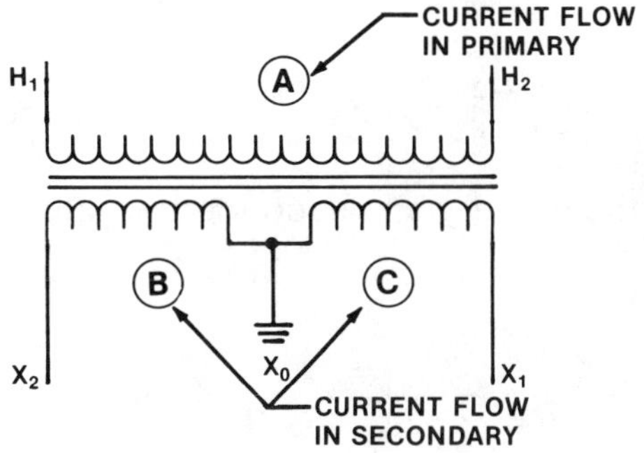

6. Label the input and output sides for the following transformers.

A. ____________

B. ____________

C. ____________

D. ____________

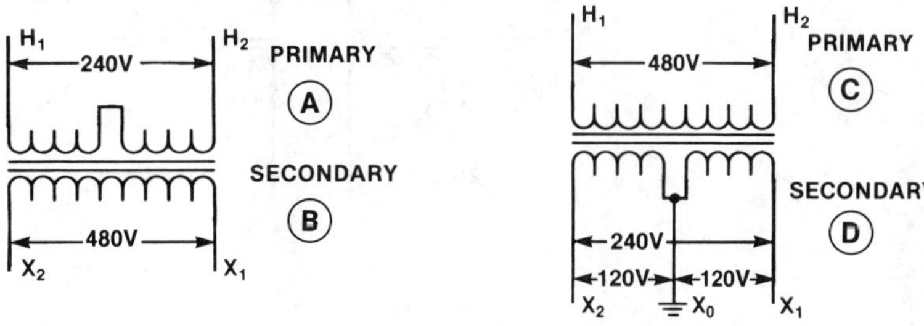

7. If the power supply to the primary side of a T- or Scott-connected bank of transformers is 480 volts, what is the reduced voltage using a 50% tap?

8. Connect the conductors from the transformer to the service equipment.

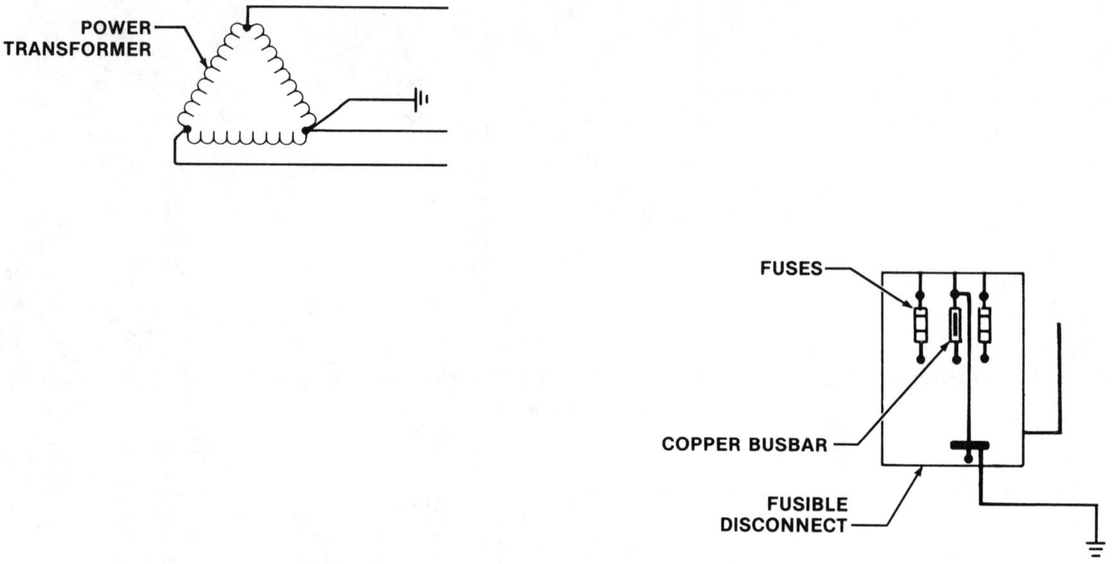

9. Connect the conductors to the primary and secondary windings for T- or Scott-connected windings.

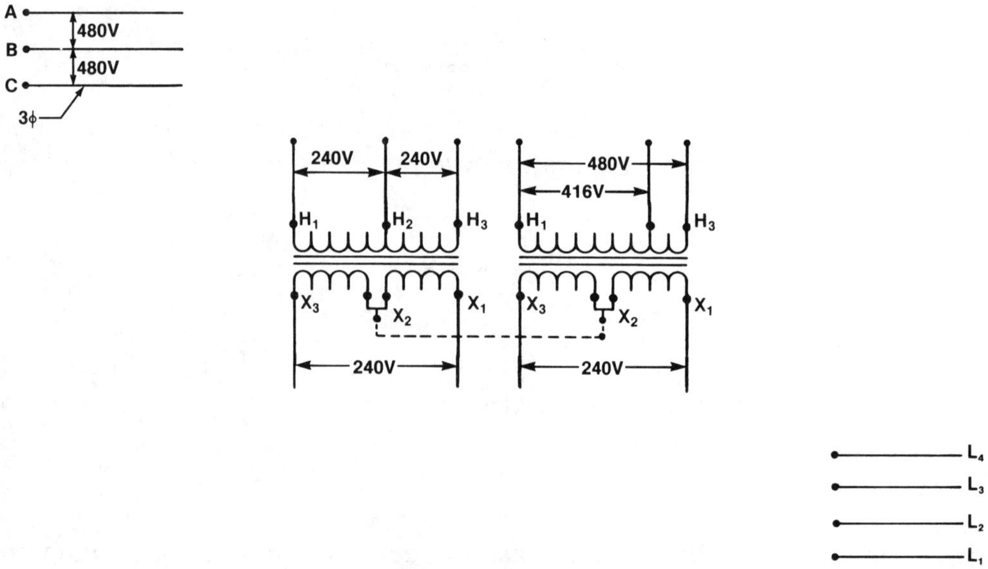

10. A bank of transformers supplies a 40-kVA load. What is the required kVA rating using Scott-connected transformers with an 86.6% tap?

AUTOTRANSFORMERS AND SECONDARY TIES

Chapter 14

Autotransformers can be used to step up or step down the voltage in a circuit or derive a ground for an ungrounded three-phase system. Autotransformers utilize one winding to buck and boost voltage. Secondary ties are used to eliminate the possibility of a complete shut-down of a feeder section in a loop system. They are derived from loop or radial systems in industrial plants.

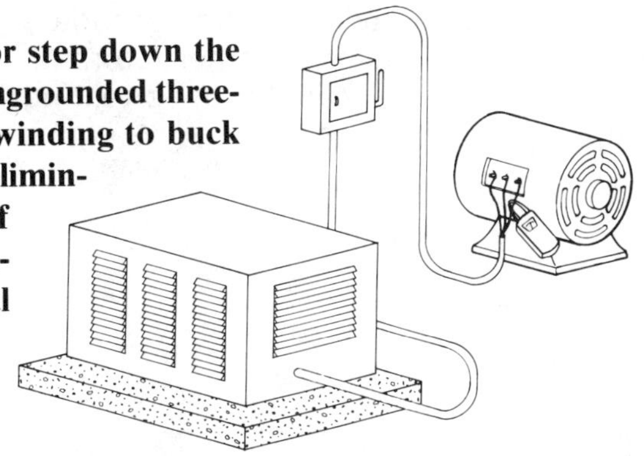

AUTOTRANSFORMERS

Autotransformers are single-winding transformers with the single winding common to the primary and secondary circuits. By connecting the primary across the full winding and tapping the secondary across part of the winding, the voltage can be reduced. The circuit to the primary is connected across part of the winding and the circuit to the secondary is connected across the entire winding to boost voltage. See Figure 14-1.

BASIC CONNECTIONS OF AN AUTOTRANSFORMER

One winding or any combination of windings of an autotransformer can be used to buck or boost the supply voltage. The windings can be connected to operate on a voltage equal to any one winding or combination of windings. See Figure 14-2.

DETERMINING REDUCED RATING OF AN AUTOTRANSFORMER

The power or kVA rating of an autotransformer is reduced from the full rating by an amount equal to the ratio of input and output voltages. For example, the input voltage is 480 volts and the output is 720 volts. The ratio is 67%, or 2 to 3; therefore, the power rating is reduced two-thirds:

$$\frac{480 \text{ V}}{720 \text{ V}} = 67\%$$

If the power load rating is 45 kVA, the rating is reduced to 15 kVA (45 × ⅔ = 30 kVA; 45 kVA − 30 kVA = 15 kVA). See Figure 14-3 for a step-

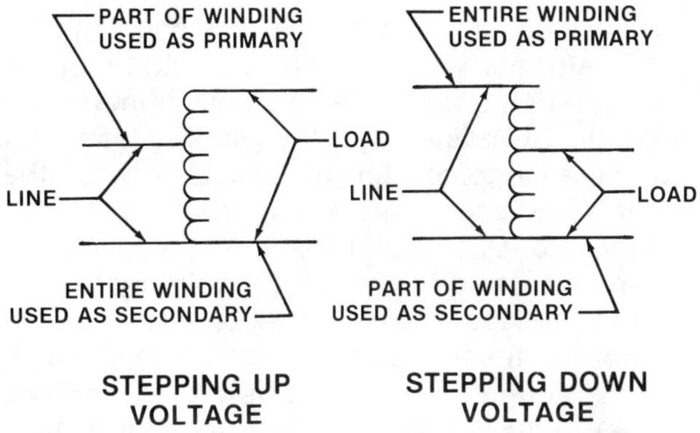

Figure 14-1. Voltage is stepped up or stepped down by utilizing all or part of the winding of an autotransformer.

by-step procedure of how an autotransformer can operate at a reduced rating.

BUCK-BOOST TRANSFORMER

A buck-boost transformer provides an economical means of lowering or raising a supplying line voltage by a small amount. The autotransformer is an insulating transformer with two primary (input) windings rated at 120 or 240 volts. The two secondary (output) windings are rated at either 12, 16, or 24 volts. The primary and secondary windings can be connected together to create a buck-boost transformer capable of correcting voltage up to plus or minus 20%.

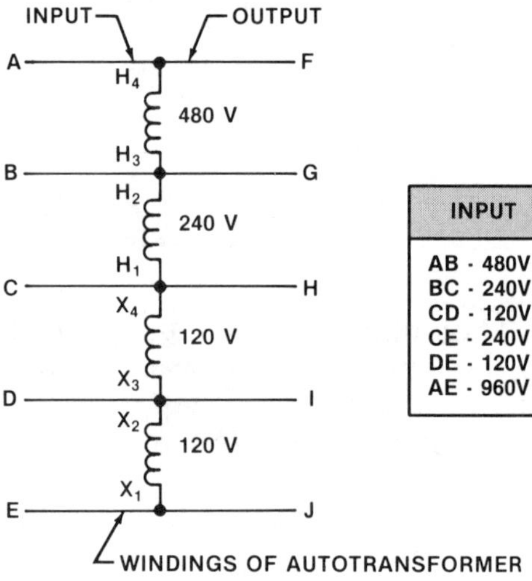

INPUT	OUTPUT
AB - 480V	FG - 480V
BC - 240V	GH - 240V
CD - 120V	HI - 120V
CE - 240V	HJ - 240V
DE - 120V	IJ - 120V
AE - 960V	FJ - 960V

Figure 14-2. Connections and voltage relation for an autotransformer.

A buck-boost transformer is an insulating type of transformer when shipped from the factory. An insulating transformer is simply primary and secondary windings with the winding leads not connected together. An autotransformer has the primary and secondary windings connected together to buck or boost the supply voltage. On the job site, a lead wire on the primary is connected to a lead wire on the secondary. The connections of the two windings change the insulating characteristics of the transformer windings to that of an autotransformer. See Figure 14-4.

Buck-boost transformers can be used to supply power to low-voltage systems such as light-

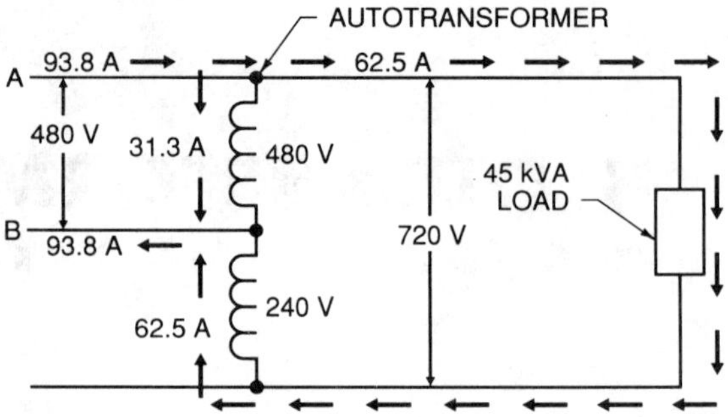

PROBLEM: What is the reduced operating current of the autotransformer?

Step 1: *Output (load current)* $= \dfrac{kVA \times 1000}{V}$

$$current = \frac{45{,}000\ VA}{720\ V} = 62.5\ A$$

Answer: **62.5 A output load current**

Step 2: *Input (supply current)* $= \dfrac{kVA \times 1000}{V}$

$$current = \frac{45{,}000\ VA}{480\ V} = 93.8\ A$$

Answer: **93.8 A input load current**

Step 3: *High voltage* $(H_1 - H_2)$ *current = input current − output current*

$$current = 93.8\ A - 62.5\ A = 31.3\ A$$

Step 4: *High voltage kVA* $= \dfrac{A \times V}{1000}$

$$kVA = \frac{31.3\ A \times 480\ V}{1000} = 15\ kVA$$

Answer: **15 kVA high voltage output**

Step 5: *Low voltage kVA* $= \dfrac{A \times V}{1000}$

$$kVA = \frac{62.5\ A \times 240\ V}{1000} = 15\ kVA$$

Answer: **15 kVA low voltage output**

NOTE: The current ratings in amps prove the reduced operation of the autotransformer.

Figure 14-3. Calculations for an autotransformer operating at a reduced rating.

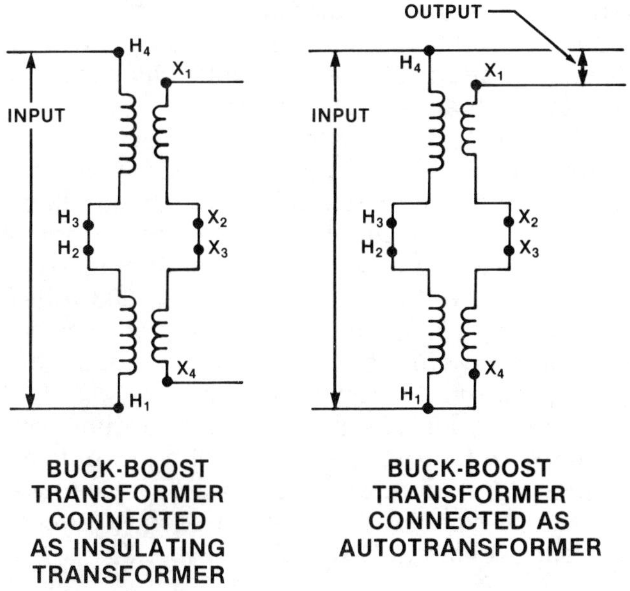

Figure 14-4. Connecting the primary and secondary windings to change the electrical characteristics of an insulating transformer to an autotransformer. The connection from H_1 to X_4 converts the windings to an autotransformer.

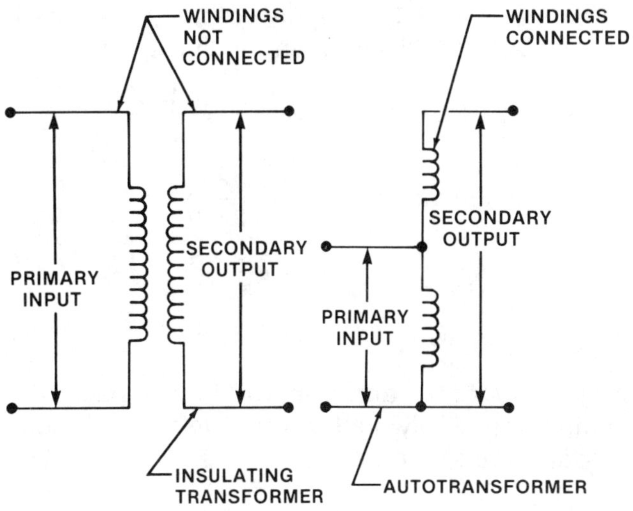

Figure 14-5. Comparison of insulating transformer to an autotransformer.

ing or control circuits. Buck-boost transformers are available in applications requiring 12, 16, 24, 32, or 48 volts.

CONNECTION OF WINDINGS

Buck-boost insulating transformers are available with two windings (one primary and one secondary) and can be connected two different ways to step up or step down the supply voltage to equipment, etc. Buck-boost insulating transformers are also available with four windings (two primaries and two secondaries) and can be connected in eight different ways to increase or decrease the supply voltage. See Figure 14-5 for a comparison of an insulating transformer and an autotransformer. Refer to Figure 14-2.

VOLTAGE APPLICATIONS

The secondary windings are the only parts of an autotransformer that do the work of transforming voltage and current. The most common application for buck-boost transformers is boosting 208 volts to 230 or 240 volts, or bucking 230 or 240 volts to 208 volts. This bucking or boosting of supply voltage is applied mostly in commercial and industrial locations on circuits used to supply air conditioning or motor loads.

Circuits supplying lighting loads can be bucked or boosted from 110 to 220 volts or 240 to 277 volts. The most important point to remember is to select the buck or boost transformer based on 12, 16, or 24 volts. In other words, the voltage supply to a motor is 209 volts, and the motor requires at least 230 volts to its terminals. A 12-, 16-, or 24-volt boost-type insulating transformer with the windings connected is required to form an autotransformer that will increase the circuit voltage. By subtracting 209 volts from 230 volts (230 V − 209 V = 21 V), a boost-type connected autotransformer increasing the voltage to 21 volts is required. Selecting a 24-volt rated insulating transformer and connecting it as an autotransformer to boost the voltage is required.

SELECTING AUTOTRANSFORMERS

The autotransformer must have a kVA, amperage, and secondary voltage rating with enough capacity to supply the requirements of the load served. The secondary current in amps can be found by multiplying the nameplate kVA by 1,000, then dividing by the secondary voltage.

Example: What are the secondary amps of a 1.5 kVA autotransformer with a secondary voltage of 24 volts? The supply voltage is boosted from a 208- to 230-volt, single-phase system.

Step 1: $sec.\ amps = \dfrac{kVA \times 1000}{sec.\ V}$

$sec.\ amps = \dfrac{1.5 \times 1000}{24\ V}$

$sec.\ amps = 62.5$

Answer: **62.5 A**

The kVA of the autotransformer can be determined by multiplying the output volts by the secondary amps, then dividing by 1,000.

Step 1: $kVA = \dfrac{output\ V \times sec.\ amps}{1000}$

$kVA = \dfrac{230\ V \times 62.5\ A}{1000}$

$kVA = 14.4$

Answer: **14.4 kVA**

NOTE: The autotransformer is capable of handling a 14.4 kVA load.

The size and rating of an autotransformer with a secondary voltage of 24 volts serving a 230-volt motor with a connected load of 14,000 volt-amps can be determined as follows:

Step 1: The load served is 14,000 VA.

Step 2: $amps = \dfrac{load\ served}{supply\ voltage}$

$amps = \dfrac{14{,}000\ VA}{230\ V}$

$amps = 61\ A$

Answer: **61 A**

Step 3: *size of autotransformer* = *amps* × *sec. volts*

size of autotransformer = 61 A × 24 V

size of autotransformer = 1464 VA

Answer: **1464 VA** or **1.46 kVA** (1464/1000 = 1.46)

Trade Tip. The calculations indicate that an autotransformer rated at least 1464 VA with a transformation voltage of 24 volts will supply a 14,000 VA load to a 230-volt motor.

Figure 14-6 shows a step-by-step procedure for determining the size autotransformer needed to supply a motor circuit in a commercial building. In Figure 14-6, the supply is too low for the motor to operate properly. An autotransformer is used to boost the voltage from 190 volts to 208 volts. If the supply voltage to a load is too high, a buck-type autotransformer can be used to reduce the voltage to the desired operating level.

Certain types of equipment cannot function properly without the correct voltage applied. Figure 14-7 illustrates a step-by-step procedure for a buck-type autotransformer used to reduce the supply voltage.

SELECTING THE OVERCURRENT PROTECTION DEVICE

The overcurrent protection device for an autotransformer is selected per 450-4 of the NEC®. The percentage times the full-load current rating of the autotransformer is based on the nameplate rating of the autotransformer. If the rated full-load input current of the autotransformer is 9 amps or more, the amperage is multiplied by 125% and the next size overcurrent protection device can be used.

For example, if the full-load input current rating is 42 amps, a 60-amp overcurrent protection device can be used per 450-4(a), Exception (42 A × 125% = 52.5 A). If the full-load input current rating is less than 9 amps, the amperage is multiplied by 167%. The next rating lower must be used per 450-4(a), Exception. For example, if the full-load input current rating is 8.3 amps, a 10-amp overcurrent protection device can be used (8.3 A × 167% = 13.9 A) per 240-6, Exception. Figure 14-8 illustrates a step-by-step procedure for finding the overcurrent protection device of an autotransformer.

GROUNDING AUTOTRANSFORMERS—450-5

It is permissible to supply branch circuits with an autotransformer if one of the secondary conductors is grounded. Note that the grounded secondary conductor must be connected to the

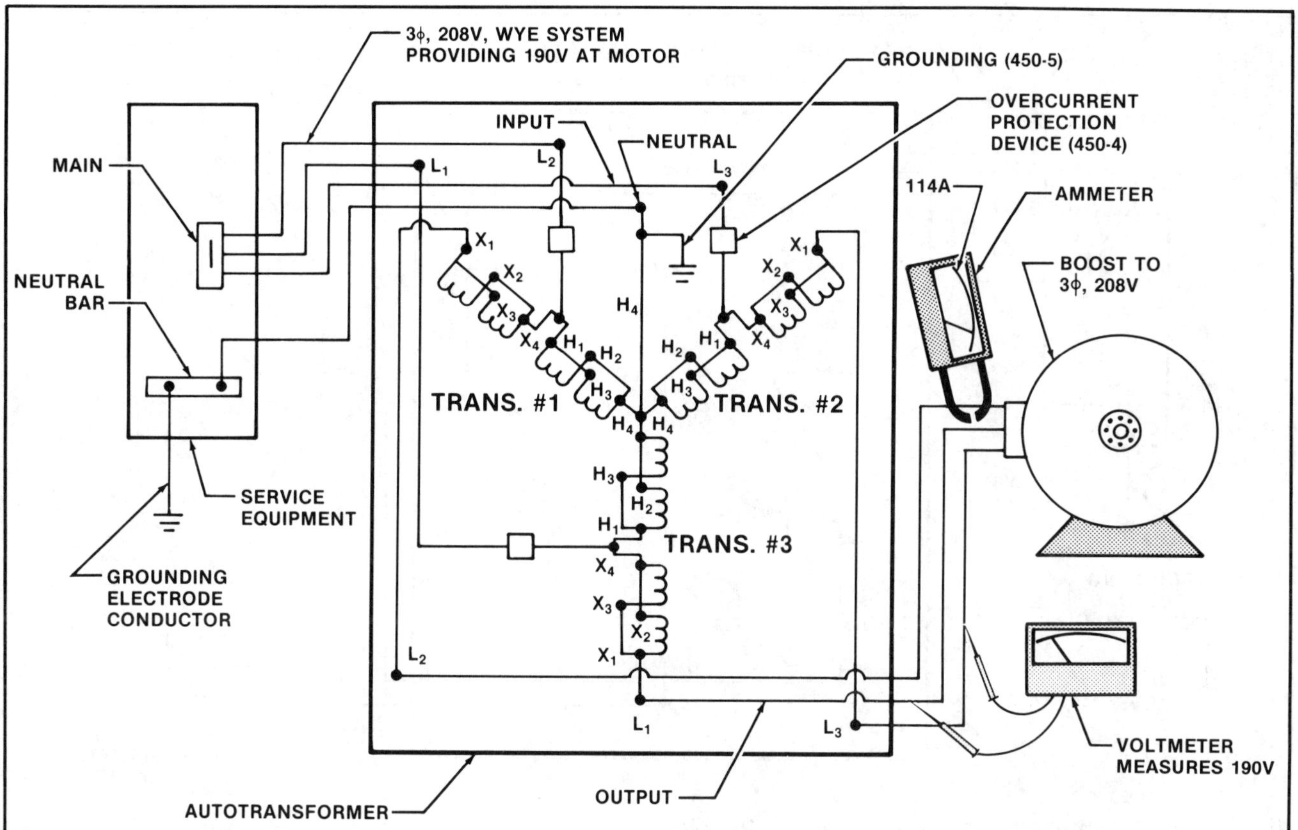

PROBLEM: What is the size autotransformer required to supply a motor load of 41,040 volt-amps (114 A) that will boost the supply voltage from 190 volts to 208 volts?

Step 1: Select autotransformer and correct transformation voltage
(208 V − 190 V = 18 V)

Step 2: Voltage selections
12, 16, or 24 V
18 V needed
24 V rated autotransformers required

Step 3: Size autotransformer at 230 V

$$kVA = \frac{VA\ of\ load}{1000}$$

$$kVA = \frac{41{,}040\ \text{VA}}{1000}$$

$$kVA = 41.04\ \text{kVA}$$

Step 4: Size autotransformer at 24 V

$$kVA = \frac{load \times transformation\ voltage}{1000}$$

$$kVA = \frac{114\ \text{A} \times 24\ \text{V} \times 1.732}{1000}$$

$$kVA = 4.739\ \text{kVA}$$

Step 5: Secondary amps at 24 V

$$sec.\ amps = \frac{kVA \times 1000}{V \times 1.732}$$

$$sec.\ amps = \frac{4.739\ \text{VA} \times 1000}{24\ \text{V} \times 1.732}$$

$$sec.\ amps = 114\ \text{A}$$

Step 6: Size of load that can be handled

$$kVA = \frac{output\ volts \times sec.\ amps}{1000}$$

$$kVA = \frac{208\ \text{V} \times 1.732 \times 114}{1000}$$

$$kVA = 41.04\ \text{kVA}$$

Answer: **Autotransformer selected must be capable of handling 41.04 kVA and increasing supply voltage up to 18 V.**

Figure 14-6. Selecting and sizing autotransformer to boost supply voltage.

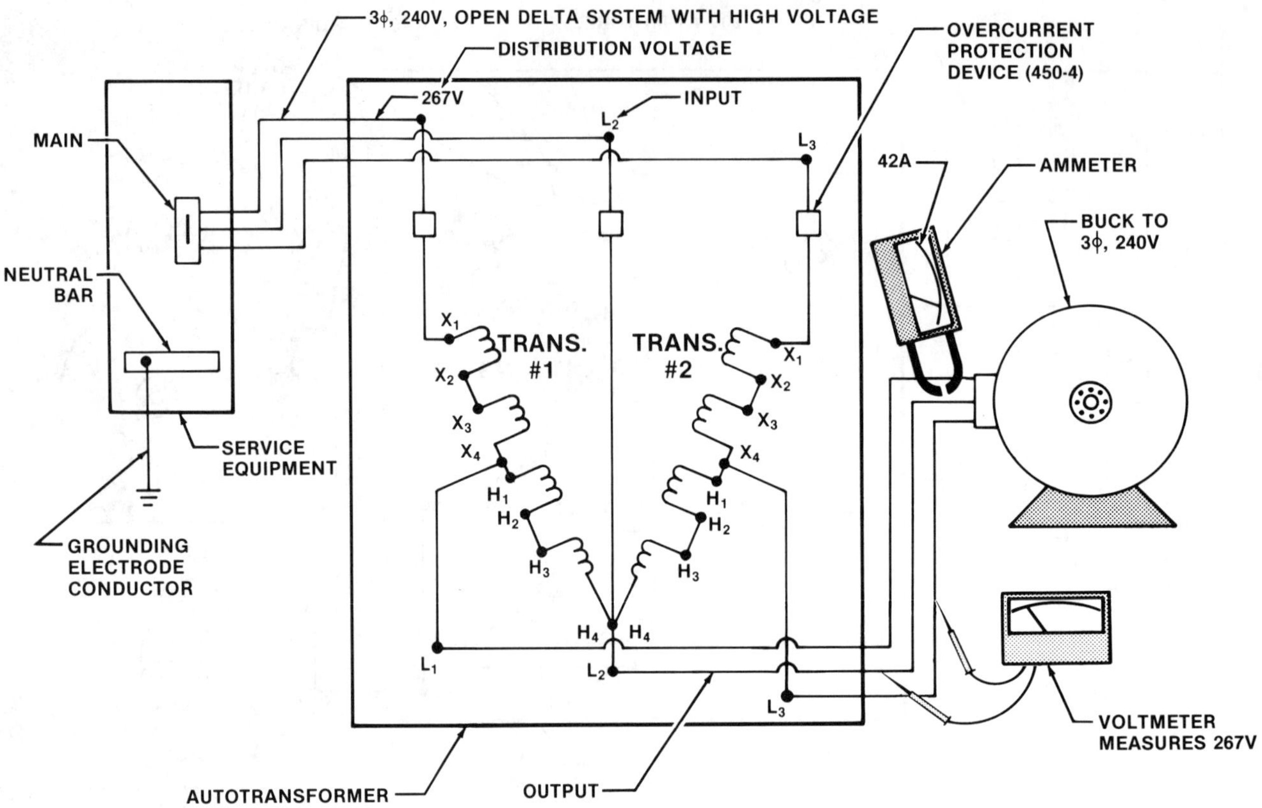

PROBLEM: What is the size autotransformer required to supply a motor load of 17,472 volt-amps (42 A) that will buck the supply voltage from 267 to 240 volts?

Step 1: Select autotransformer and correct transformation voltage
(267 V − 240 V = 27 V)

Step 2: Voltage selections
12, 16, 24, 32, or 48 V
27 V needed
32 V rated autotransformers required

Step 3: Size autotransformer at 230 V

$$kVA = \frac{VA\ of\ load}{1000}$$

$$kVA = \frac{17{,}472\ \text{VA}}{1000}$$

$$kVA = 17.472\ \text{kVA}$$

Step 4: Size autotransformer at 32 V

$$kVA = \frac{load \times transformation\ voltage}{1000}$$

$$kVA = \frac{42\ \text{A} \times 32\ \text{V} \times 1.732}{1000}$$

$$kVA = 2.328\ \text{kVA}$$

Step 5: Secondary amps at 32 V

$$sec.\ amps = \frac{kVA \times 1000}{V \times 1.732}$$

$$sec.\ amps = \frac{2.328\ \text{kVA} \times 1000}{32\ \text{V} \times 1.732}$$

$$sec.\ amps = 42\ \text{A}$$

Step 6: Size of load that can be handled

$$kVA = \frac{output\ volts \times sec.\ amps}{1000}$$

$$kVA = \frac{240\ \text{V} \times 1.732 \times 42\ \text{A}}{1000}$$

$$kVA = 17.472\ \text{kVA}$$

Answer: **Autotransformer selected must be capable of handling 17.472 kVA and decreasing supply voltage to 27 V.**

Figure 14-7. Selecting and sizing autotransformer to buck supply voltage.

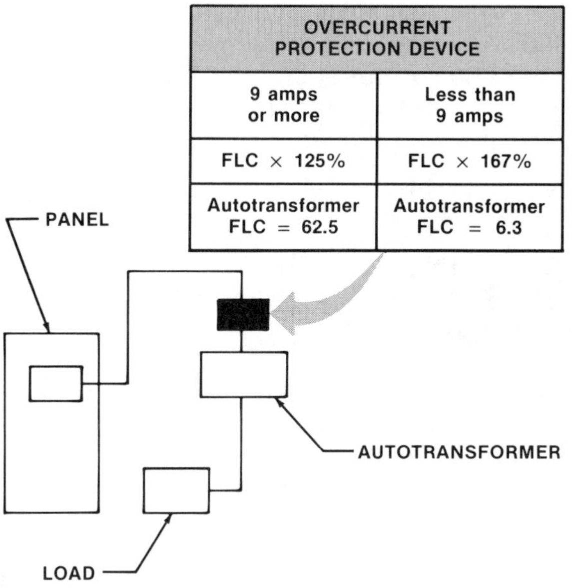

OVERCURRENT PROTECTION DEVICE	
9 amps or more	Less than 9 amps
FLC × 125%	FLC × 167%
Autotransformer FLC = 62.5	Autotransformer FLC = 6.3

PROBLEM: What size overcurrent protection device is required to protect an autotransformer with an input full-load current rating of 62.5 or 6.3 amps.

Step 1: *Input FLC rating* = 62.5 A

Step 2: *450-4(a), Ex.*

62.5 A × 125% = 78 A

Step 3: *240-6.*

80 A device

Answer: **80 A overcurrent protection device for 62.5 A FLC**

NOTE: Where the amperage does not correspond to a standard overcurrent protection device, the next higher size may be used (input FLC 9 A or more).

Step 1: *Input FLC rating* = 6.3 A

Step 2: *450-4(a), Ex.*

6.3 A × 167% = 10.5 A

Step 3: *240-6, Ex.*

10 A device

Answer: **10 A overcurrent protection device for 6.3 A FLC**

Figure 14-8. Determining the overcurrent protection device for an autotransformer.

grounded conductor of the primary circuit. Grounding is not required when the voltage is transformed from 240 volts to 208 volts or from 208 volts to 240 volts. Grounding is not required for autotransformers supplying feeder circuits or services. See Figure 14-9.

Autotransformers are used to derive a three-phase, four-wire, ungrounded distribution system. This may be done by connecting three autotransformers in a star (wye) configuration to the three-phase ungrounded system. Many electrical systems in use today are not grounded. Grounding autotransformers may be used to derive a neutral when it is necessary to ground existing ungrounded delta systems. The type generally installed for this purpose is the three-phase zigzag transformer.

ZIGZAG TRANSFORMER

When there is no ground fault on any one of the phase legs of the three-phase ungrounded system, the current flow in the transformer windings is balanced. This is because equal impedances are connected across each set of phase legs. If a ground fault develops on any leg of the three-phase system, the windings of the transformer will have a very low impedance path for the return of the fault current. This low impedance path will permit a great amount of fault current to flow and trip open the overcurrent protection device. See Figure 14-10.

TRANSFORMING THREE-WIRE CIRCUIT TO THREE-PHASE, FOUR-WIRE CIRCUIT—450-5(a)

When autotransformers are connected to derive a neutral to ground an ungrounded system, the following rules apply:

(1) Connections: Transformers must be connected directly to the ungrounded system with no switches or overcurrent protection devices between the connection and the autotransformer.

(2) Overcurrent protection: Overcurrent protection sensing devices must be provided to trip at 125% of the rated current in the phase wire or neutral conductor. If the input current is 9 amps or more, computed at 125%, the next higher rating may be used. Input current of less than 9 amps must not exceed 167%.

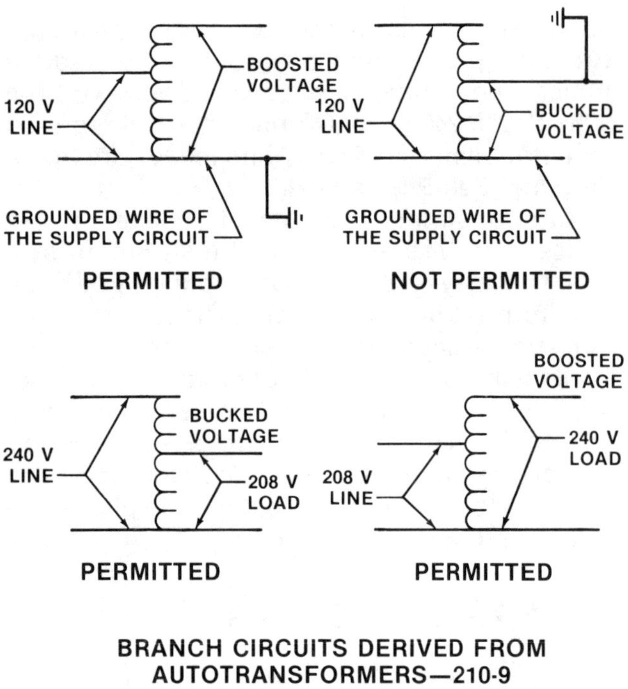

Figure 14-9. Using autotransformers to step up or step down voltage.

(3) Transformer fault sensing: Internal faults or single-phasing of the transformer must be provided with a sensing system.

(4) Rating: Autotransformers must have a continuous neutral rating sufficient to handle the specified ground-fault current.

See Figure 14-11.

DETECTING GROUNDS ON THREE-PHASE, THREE-WIRE SYSTEMS—450-5(b)(2)

When autotransformers are used to detect grounds on three-phase, three-wire systems, the following rules apply:

(1) Transformers are wye-connected to the three-phase, three-wire system. The common point of the wye connection is grounded.

(2) Rating: Autotransformers must have a continuous neutral current rating sufficient to handle the maximum ground-fault current that it may have to carry.

(3) Overcurrent protection: The overcurrent protection device must be set to trip at not more than 125% of the rated phase current of the

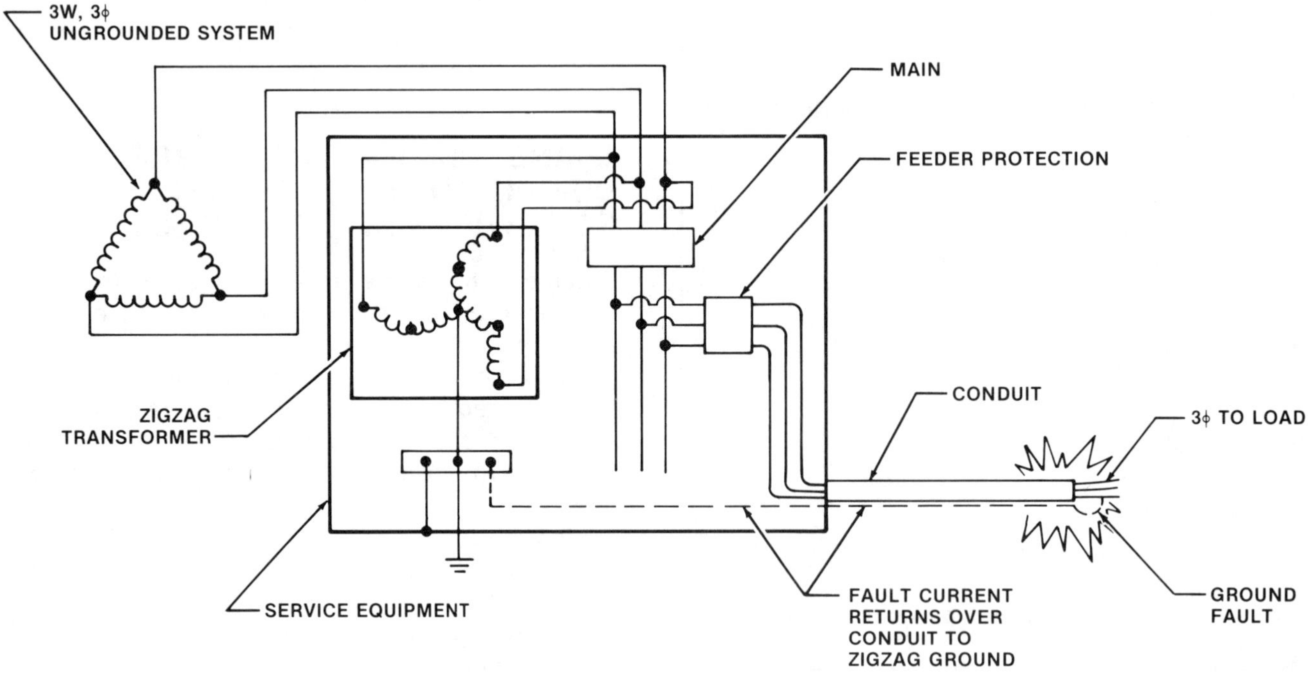

Figure 14-10. Using a zigzag transformer to convert an ungrounded system to a grounded operation.

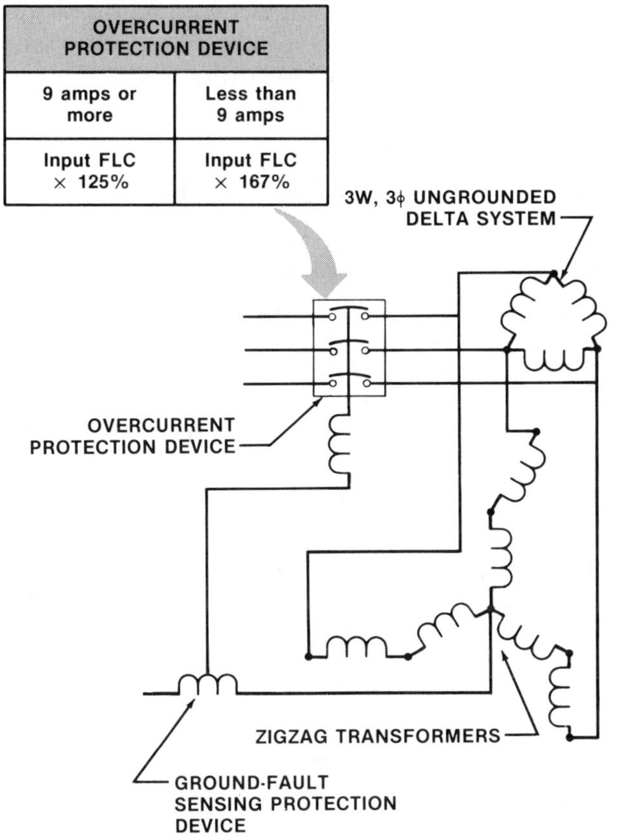

450-5(a)(2)
450-4(a), Ex.

Figure 14-11. Selecting the overcurrent protection sensing device.

transformer. If the overcurrent protection device is connected in the autotransformers' connection, 42% of the device's rating may be used. See Figure 14-12.

SECONDARY TIES—450-6

A secondary tie is a circuit of not over 600 volts between phases that connects sources of power or power supply points, such as two transformer secondaries. The tie may consist of one or more conductors per phase. In a network distribution system for industrial plants, three-phase banks of transformers are installed at various locations in the plant with two high-tension primary feed circuits supplying the transformers. At each transformer bank a double-throw switch is used to supply either primary feeder circuit to any bank of transformers. Secondary voltage must be 600 volts or less, and the primary conductors must be sized with enough capacity that either circuit can carry the total load should a fault occur in one of the circuit feeders.

Overcurrent protection must be provided for secondary ties at both ends. Fuses may provide this protection if they comply to the current-carrying ampacity of the conductors. See 450-6(a).

Limiters are generally used to protect the ties at both ends. See 450-6(a)(1). Limiters protect against short circuit but not against overloads. When protection is provided by limiters, the following rules apply:

(1) Ampacity of the tie must not be less than 67% of the rated secondary current of the largest transformer connected to the tie. All loads are at the transformer supply points.

(2) Ampacity of the tie must not be less than 100% of the rated secondary current of the largest transformer connected to the secondary tie. All loads are connected between the transformer supply points.

The secondary conductors between the transformer and the load must have a circuit breaker to protect the transformer. The circuit breaker may be set up to 250% of the transformer sec-

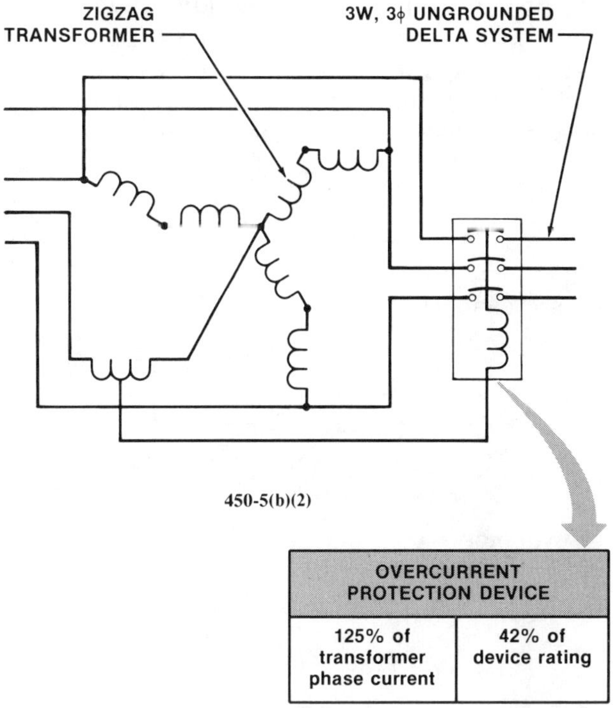

Figure 14-12. Selecting the rating of the overcurrent protection device.

ondary full-load current rating. A reverse-circuit relay must be installed to open the circuit in case the transformer should lose power for any reason. If the voltage is over 150 volts-to-ground, a disconnecting switch is required at each end.

Corresponding phase conductors must be joined together at the tap points. However, the exception to 450-6(a)(4) permits loads to be connected to individual conductors and to be without limiters where the total ampacity of the ties is at least 133% of the secondary rated current of the largest transformer. The total load of the taps must not exceed this value.

USING RADIAL OR LOOP SYSTEMS FOR INDUSTRIAL PLANT DISTRIBUTION

Power to industrial plants can be distributed in many types of voltage and hook-up arrangements. There are four basic types utilized in the industry. These are radial low-voltage systems, radial high-voltage systems, distribution loop systems, and network power systems.

Radial Low-voltage Systems

The power company provides high voltage to the plant's transformer and service equipment. Reduced voltage feeders are run to switchboards and load centers. This method of distributed power is costly due to larger conductors and conduits. The elements and components of the load centers are more expensive due to the lower voltage and higher current ratings. See Figure 14-13.

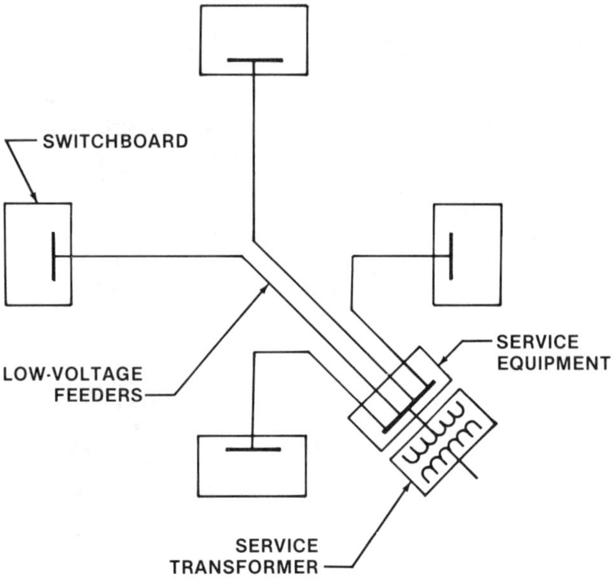

Figure 14-13. A radial low-voltage system.

Radial High-voltage Systems

The power company provides high voltage to the plant's service equipment. High-voltage feeder circuits are run to transformers at each distribution point and the voltage stepped down to the desired operating value.

The feeders with the higher voltage require smaller conduits, conductors, and equipment. The total cost of such an arrangement of power distribution is less expensive due to the higher voltage. The concept is to raise the voltage on a feeder circuit and lower the current ratings for all the elements and components for the feeder circuit. See Figure 14-14.

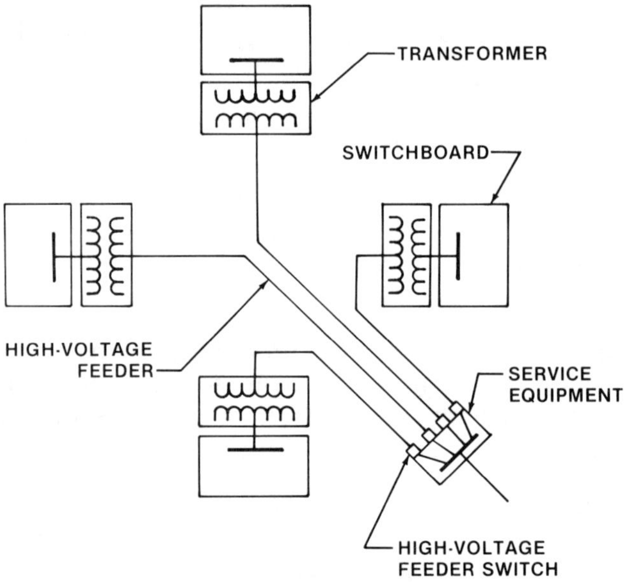

Figure 14-14. A radial high-voltage system.

Distribution Loop System

The distribution loop system is superior to the radial low- or high-voltage system. The power company delivers power to the loop where it connects to a pair of circuit breakers and branches out to form a loop or circle. Circuit breakers are installed between each transformer and load center connected to the loop. The circuit breakers provide protection for any section of the feeders connected to the loop and also isolates in case of a fault or disorderly shutdown.

If trouble developed between switchboard D and switchboard C, the circuit breaker would remove the damaged load center. If switchboard D was damaged in some way and created a fault condition, one of the circuit breakers installed from the service tap or in the loop system would disconnect switchboard D and service would continue to the rest of the switchboards connected to the loop. See Figure 14-15.

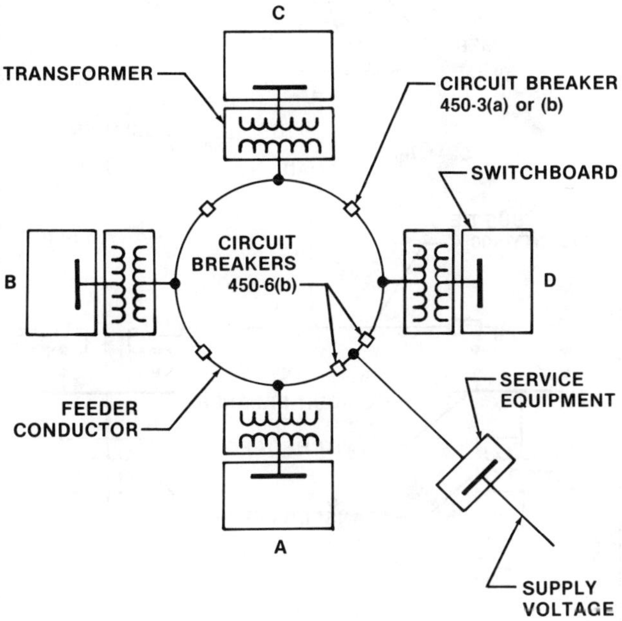

Figure 14-15. A distribution loop.

Loop Systems Using Bus Ties. This type of loop system is designed to eliminate a complete shutdown of a feeder section that has been damaged. The damage could be the transformer, the service equipment, or any of the elements that make up the feeder system. The loop feeder has, in addition, a second loop that is connected between the transformers forming the loop.

These secondary connections installed at each switchboard (load center) location is protected with circuit breakers. These loops from each switchboard form a continuous circle from A through D and back to A. A fault in any section of the feeder will not isolate any part of the plant from a source of power. A transformer in any section could have a fault or develop trouble, and the service of the section would continue to receive power from the continuous loop. See Figure 14-16.

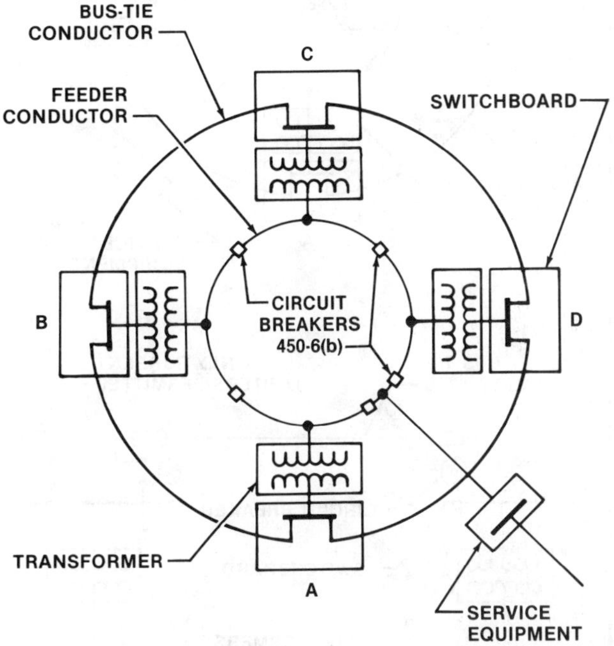

Figure 14-16. Bus ties used in a loop system.

Requirements for Bus-tie Conductors. Bus-tie conductors or secondary ties are low-voltage secondary loop connections. They are rated at 600 volts or less between phases per 450-6 of the NEC®. Bus ties are used to connect two power sources to the secondaries of two transformers. Loads can be connected at the transformer supply points and overcurrent protection devices set to limit the maximum current to 150% of the capacity of the conductor. If protection of conductors is not provided at 150% or less, the current rating of the bus tie must be at least 67% of the full-load current rating of the largest transformer. See Figure 14-17.

If loads are tapped from the secondary bus tie and not at the transformer location, the current-

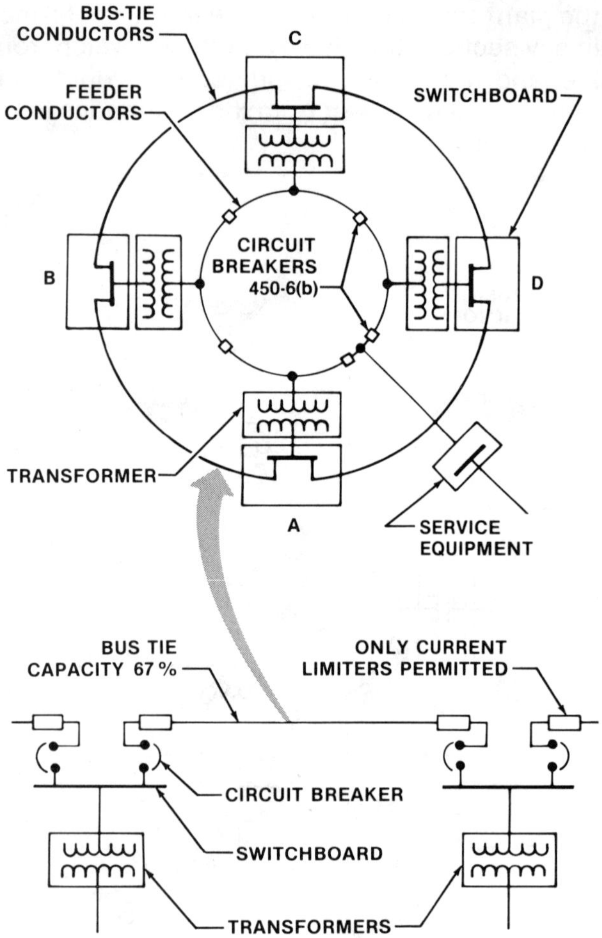

PROBLEM: The largest transformer is rated at 750 kVA and connected for a 4-wire, three-phase, 480/277-volt system. What size bus tie is required?

Step 1: $FLC = \dfrac{kVA \times 1000}{V \times 1.732}$

$FLC = \dfrac{750 \text{ kVA} \times 1000}{480 \times 1.732}$

$FLC = 903 \text{ A}$

Step 2: 67% × *FLC of transformer*

$.67 \times 903 \text{ A} = 605 \text{ A}$

Answer: **minimum 605 A bus tie**

Figure 14-17. Bus ties calculated at 67%.

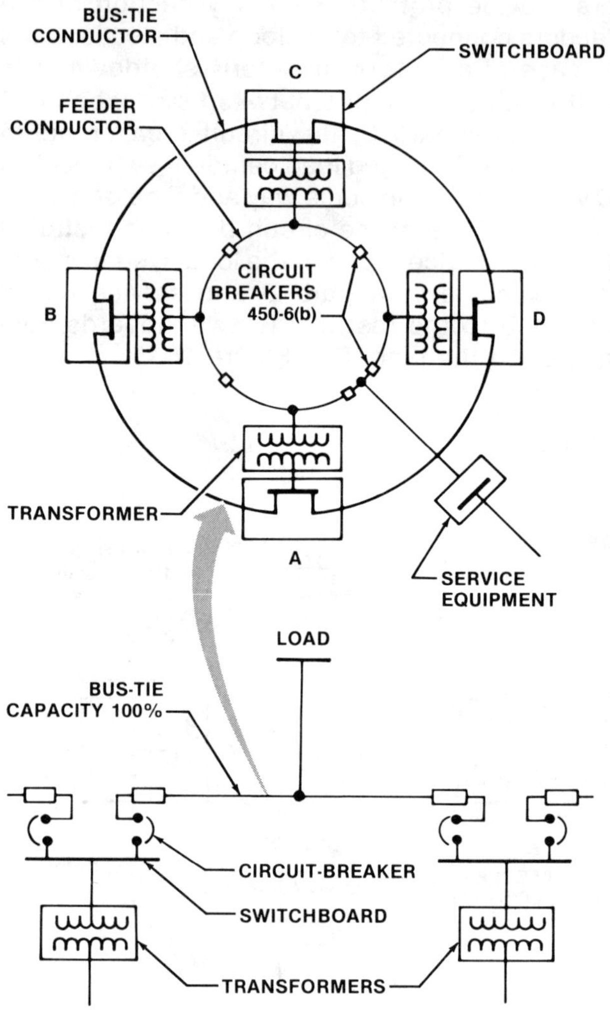

PROBLEM: The largest transformer is rated at 500 kVA and is connected to a 4-wire, three-phase, 208/120-volt system. What size bus tie is required?

Step 1: $FLC = \dfrac{kVA \times 1000}{V \times 1.732}$

$FLC = \dfrac{500 \text{ kVA} \times 1000}{208 \text{ V} \times 1.732}$

$FLC = 1389 \text{ A}$

Step 2: 100% × *FLC of transformer*

$1.00 \times 1389 \text{ A} = 1389 \text{ A}$

Answer: **minimum 1389 A bus tie**

Figure 14-18. Bus ties calculated at 100%.

carrying capacity of the bus tie must be 100% of the full-load current rating of the largest transformer. See Figure 14-18.

The secondary tie can consist of a number of conductors parallel per phase with loads tapped to individual conductors between the locations of the transformers. If these loads are not tapped to every one of the tie conductors, the combined total rating of the conductors between stations must be at least 133% of the full-load secondary current rating of the largest transformer. See Figure 14-19.

The loads tapped must be equally divided on each phase and on the individual conductors of each phase as close as possible. See Figure 14-20.

Protection of Bus Ties. A current limiter or automatic circuit breaker must be installed at both ends of each tie to protect the conductors from short circuit conditions. If the operating voltage is above 150 volts-to-ground and current limiters are used, a switch must be provided at either end of the tie. In addition, an overcurrent protection device set at 250% or less of the rated full-load current must be installed in the secondary circuit of each transformer to protect the tie conductors. See Figure 14-21.

If reverse current exceeding the full-load current of a transformer tries to flow into the unit, a reverse-current relay must be installed to actuate a circuit breaker. The circuit breaker actuated by the reverse-current relay will disconnect the secondary windings of the transformer due to faults or current feedback.

Network Power Systems

A simple network system consists of two power sources fed into a loop and connected to transformers. Power source No. S1 is connected to transformers A and C. Power source No. S2 is connected to transformers B and D. Switches are installed to provide transfer means in case one of the transformers develops trouble. See Figure 14-22.

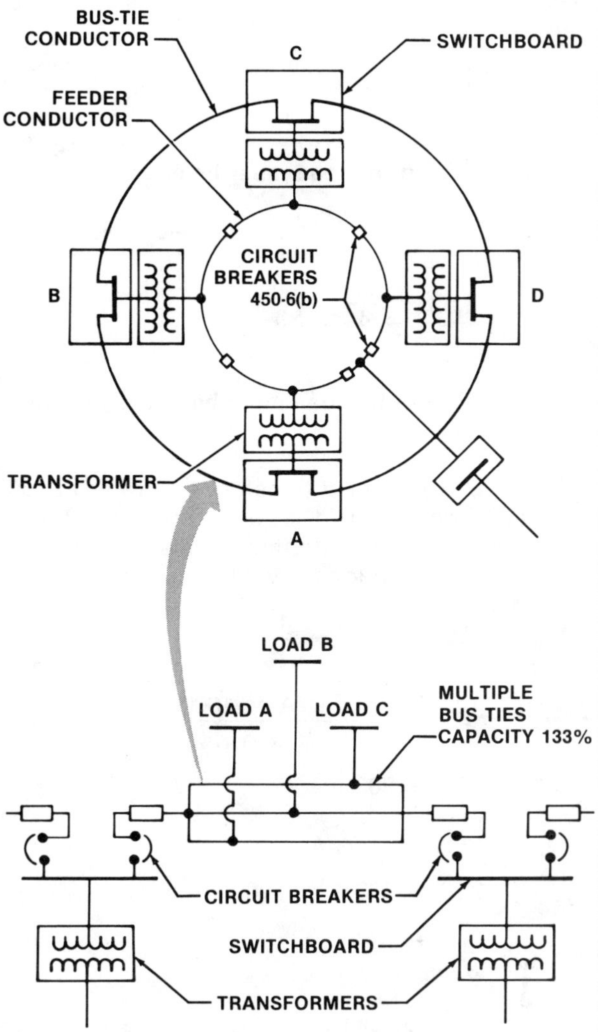

PROBLEM: The largest transformer is rated at 400 kVA and connected for a 4-wire, three-phase, 208/120-volt system. What size bus tie is required?

Step 1: $FLC = \dfrac{kVA \times 1000}{V \times 1.732}$

$FLC = \dfrac{400 \text{ kVA} \times 1000}{208 \text{ V} \times 1.732}$

$FLC = 1111 \text{ A}$

Step 2: 133% × *FLC of transformer*

$1.33 \times 1111 \text{ A} = 1478 \text{ A}$

Answer: **minimum 1478 A bus tie**

Figure 14-19. Bus ties calculated at 133%.

PROBLEM: Two loads are tapped from a tie. Divide the loads evenly.

Step 1: $\dfrac{1.33}{2} = 67\%$

Answer: **67% of the load must be equally divided**

PROBLEM: Three loads are tapped from a tie. Divide the loads evenly.

Step 1: $\dfrac{1.33}{3} = 44\%$

Answer: **44% of the load must be equally divided**

PROBLEM: The largest transformer is rated at 750 kVA and connected for a four-wire, three-phase, 480/277-volt system with three loads tapped from the tie. Divide the loads evenly.

Step 1: $FLC = \dfrac{kVA \times 1000}{V \times 1.732}$

$FLC = \dfrac{750 \text{ kVA} \times 1000}{480 \text{ V} \times 1.732}$

$FLC = 903 \text{ A}$

Step 2: $133\% \times 903 \text{ A}$

$1.33 \times 903 \text{ A} = 1200.9 \text{ A}$

Step 3: $\dfrac{percent}{number\ of\ loads}$

$\dfrac{1.33}{3} = 44\%$

Step 4: $FLC \times 44\%$

$1200.9 \times .44 = 528 \text{ A}$

Answer: **Load must be divided as close as possible to 528 A**

Figure 14-20. Loads divided evenly on ties.

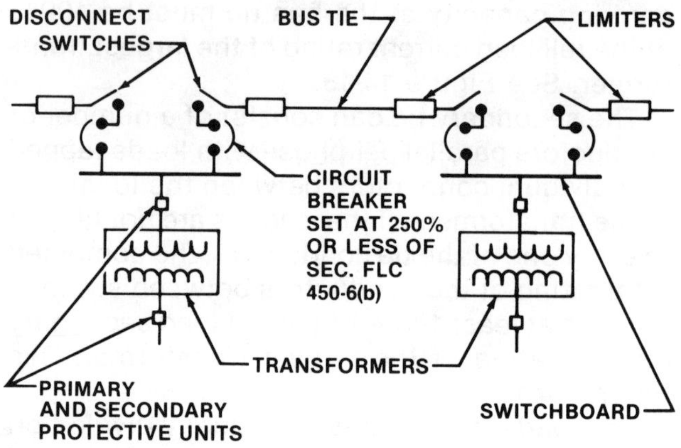

Figure 14-21. Protection of bus ties.

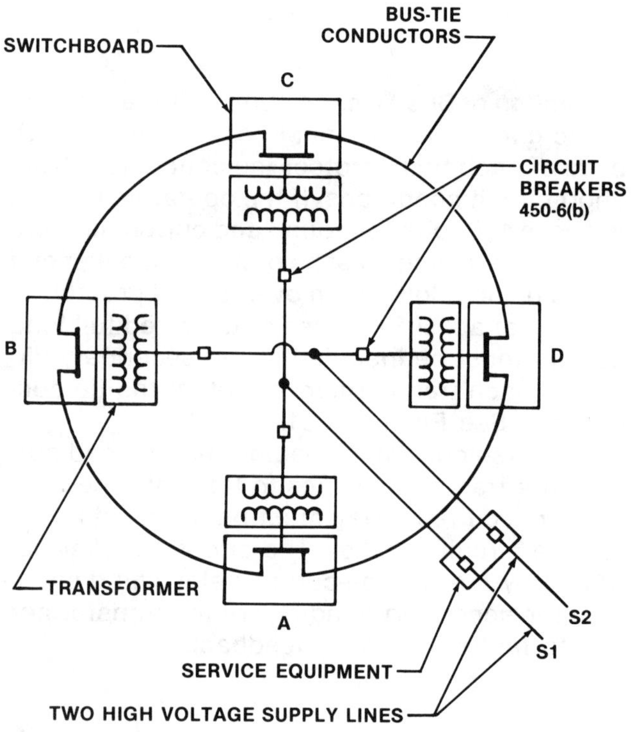

Figure 14-22. Simple network system.

REVIEW—CHAPTER 14

Name ______________________________ Date ______________

True-False

T F 1. Autotransformers are single-winding transformers with the single winding common to the primary and secondary circuits.

T F 2. If the primary is connected across the full winding and the secondary across part of the winding, the voltage supply would be increased.

T F 3. If the entire winding of the secondary is used, the voltage will be boosted.

T F 4. Any one winding or combination of windings of autotransformers can be used to boost or buck the supply voltage.

T F 5. Buck and boost insulating transformers are only available with two windings.

T F 6. The secondary windings are the only parts of autotransformers that perform the work of transforming voltage and current.

T F 7. Autotransformers can be used to supply branch circuits if one of the secondary conductors is grounded.

T F 8. Autotransformers must not be used to derive a neutral where there is an existing ungrounded delta system.

T F 9. Autotransformers must have a continuous neutral rating sufficient to handle the amount of ground-fault current that can flow.

T F 10. Both ends of secondary ties are usually protected by limiters.

T F 11. The secondary conductors between the transformer and the load served must have a circuit breaker to protect the transformer.

T F 12. The elements and components of load centers are more expensive where lower voltage systems are used.

T F 13. Loop distribution systems are superior to low- or high-voltage radial systems.

T F 14. High voltage systems are utilized in industrial locations to save on copper, conduits, and size of equipment.

T F 15. The current is high on high-voltage systems and low on low-voltage systems.

Completion

__________ 1. Radial low-voltage systems usually have __________ bank(s) of transformers that supply the service equipment.

__________ 2. Radial high-voltage systems usually have one bank of transformers that supplies the service equipment where additional transformers are installed to step the voltage __________.

__________ 3. Loop systems use __________ to eliminate a complete shutdown of a feeder section that has been damaged.

_______ **4.** Bus ties are used to connect two power sources to the _______ of two transformers.

_______ **5.** Bus ties must be protected at both ends of each tie to protect the _______.

_______ **6.** A simple network system consists of two power sources fed into a loop and connected to _______.

_______ **7.** A(n) _______ current relay must be installed to disconnect the secondary conductors from the tie due to faults or current feedback.

_______ **8.** Switches must be installed in the network systems to _______ in case trouble develops in one of the transformers.

_______ **9.** Where bus ties are used, protection must be provided against _______ current flow.

_______ **10.** In a network system, circuit breakers are provided in the _______ to each transformer.

_______ **11.** Limiters must be provided between each _______ and bus tie.

_______ **12.** Disconnecting means must be provided between each transformer and _______.

_______ **13.** Secondary ties can consist of a number of conductors _______ per phase.

_______ **14.** Loads can be _______ at locations along the tie if proper protection is provided.

_______ **15.** Delta systems that are not grounded can be grounded by using three-phase _______ transformers.

Multiple Choice

_______ **1.** Autotransformers can be connected together to buck or boost the voltage to plus or minus _______%.

A. 20
B. 25
C. 30
D. 35

_______ **2.** The two secondary output windings of autotransformers are rated at _______ volts.

A. 6-8-10
B. 12-16-24
C. 25-30-40
D. 40-60-80

_______ **3.** The secondary current of autotransformers can be determined by dividing the kVA rating by the _______.

A. primary turns
B. primary voltage
C. secondary turns
D. secondary voltage

______ **4.** If the voltage to a load is too high, the voltage can be reduced by using a(n) ______.

A. boost autotransformer
B. buck autotransformer
C. isolation transformer
D. separately derived system

______ **5.** Autotransformers with a full-load current rating of 9 amps or more can be protected by multiplying the full-load current by ______%.

A. 125
B. 167
C. 250
D. 300

______ **6.** Autotransformers with full-load current ratings of less than 9 amps can be protected by multiplying the full-load current by ______%.

A. 125
B. 167
C. 250
D. 300

______ **7.** Autotransformers are not required to be grounded where the voltage transformed is from 240 to ______ volts.

A. 120
B. 208
C. 277
D. 480

______ **8.** Each secondary circuit of transformers that supplies power to a bus tie must be protected at ______% of the full-load current.

A. 110%
B. 115%
C. 200%
D. 250%

______ **9.** Where multiple bus ties are made, the capacity of the conductors must be at least ______% of the largest transformer.

A. 125
B. 133
C. 150
D. 175

______ **10.** Where loads are tapped from a bus tie, the capacity of the conductors must be at least ______% of the largest transformer.

A. 100
B. 110
C. 125
D. 133

Problems

1. Draw an autotransformer with a boost connection.

2. Draw an autotransformer with a buck connection.

3. Tap the autotransformer for 480-volt, single-phase operation on the secondary side.

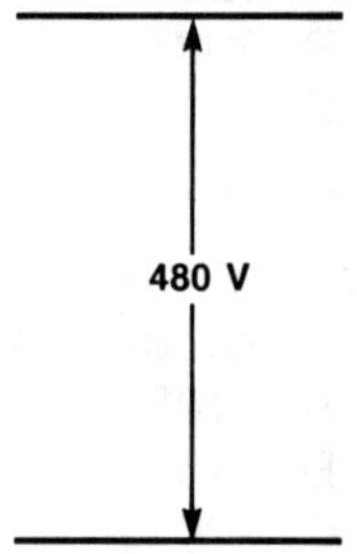

4. What is the reduced operating kVA and current rating of the following autotransformer?

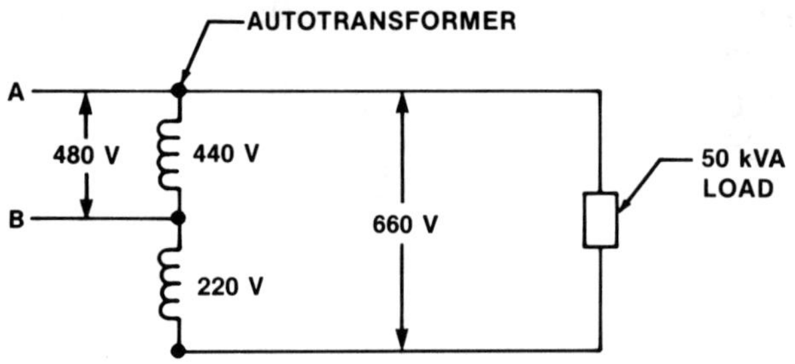

5. What size kVA autotransformer is required to supply a load of 48,000 volt-amps that will boost the supply voltage from 208 to 230 volts?

6. What is the size of the overcurrent protection device required to protect an autotransformer with a primary current of 42 amps?

7. What size bus tie is required for the tie between transformers with the largest transformer rated at 1200 kVA and three-phase, 480 volts?

8. What size bus tie is required for the tie and tap tie between the transformers with the largest transformer rated 800 kVA and three-phase, 208 volts?

9. What size bus tie is required for multiple ties between transformers where the largest transformer is rated 600 kVA and operating at three-phase, 240 volts?

10. The largest transformer of a loop system is rated 650 kVA and operates at three-phase, 460 volts. How must the load be divided in using four multiple taps from the main tie?

TRANSFORMER EXAM

Chapter 15

The four tests in this chapter provide a comprehensive review of the material presented in chapters 10 through 14. Each test contains a variety of True-False, Completion, and Multiple Choice questions. Record your answers in the spaces provided. Each test also contains Problems. Show your work in the spaces provided. Your instructor may require Code validation for specific questions and/or problems.

Subject matter covered in these tests includes:

- **Delta-to-Delta Connected Transformers**
- **Delta-to-Wye Connected Transformers**
- **Wye-to-Wye Connected Transformers**
- **Wye-to-Delta Connected Transformers**
- **Buck and Boost Transformers**
- **Location**
- **Separation from Combustible Material**
- **Troubleshooting**
- **Low-Voltage Radial Systems**
- **High-Voltage Radial Systems**
- **Network Systems**
- **Connecting Windings**
- **Testing Windings**
- **Finding Full-Load Current**
- **Sizing Conductors**
- **Sizing Overcurrent Protection Devices**
- **Secondary Ties**
- **Secondary Taps**

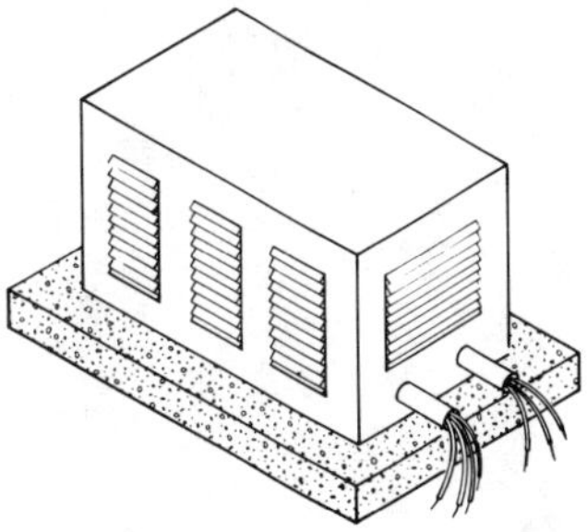

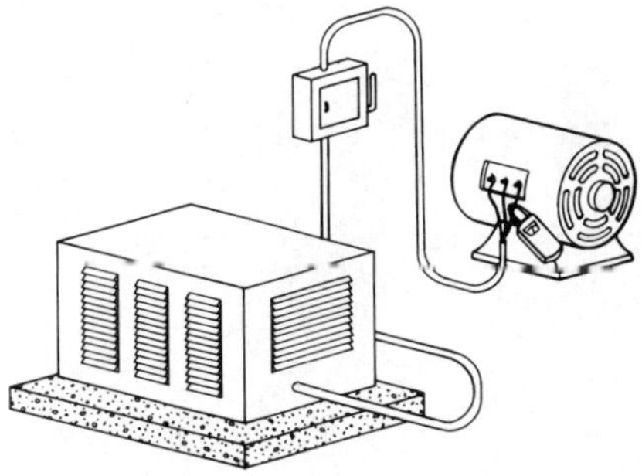

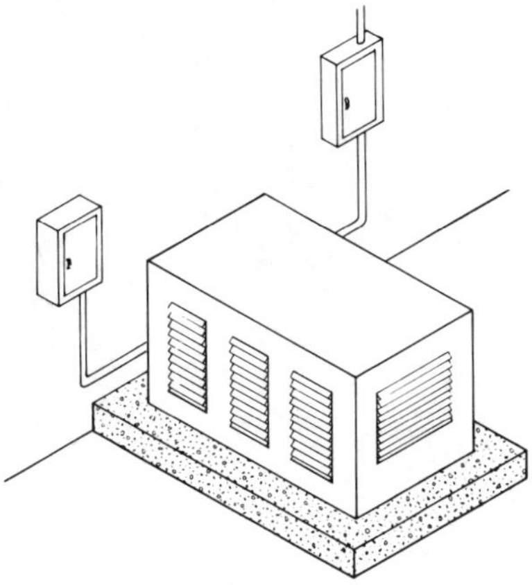

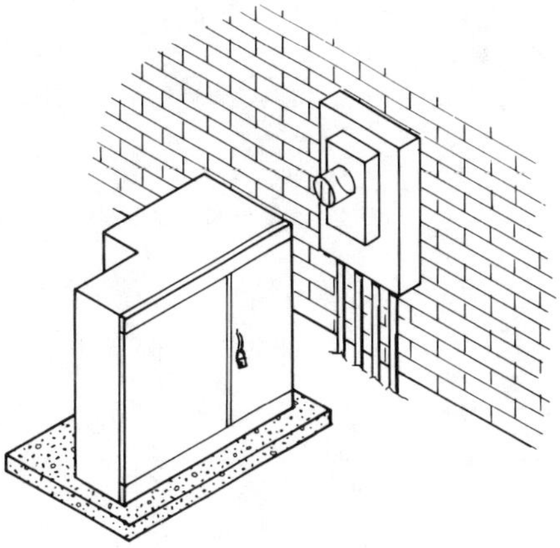

TEST 1—CHAPTER 15

Name ______________________________ **Date** __________

True-False

T F 1. When a primary side of a transformer has more turns than the secondary side, the input voltage is higher.

T F 2. When the secondary side of a transformer has less turns than the primary side, the output amperage of the secondary is stepped up.

T F 3. Transformers with high impedance have higher fault currents than those with low impedance.

T F 4. When the primary and secondary windings of a transformer have the same number of turns, the amperage on the secondary is the same as on the primary.

T F 5. Coils in transformers not equipped with soft iron cores are not as efficient because of the air gap.

T F 6. Under no condition can transformers rated 600 volts or less be mounted above suspended ceilings.

T F 7. Transformers must be installed in locations where the heat from their windings can be dissipated easily.

Completion

__________ 8. When power is applied to the primary coil of transformers, voltage is induced into the secondary by __________.

__________ 9. Counterelectromotive force is a(n) __________ voltage that opposes applied voltage.

__________ 10. When a coil is placed on a steel core and voltage is applied, the coil becomes __________.

__________ 11. Neutral conductors must be sized large enough to carry the maximum __________ current of the system.

__________ 12. If the line voltage of a delta-connected system is 480 volts, the phase voltage is __________ volts.

__________ 13. Transformers rated at 600 volts or less and with an output of __________ kVA or less can be mounted above suspended ceilings.

Multiple Choice

__________ 14. For a 240-volt coil tapped in the center and connected to ground, the voltage to ground is __________ volts.

A. 120
B. 208
C. 240
D. none of the above

__________ **15.** For a 480-volt, corner-grounded delta system, the voltage to ground is __________ volts from the grounded phase.
A. 0
B. 208
C. 240
D. 480

__________ **16.** If the current in a wye system is 88 amps per phase, the coil amperage is __________ amps.
A. 51
B. 88
C. 102
D. 176

__________ **17.** If the phase-to-phase voltage of a wye-connected transformer is 208 volts, the coil voltage is __________ volts.
A. 120
B. 200
C. 208
D. 240

__________ **18.** When phase A of a transformer carries 130 amps and phase B pulls 190 amps, the neutral draws __________ amps.
A. 20
B. 30
C. 40
D. 60

__________ **19.** The lighting transformer for an open delta system is determined by multiplying the single-phase load by 100% and the three-phase load by __________%.
A. 33
B. 42
C. 58
D. 67

Problems

20. Fill in the blanks with the correct voltage.

A. __________ volts

B. __________ volts

C. __________ volts

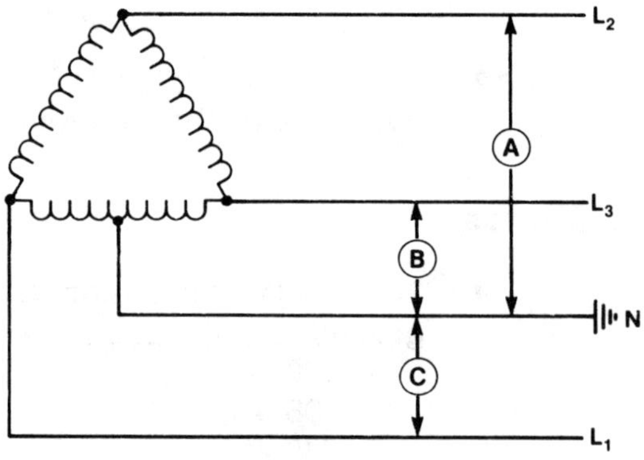

21. What is the amperage from phase to phase in a closed delta system with 74 amps of coil current?

22. What is the current in L_1 when the current in L_2 is 110 amps and the current in L_3 is 132 amps (closed delta system)?

23. Connect the phases of the corner-grounded system to the service equipment and ground the service.

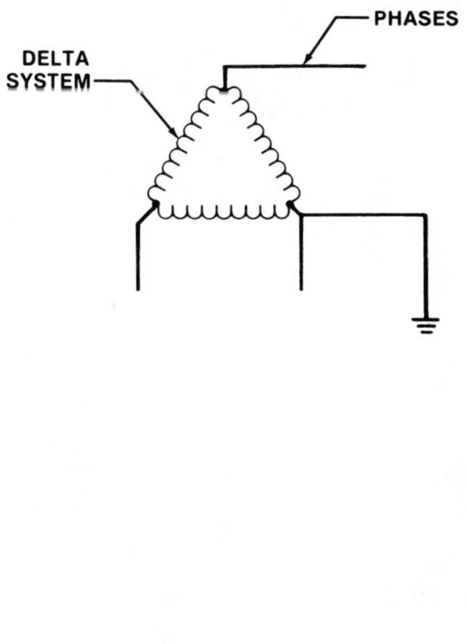

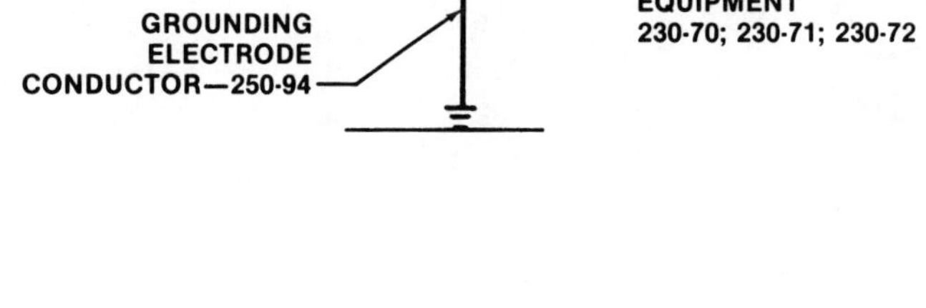

24. Fill in the blanks for the missing values.

A. Voltage stepped ____________

B. Turns ____________

C. Impedance stepped ____________

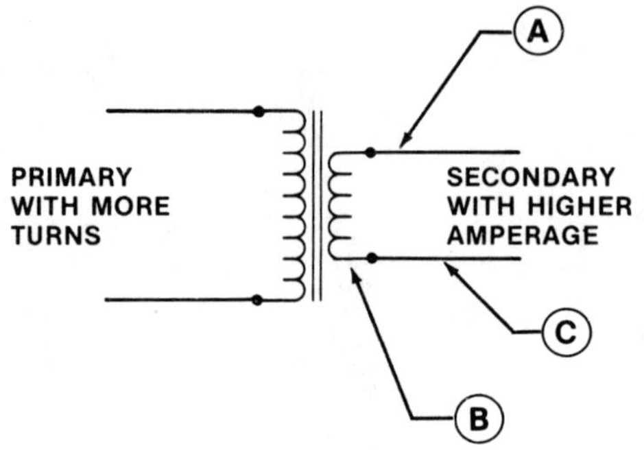

25. What is the full-load current rating of a 100 kVA transformer supplying a building that is operating on 120/240-volt, single-phase?

TEST 2—CHAPTER 15

Name ______________________________ **Date** ______________

True-False

T F **1.** Transformers attached to buildings must have safeguards to protect the building and any openings from oil leaks.

T F **2.** Transformer vaults must have a minimum two-hour rated door.

T F **3.** Ventilation openings without ducts or flues going directly to the outside must have an area of at least 2 sq. in. for each kVA of transformer rating.

T F **4.** The high leg of 120/240-volt, three-phase, four-wire systems is identified by marking it red.

T F **5.** The impedance of transformers determines the amount of fault current that transformers deliver.

T F **6.** The overcurrent protection device on the primary side of a transformer will clear the circuit if a short circuit occurs.

Completion

__________ **7.** Transformers rated at 120/240 volts, single-phase must have a neutral conductor sized to carry the unbalanced current between __________.

__________ **8.** The amperage that transformers deliver is determined by dividing the kVA by the __________.

__________ **9.** A 30 kVA, 480-volt, three-phase transformer supplies ________ amps.

__________ **10.** The power transformers on a closed delta system can be determined by multiplying the single- and three-phase loads by __________.

__________ **11.** Secondary conductors of three- to four-wire transformers must not extend more than __________′ unless overcurrent protection is installed.

__________ **12.** Overcurrent protection devices on the primary side can be increased up to __________% if the device trips open due to inrush current of motors.

__________ **13.** The __________ of transformers must be known when combination protection is provided for transformers rated over 600 volts.

Multiple Choice

__________ **14.** If the ratio of a transformer is 3 to 1 and the primary has the lesser rating of of 60 amps, the secondary current would be __________ amps.

A. 100
B. 150
C. 175
D. 180

__________ **15.** Wye-connected transformer size can be determined by multiplying the single- and three-phase loads by __________%.
A. 25
B. 33
C. 42
D. 58

__________ **16.** If the primary of a transformer is 480 volts and the secondary is 24 volts, the ratio would be __________ to 1.
A. 10
B. 15
C. 20
D. 30

__________ **17.** The amperage of three-phase transformers can be determined by dividing the total kVA by the voltage times the __________.
A. $\sqrt{2}$
B. $\sqrt{3}$
C. number of turns
D. none of the above

__________ **18.** Overcurrent protection devices for transformers rated with a primary full-load current of 9 amps or more can be selected at __________% of the full-load current.
A. 125
B. 150
C. 175
D. 200

__________ **19.** Overcurrent protection devices for transformers rated with a primary full-load current of 2 amps or more but less than 9 amps are selected at __________% of the full-load current.
A. 125
B. 167
C. 175
D. 200

Problems

20. What is the three-phase full current rating for three 50 kVA, 240-volt transformers connected for delta operation?

21. What is the minimum kVA rating needed to supply a building with a single-phase load of 20,000 VA and a three-phase load of 15,000 VA? (The supply is 120/208 volts. Use a wye system.)

22. What is the minimum kVA rating of two power transformers used to supply a single-phase load of 26 kVA and a three-phase load of 42 kVA? (The supply is 120/240 volts. Use a closed delta system.)

23. What is the minimum kVA rating of a lighting transformer used to supply a single-phase load of 242 amps and a three-phase load of 88 amps? (The supply is 115/230 volts. Use an open delta system.)

24. Connect the windings for open delta operation where one of the transformers of a closed delta system is defective.

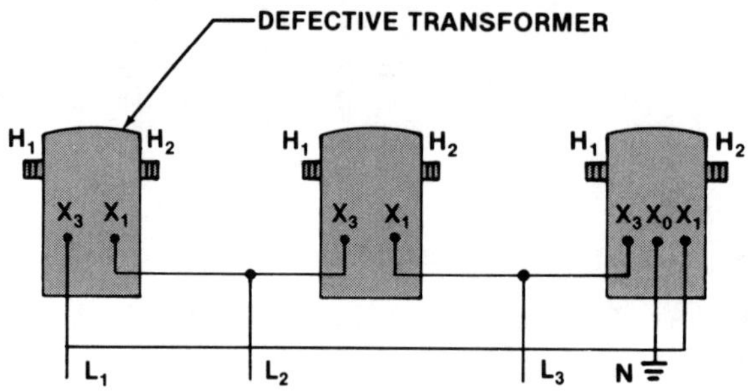

25. What is the available fault current at the terminals of a 225 kVA, 240-volt, three-phase transformer with an impedance of 2%?

TEST 3—CHAPTER 15

Name ____________________ **Date** ____________

True-False

T F 1. When the 10′ tap rule is applied, conductors are not required to be routed in conduit.

T F 2. Panelboards that are classified as power panels are required to have a main when the 10′ tap rule is applied.

T F 3. Windings wound for additive polarity must be connected additive to ensure proper voltage.

T F 4. Two subtractive windings and an additive winding can be connected provided the additive winding is connected subtractive.

T F 5. Additive connected windings will subtract when the voltage is measured.

T F 6. The building steel is the first choice for the grounding electrode when grounding a separately derived system.

Completion

__________ 7. Power panels must not have more than 10% of their single-pole devices rated at __________ amps or less with neutral loads.

__________ 8. When one transformer is defective in a closed delta hookup and each transformer is rated 75 kVA, the load available for open delta is determined by multiplying the remaining two by __________%.

__________ 9. A power transformer is usually __________ than the lighting transformer in open delta systems.

__________ 10. When three transformers are used to connect a closed delta system, the lighting transformer is __________ than the two power transformers.

__________ 11. The turns ratio of transformers is equal to one voltage __________ by the other voltage.

__________ 12. When transformer windings are connected in __________, always connect the odd-numbered terminals together and the even-numbered terminals together.

Multiple Choice

__________ 13. Overcurrent protection devices for transformers rated with a primary full-load current of less than 2 amps are selected at __________% of the full-load current.

A. 300
B. 400
C. 500
D. 600

_____ **14.** Where combination protection is used on transformers rated over 600 volts with an impedance of 1.5, the secondary side can be protected with fuses at _____% of the full-load current.
A. 125
B. 225
C. 250
D. 300

_____ **15.** Primary overcurrent protection devices can protect the secondary conductors of a two-wire to _____-wire hookup.
A. two
B. three
C. four
D. none of the above

_____ **16.** T-connected transformers must have a rating of approximately _____% above the load supplied.
A. 110
B. 115
C. 116
D. 125

_____ **17.** When transformers are reverse-connected, the voltage is less at full load than at _____.
A. reverse speed
B. no load
C. both A and B
D. neither A nor B

_____ **18.** Two transformers can be used to supply two-phase power for special equipment if _____ connections are used.
A. open delta
B. wye
C. T
D. none of the above

_____ **19.** Windings of transformers can be connected in a bank to form _____ connections.
A. wye-to-wye
B. delta-to-wye
C. wye-to-delta
D. all of the above

Problems

20. Fill in the blanks for the transformer vault.

A. __________ minimum thickness of roof and walls (in inches)
B. __________ minimum thickness of floor (in inches)
C. __________ minimum fire rating of door (in hours)

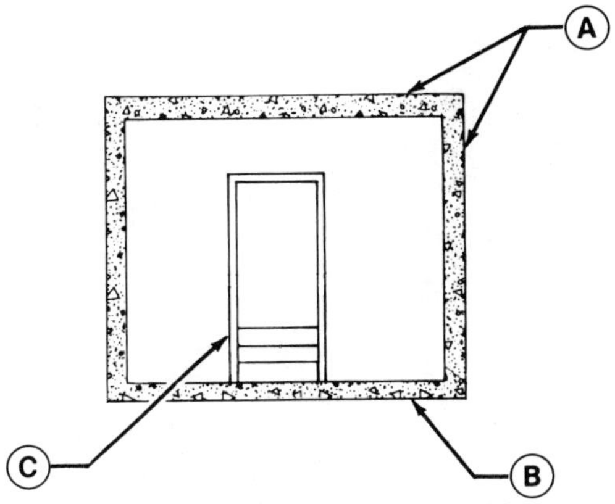

21. Fill in the blanks for the transformer above a hung ceiling.

A. __________ volts or less

B. __________ kVA or less

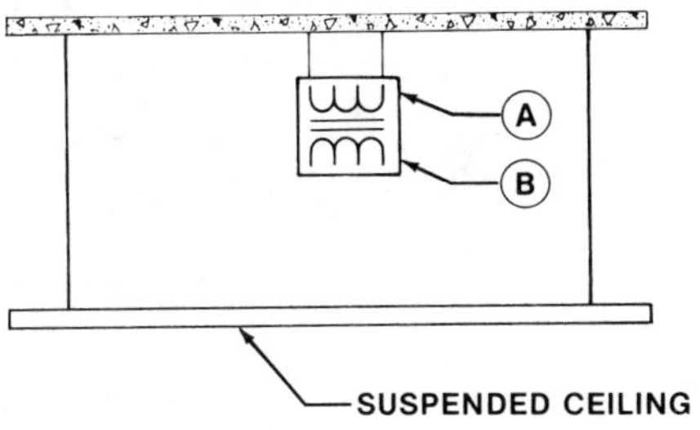

22. Fill in the blanks with the proper clearances for transformers that are not totally enclosed or ventilated, but have 80°C rise insulation.

A. __________

B. __________

C. __________

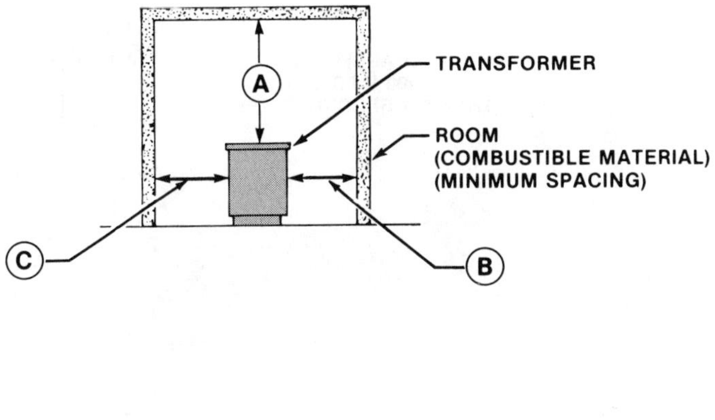

23. Connect the bank of transformers for delta-to-wye operation using subtractive polarity.

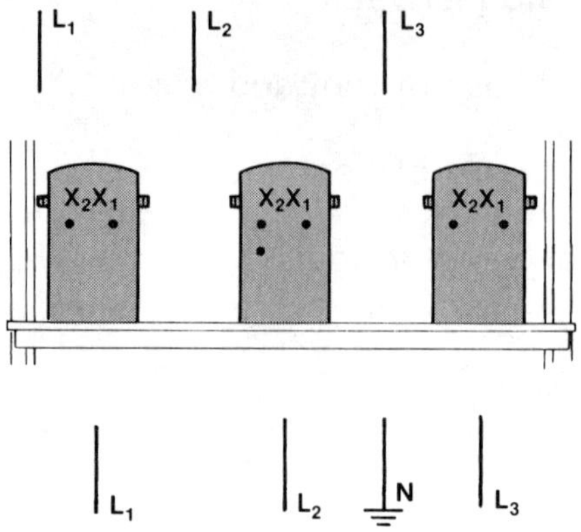

24. Connect the windings for wye-to-wye operation.

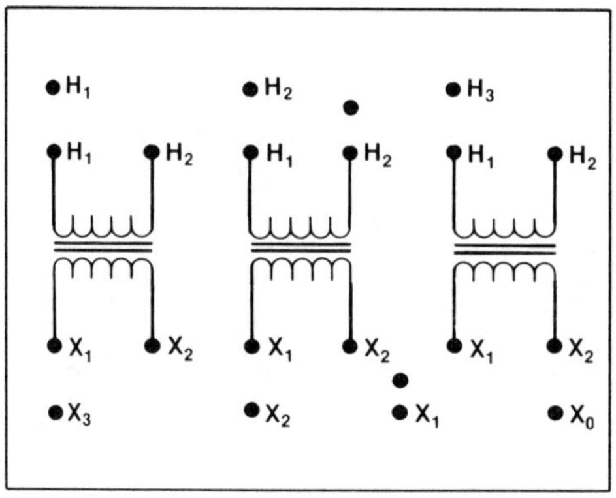

25. Connect the windings for 120/240-volt, single-phase and fill in the blanks with the proper voltage.

A. ______________ volts

B. ______________ volts

C. ______________ volts

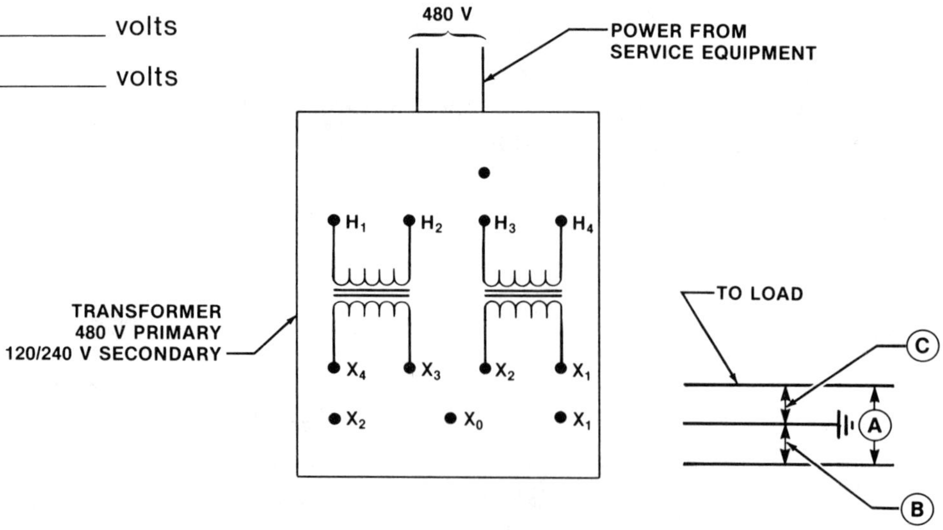

TEST 4—CHAPTER 15

Name ____________________ Date ____________

True-False

T F 1. The size of the grounding electrode conductor used to ground separately derived systems is selected from the size main in the equipment served.

T F 2. When the full winding of an autotransformer supplies the load, the voltage is stepped up.

T F 3. If only part of the winding is used on the line side of an autotransformer, the voltage on the load side would be reduced.

T F 4. Autotransformers are available with four windings.

T F 5. The overcurrent protection for autotransformers is selected by multiplying the full-load current rating by a percentage based on the amperage of the primary.

T F 6. When voltage is too high to supply a load, a boost-type autotransformer may be used.

Completion

__________ 7. The maximum voltages available from an autotransformer are equal to the __________ of the winding voltages.

__________ 8. Branch circuits of __________ volts that are grounded can be supplied with autotransformers.

__________ 9. Grounding of autotransformers is not required where the branch circuit ______ the voltage from 240 volts to 208 volts.

__________ 10. Autotransformers can be connected to derive a neutral for __________ systems in industrial plants.

__________ 11. Autotransformers are provided with ground-fault __________ protection devices.

__________ 12. The neutral of an autotransformer must be sized to carry the __________ fault current of the system.

Multiple Choice

__________ 13. The input terminals of transformers are marked with the letter __________.

A. H
B. X
C. O
D. Z

________ **14.** For an overcurrent protection device connected in the neutral of an autotransformer, the rating of the device is selected at ________%.
A. 50
B. 67
C. 75
D. 80

________ **15.** Where secondary ties are made, a reverse-________ relay must be used to open the circuit in case of fault currents.
A. current
B. voltage
C. both A and B
D. neither A nor B

________ **16.** Secondary ties made at supply points only must be rated at least ________% of the largest transformer.
A. 42
B. 50
C. 67
D. 70

________ **17.** Secondary ties made between transformer supply points must be at least ________% of the largest transformer.
A. 42
B. 67
C. 80
D. 100

________ **18.** When voltage exceeds 150 volts-to-ground, the secondary ties with limiters must have a switch at each ________.
A. end
B. corner
C. both A and B
D. neither A nor B

Problems

19. Connect the windings for delta-to-delta operation using the terminals.

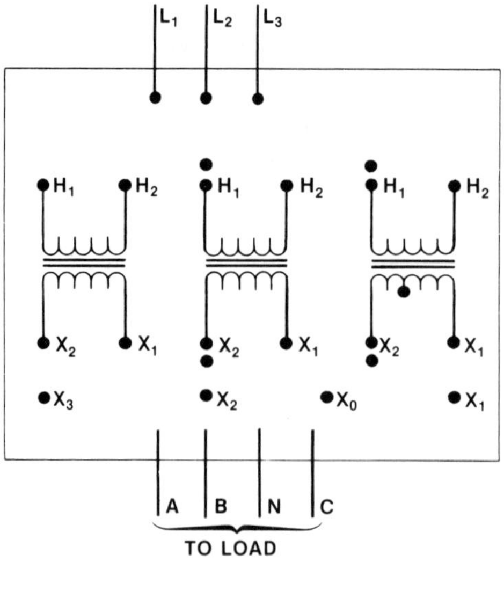

20. Connect the windings for open delta operation.

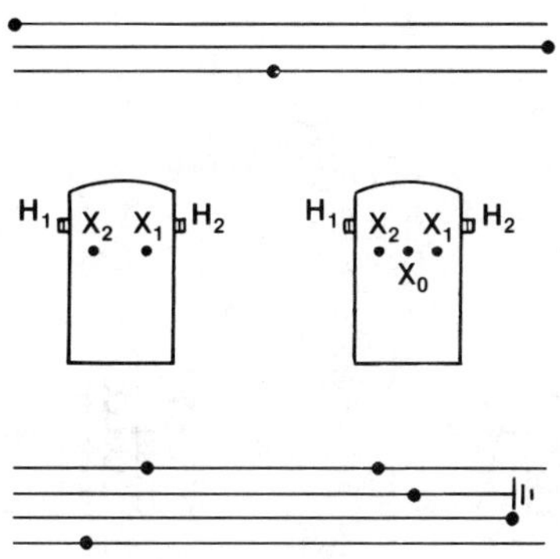

__________ **21.** This is a(n) ____________ voltage radial system.

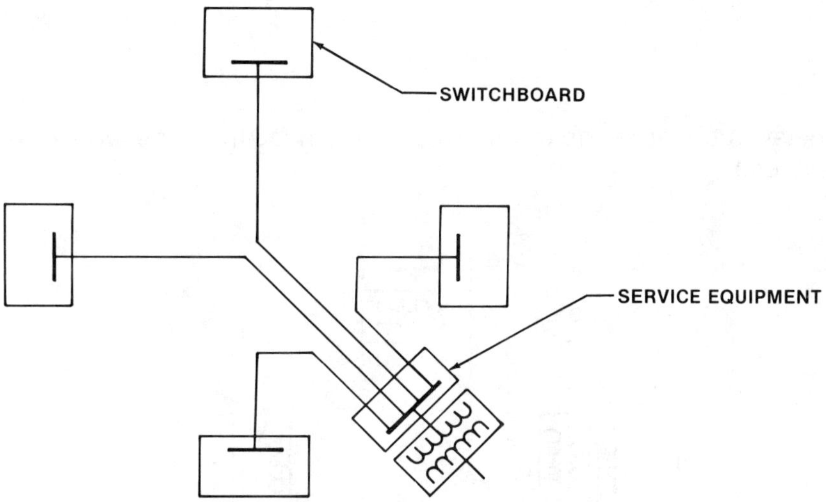

__________ **22.** This is a(n) ____________ loop system.

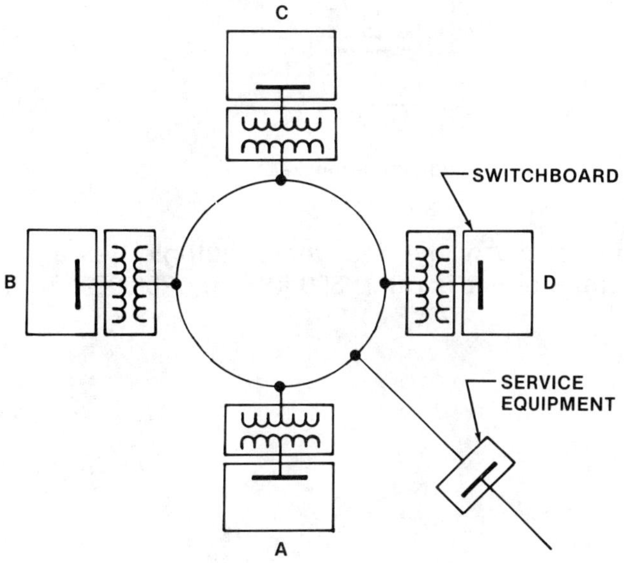

23. Complete the connections for the secondary ties. (Do not show protection devices.)

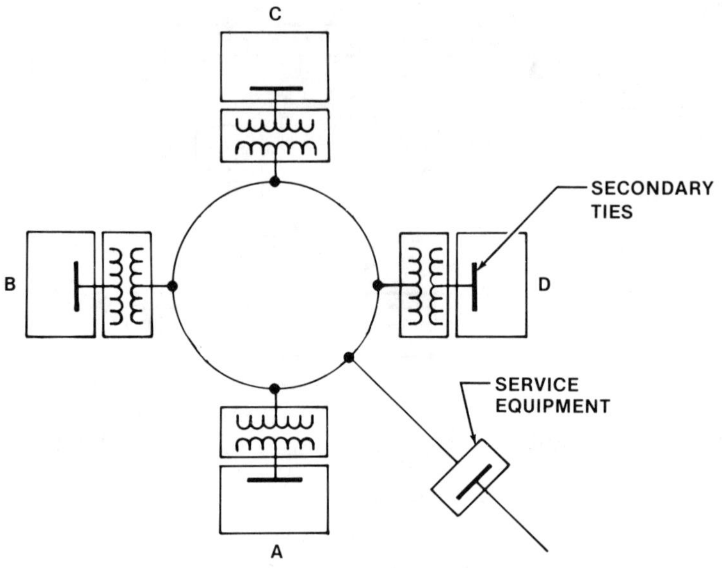

24. Connect the switchboards and transformers to form a simple network system. (Do not show protection devices.)

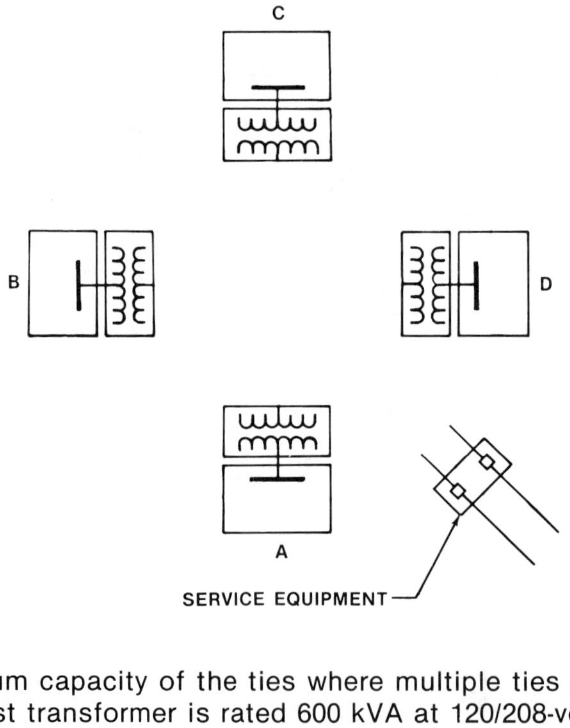

25. What is the minimum capacity of the ties where multiple ties are made between two transformers? The largest transformer is rated 600 kVA at 120/208-volt, three-phase.

MOTORS AND TRANSFORMERS FINAL EXAM

Chapter 16

The eight tests in this chapter provide a comprehensive review of the material presented in chapters 1 through 14. Each test contains a variety of True-False, Completion, and Multiple Choice questions. Record your answers in the spaces provided. Each test also contains Problems. Show your work in the spaces provided. Your instructor may require Code validation for specific questions and/or problems.

Subject matter covered in these tests includes:

- Motor Theory and Identification
- Motor Nameplates
- Troubleshooting Procedures
- Testing Connections and Windings
- Finding Starting Torque
- Finding Full-Load Torque
- Finding Locked-Rotor Current
- Finding Horsepower
- Finding Full-Load Current
- Sizing Conductors
- Sizing Overcurrent Protection Devices
- Sizing Overload Protection Devices
- Sizing Disconnects
- Sizing Reduced Starters
- Sizing Controllers
- Selecting Control Circuits
- Delta-to-Delta Connected Transformers
- Delta-to-Wye Connected Transformers
- Wye-to-Wye Connected Transformers
- Wye-to-Delta Connected Transformers
- Buck and Boost Transformers
- Location
- Separation from Combustible Material
- Troubleshooting
- Low-Voltage and High-Voltage Radial Systems
- Network Systems
- Connecting and Testing Windings
- Finding Full-Load Current
- Sizing Conductors
- Sizing Overcurrent Protection Devices
- Secondary Ties and Taps

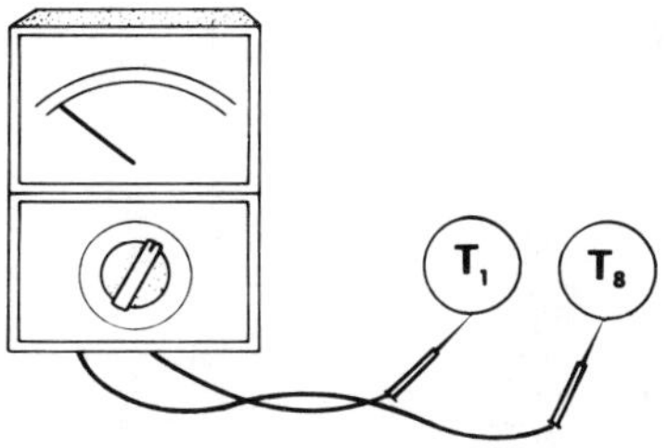

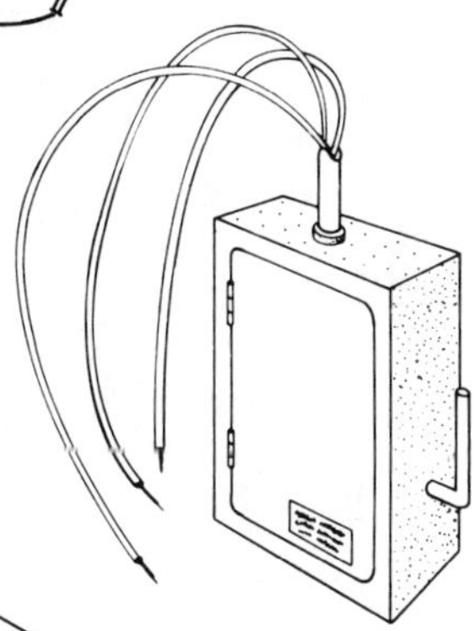

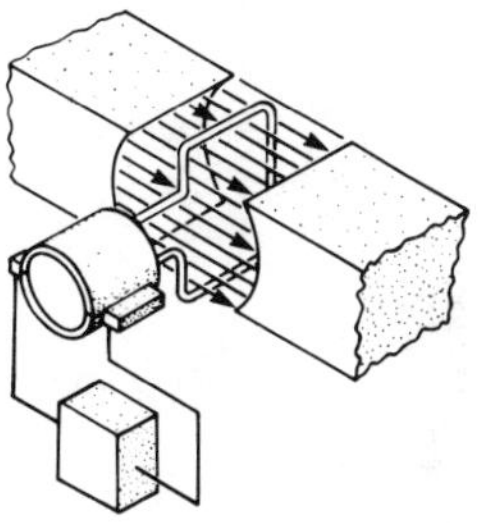

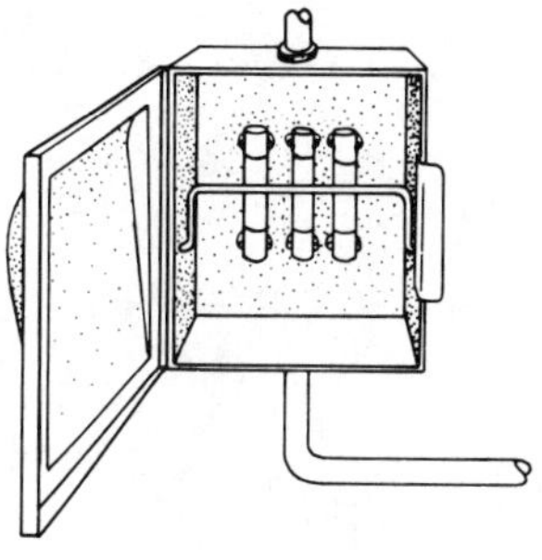

TEST 1—CHAPTER 16

Name ______________________________ Date ______________

True-False

T F 1. Windings of induction motors are connected in series for the higher operating voltage.

T F 2. Field pole windings of induction motors switch from north to south pole with alternating current.

T F 3. North field poles repel north poles on the rotor of induction motors.

T F 4. When the secondary of transformers has more turns than the primary, the amperage of the secondary is stepped up.

T F 5. Two 120-volt coils can be connected in series to obtain 240 volts.

T F 6. The midpoint of two 120-volt coils must be tapped and connected to ground to obtain 120/240-volt, single-phase.

T F 7. The neutral carries all the balanced current between two ungrounded conductors.

Completion

__________ 8. For the 25′ tap rule to be applied, a(n) __________ is always required in the panelboards.

__________ 9. Where the 10′ tap rule is applied, a main is not required in __________ panels.

__________ 10. The grounded conductor in corner-grounded systems can be __________ or gray in color.

__________ 11. The secondary side of transformers must be protected from __________.

__________ 12. Circuits feeding the primary side of transformers must be provided with __________ grounding.

__________ 13. A gas pipe feeding through a transformer __________ is not permitted.

Multiple Choice

__________ 14. The input side of a transformer is the high voltage __________ side.

A. secondary
B. primary
C. output
D. none of the above

_____ **15.** Three 120-volt circuits to ground can be obtained from a _____ system.
A. wye
B. delta
C. both A and B
D. neither A nor B

_____ **16.** The high leg of closed-delta systems is served with a voltage drop of 10% and is measured at _____ volts.
A. 168
B. 187
C. 194
D. 208

_____ **17.** The grounded conductor on the line side of the service must be calculated at _____% of the largest phase conductor.
A. 10
B. 12½
C. 15
D. 30

_____ **18.** Open-delta connected systems use only _____ transformers.
A. two
B. three
C. none of the above
D. all of the above

_____ **19.** The service disconnecting means can serve as the disconnect for _____ motor(s).
A. one
B. two
C. three
D. any number of

Problems

20. Which of the transformers is connected for additive polarity?

Ⓐ

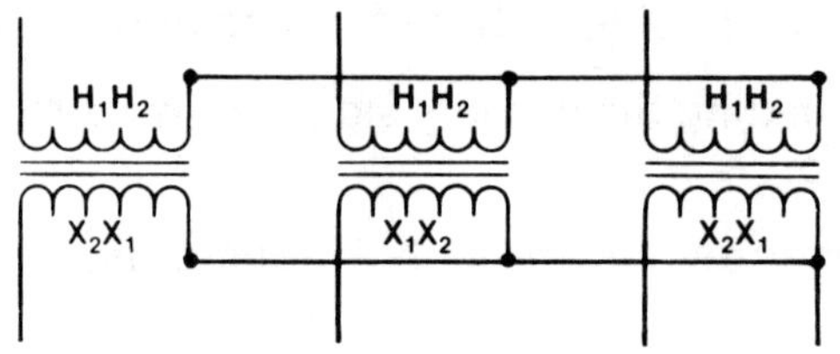

Ⓑ

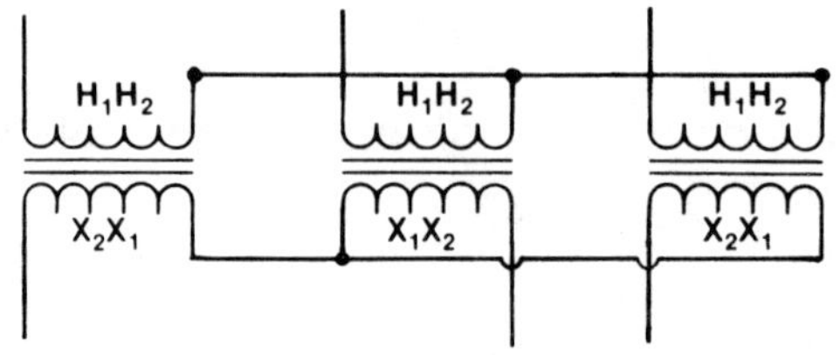

21. What size overloads and maximum size fuses are required to protect the motor from overloaded conditions (size the fuses for backup overload protection)?

A. __________ size overloads

B. __________ maximum size fuses

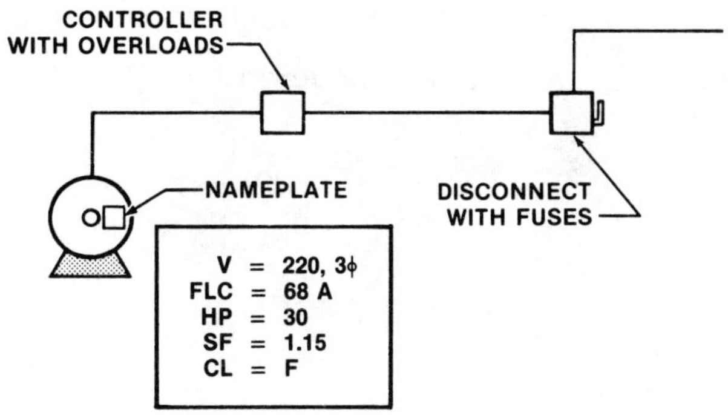

22. What size nonfusible disconnect and THWN copper conductors are required to supply the motor?

A. __________ size nonfusible disconnect

B. __________ size THWN copper conductors

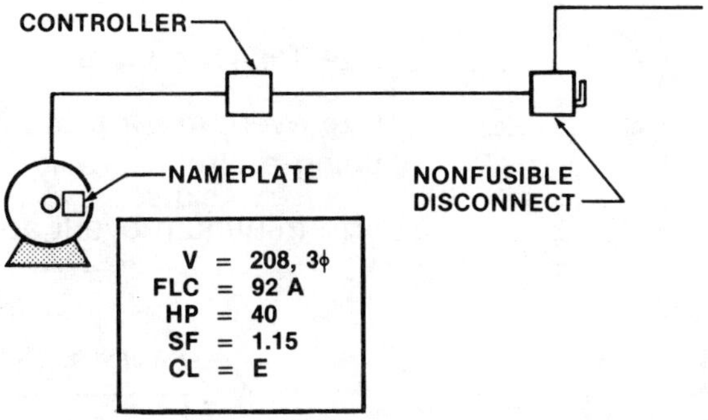

23. What is the running and starting torque of the motor?

A. __________ running torque

B. __________ starting torque

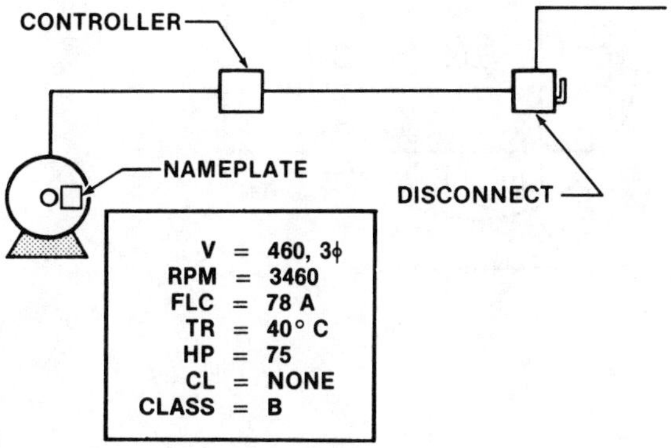

24. What size overcurrent protection device and conductors are required for the primary of the transformer used at continuous duty?

A. __________ size overcurrent protection device

B. __________ size conductors

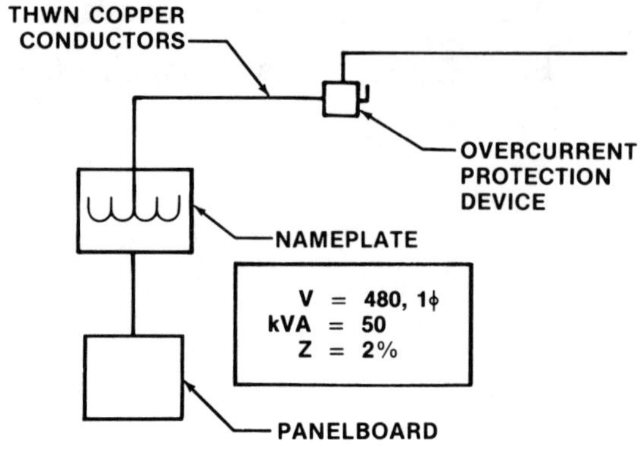

25. Fill in the blanks for the split-phase motor and connect the windings to the 240-volt line.

A. __________ size THWN copper conductors

B. __________ size overcurrent protection device using a circuit breaker

C. __________ size minimum overloads

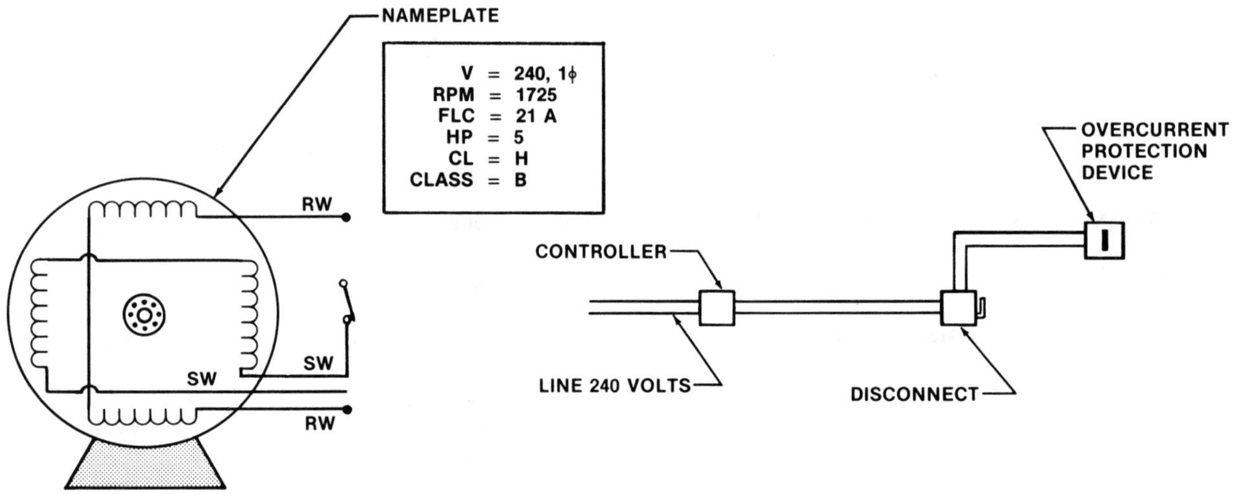

TEST 2—CHAPTER 16

Name ______________________________ **Date** ____________

True-False

T F 1. Transformers connected for closed-delta operation have the same line voltage and phase voltage.

T F 2. Transformers connected for wye operation do not have the same amount of line and phase current.

T F 3. The first 200 amps of neutral current must be calculated at 100% for incandescent lighting loads.

T F 4. The holding coil of three-wire control circuits holds the power to the coil until the stop button is pressed.

T F 5. Field windings of split-phase motors must be series-connected to operate on lower voltage.

T F 6. Centrifugal switches are connected in parallel with the starting winding in split-phase motors.

Completion

__________ 7. Sprinkler __________ are permitted in transformer vaults to provide fire protection.

__________ 8. The voltage-to-ground on the high leg of delta systems is __________ volts.

__________ 9. Capacitors used to correct power factor for induction motors must have a means of __________.

__________ 10. Capacitors supplying motors rated at 600 volts or less must discharge in ______ minute(s) after the motor is disconnected.

__________ 11. An additional __________ is not required for capacitors installed on the load side of magnetic starters.

__________ 12. Equipment grounding connections must be provided either inside or outside of the motor __________.

__________ 13. The overcurrent protection device for the __________ side of transformers is selected and based on the full-load current.

Multiple Choice

__________ 14. The general rule is that each motor be provided with a disconnecting ________.
A. start button
B. means
C. stop button
D. limit switch

______ **15.** Overcurrent protection devices for transformers can be selected at 125% of the primary full-load current when the amperage is ______.
A. 2 amps or less
B. 9 amps or less
C. 9 amps or more
D. none of the above

______ **16.** An 80% tap on an autotransformer reduces 480 volts to a value of ______ volts.
A. 240
B. 277
C. 375
D. 384

______ **17.** An autotransformer with a 65% tap reduces the inrush current of a motor pulling 1250 amps to ______ amps.
A. 312.5
B. 525
C. 650
D. 1000

______ **18.** A three-phase, 480-volt, 50-horsepower motor must be supplied with conductors having a rating of ______ amps.
A. 68
B. 75
C. 81
D. 90

______ **19.** A three-phase load of 722 amps requires a 480 V transformer rated at least ______ kVA.
A. 650
B. 680
C. 700
D. 750

Problems

20. What size overcurrent protection device and THHN copper conductors are required to supply the panelboard? Terminals are 75°C.

A. ______ size conductors
B. ______ size main

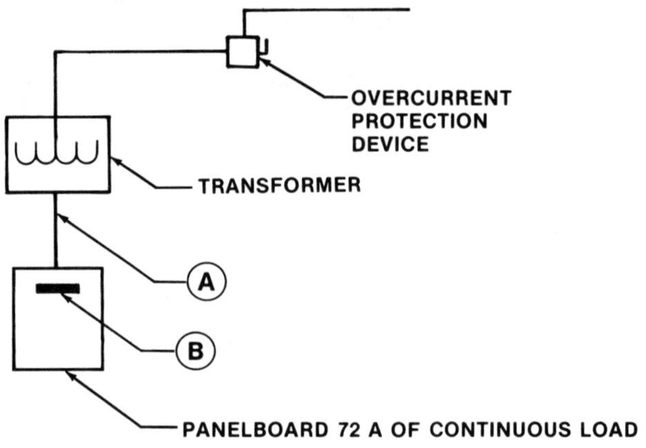

21. What size transformer is required to supply a load of 203 amps? Connect the secondary side for a 120/208-volt, four-wire system.

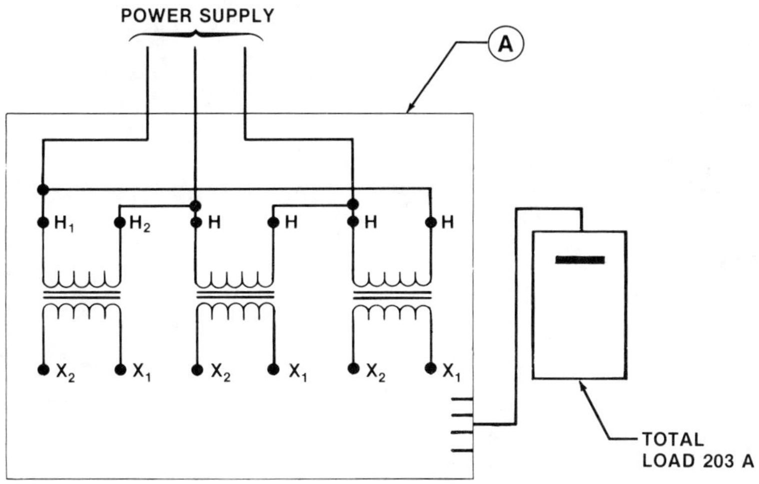

22. Connect the leads from the terminals of the resistance banks to the slip rings of the wound-rotor motor.

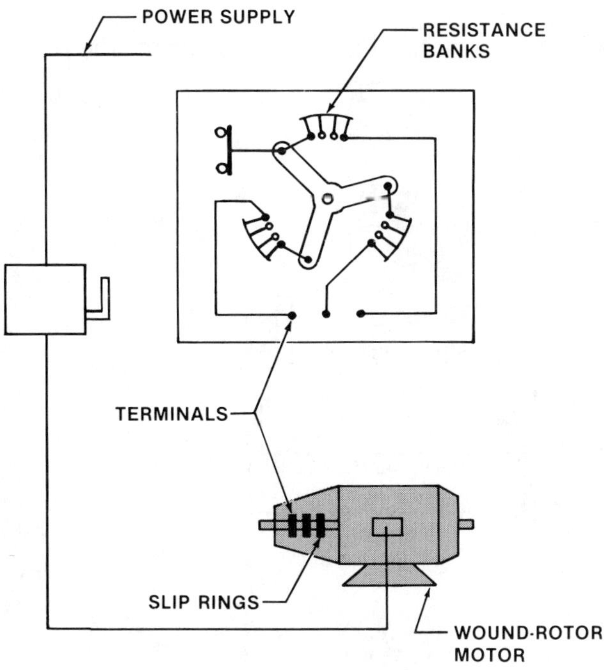

23. Connect the leads to the terminals and from the magnetic starter to the windings of the part-winding motor.

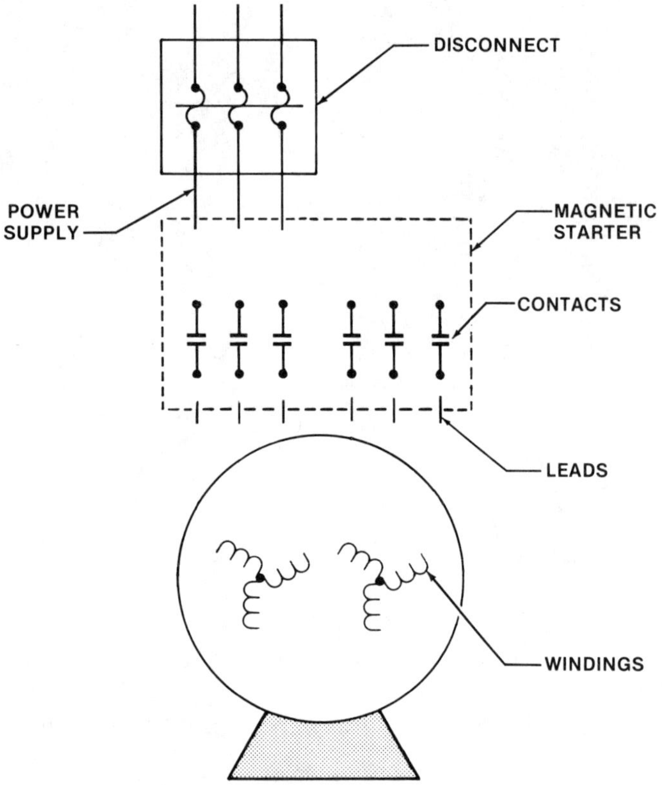

24. Connect the resistors to reduce the starting current to the motor. Fill in the blanks for the proper operation. To start, close:

A. ____________ (step 1)

B. ____________ (step 2)

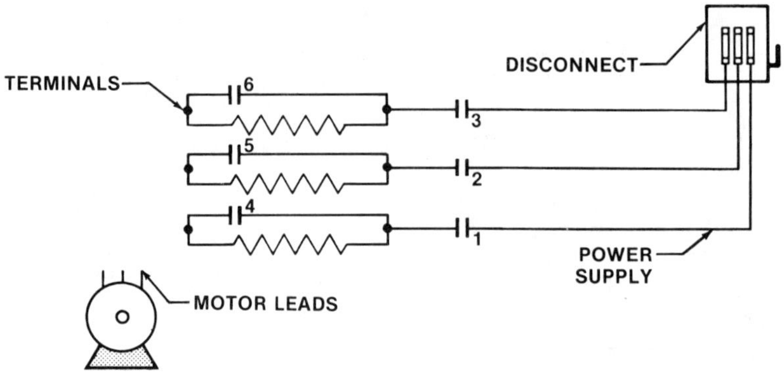

25. Connect the autotransformer using a 50% tap to reduce the starting current and allow the motor to start. Connect the leads to the motor windings for wye operation.

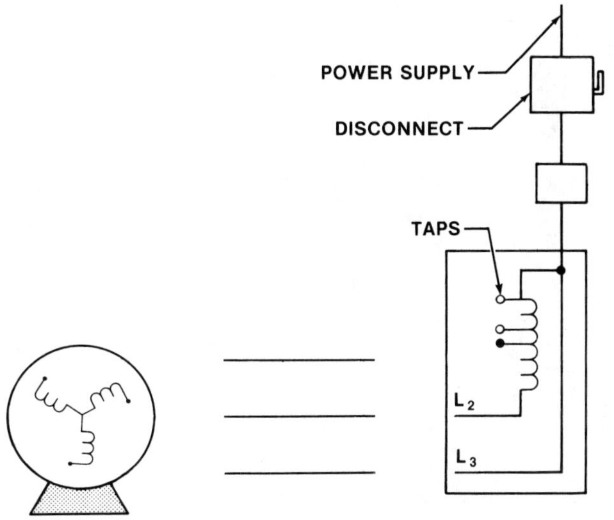

TEST 3—CHAPTER 16

Name ______________________________ Date ______________

True-False

T F 1. If the rotor of a shaded-pole motor is installed in the opposite direction, the rotor will rotate in the opposite direction.

T F 2. Synchronous motors run at synchronous speed once the motor is started.

T F 3. If the slip rings are not set against the rotor, wound-rotor motors will not run properly.

T F 4. Rotors of repulsion motors are built in the same manner as rotors of split-phase motors.

T F 5. Standard repulsion motors are equipped with a short-circuiter and brush-lifting mechanism.

T F 6. The direction repulsion motors rotate can be reversed by interchanging any two of the leads to the windings.

Completion

__________ 7. Motors rated above 1500 horsepower must be provided with __________ temperature detectors.

__________ 8. Motors with integral thermal protection and full-load current ratings not exceeding 9 amps can be set at __________% or less.

__________ 9. Motors that can be restarted __________ after shutdown must not be installed where personnel can be injured.

__________ 10. Fusible disconnects supplying motors must be selected by the minimum size __________ required to start the motor.

__________ 11. Control circuits that are derived from the service equipment and not the motor circuit must be __________ by the motor control disconnect.

__________ 12. Remote control circuits must be arranged in such a manner that a(n) __________ will not start the motor.

Multiple Choice

__________ 13. Power conductors and Class I control circuits can occupy the same __________ supplying a motor.

A. power circuit
B. conduit system
C. either A or B
D. neither A nor B

_______ **14.** The disconnecting means for controllers, located out of sight, to motors rated over 600 volts must be capable of being _______ in the open position.
A. locked
B. closed
C. either A or B
D. neither A nor B

_______ **15.** A 2% impedance transformer with a full-load current rating of 624 amps has a fault current rating of _______ amps.
A. 26,800
B. 27,950
C. 31,200
D. 36,500

_______ **16.** A three-phase, 208-volt, 25-horsepower motor requires a _______-amp circuit breaker to allow the motor to start.
A. 150
B. 175
C. 200
D. 225

_______ **17.** Three-phase, three-wire, corner-grounded circuits that supply motors must have fuses in each _______.
A. phase conductor
B. ungrounded leg only
C. both A and B
D. neither A nor B

_______ **18.** If the line current of a closed delta system is 258 amps, the phase current in the coils is _______ amps.
A. 125
B. 130
C. 140
D. 150

_______ **19.** A 70-amp overcurrent protection device can protect a remote motor control circuit with No. _______ aluminum conductors.
A. 14
B. 12
C. 10
D. 8

Problems

20. Fill in the blanks concerning the part-winding motor.

A. To start, close contacts ________________.

B. To run, close contacts ________________.

C. Reduced starting current = ________________ A

D. Reduced starting torque = ________________ ft lb

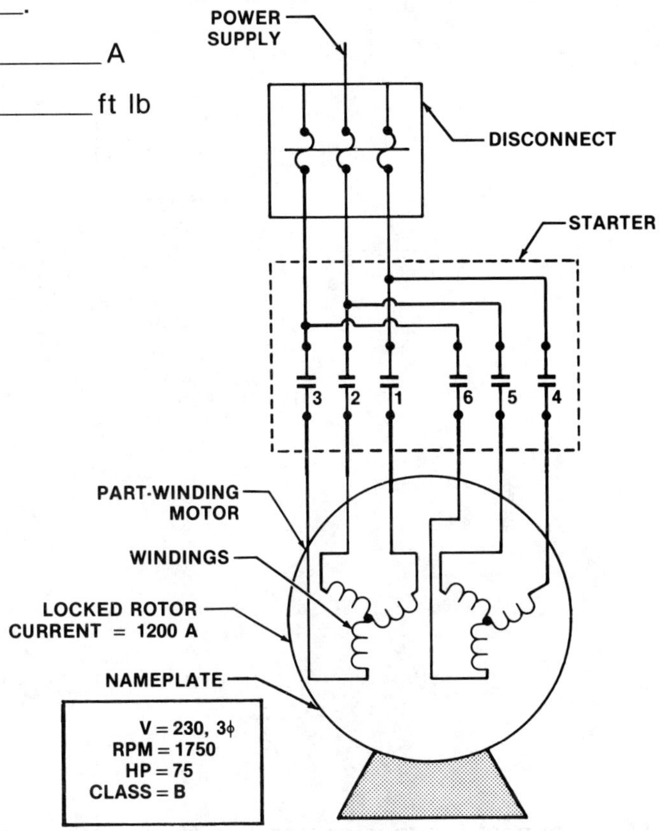

21. Connect the start and stop buttons to start and run the motor. Connect the overloads in the control circuit.

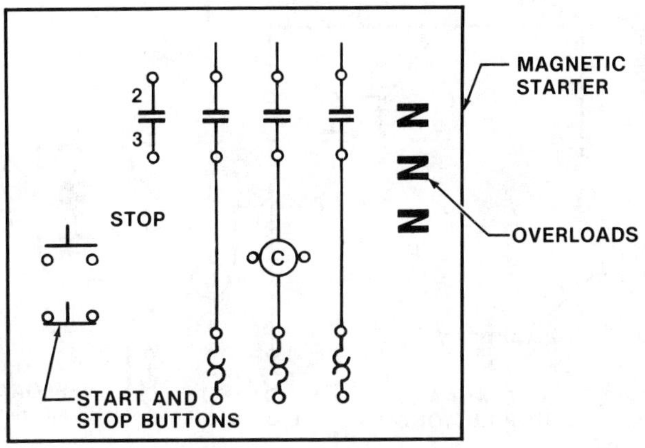

22. Connect the start, stop, and pilot light to start and stop the motor, and indicate when it is running. Connect the leads to the motor for the higher voltage.

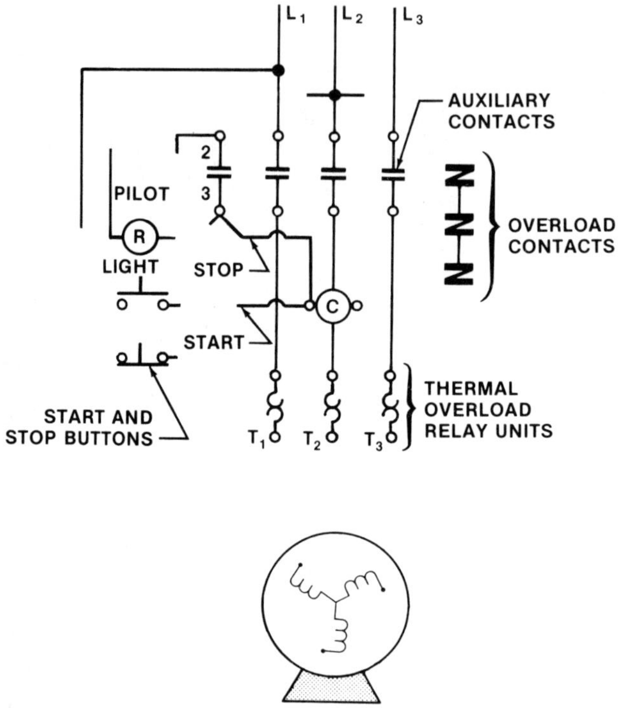

23. Connect the master stop button to prevent the motor from being started with the start button. Make the other necessary connections to start and stop the motor.

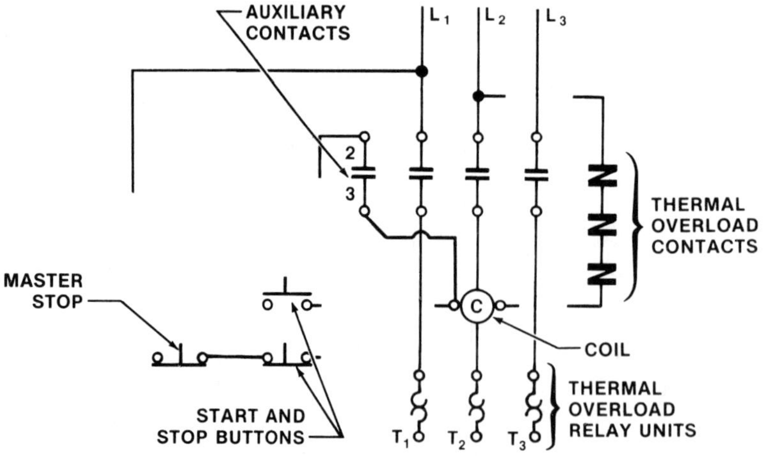

24. Connect the hand-off-automatic to the control circuits.

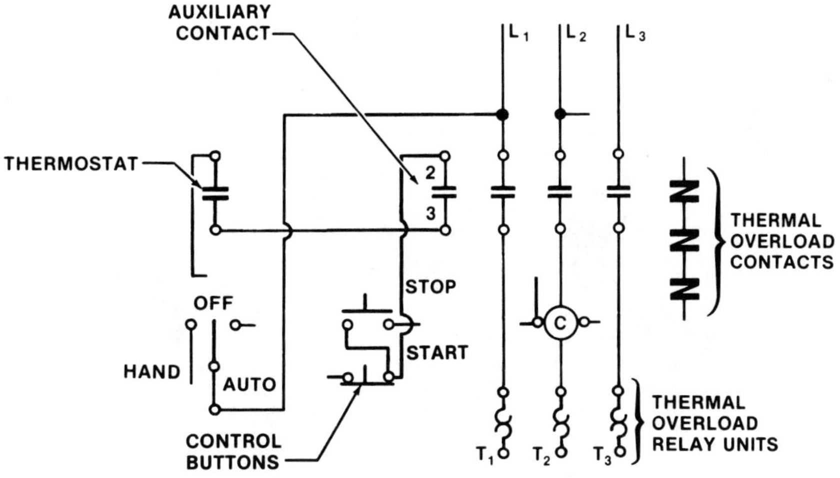

25. Connect the extra set of auxiliary contacts to energize another starter. Connect all control circuits.

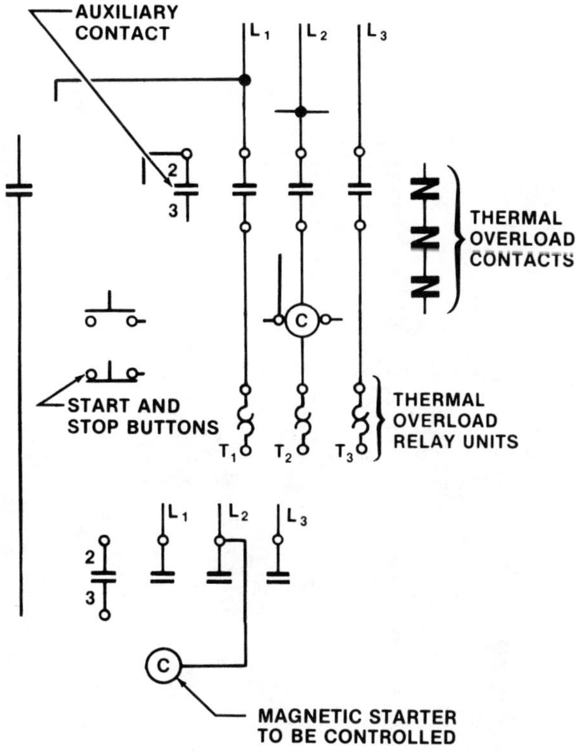

TEST 4—CHAPTER 16

Name ______________________________ Date ______________

True-False

T F **1.** When synchronous motors are used to lead or lag the current, the full-load current rating must be adjusted for three-phase.

T F **2.** Excessive greasing of a motor's ball bearings may cause the motor to overheat.

T F **3.** Dirt on the slip rings of wound-rotor motors do not affect the operation or speed control of a motor.

T F **4.** To check for a defective exciter in synchronous motors, turn the rotor by hand and check the output.

T F **5.** Where 10′ taps are made from the secondary of transformers, the length of the conductors in conduit can be routed over 10′.

T F **6.** Secondary ties are used to ensure power in the event of trouble on the network.

Completion

__________ **7.** Control transformers installed in motor control centers to reduce control voltage must be connected on the __________ side of the disconnect.

__________ **8.** Controllers must be capable of __________ the stall current of the motor they start and stop.

__________ **9.** Disconnects used to disconnect motors shall be plainly marked to __________ the ON and OFF position.

__________ **10.** One of the __________ for motor circuits must be readily accessible.

__________ **11.** The __________ winding is wound on top of the running winding.

__________ **12.** The running winding is wound with __________ wire than the starting winding.

Multiple Choice

__________ **13.** With a single-phase load of 67 amps and a three-phase load of 82 amps, a power transformer is rated at __________ kVA on open delta.

A. 16.56
B. 19.8
C. 20
D. 22.12

__________ **14.** A single-phase, 480-volt, 37½ kVA transformer can be protected with an overcurrent protection device rated at __________ amps.

A. 80
B. 90
C. 100
D. 110

__________ **15.** Loads are tapped at transformer supply points using secondary ties and the largest transformer is a 200 kVA, 208-volt, three-phase transformer. The bus ties must be rated at least __________ amps.
A. 250
B. 372
C. 450
D. 500

__________ **16.** The actual speed of an 1800 rpm motor with 13% slip is __________ rpm.
A. 1650
B. 1750
C. 3450
D. none of the above

__________ **17.** Heavy loads that are hard to start (such as elevators) should be started by using a class __________ design.
A. A
B. B
C. C
D. none of the above

__________ **18.** The torque required to drive a load of 148 foot-pounds is __________ horsepower with the rotor turning at 1725 rpm.
A. 30
B. 40
C. 50
D. 60

Problems

19. Connect the control circuits to jog the motor.

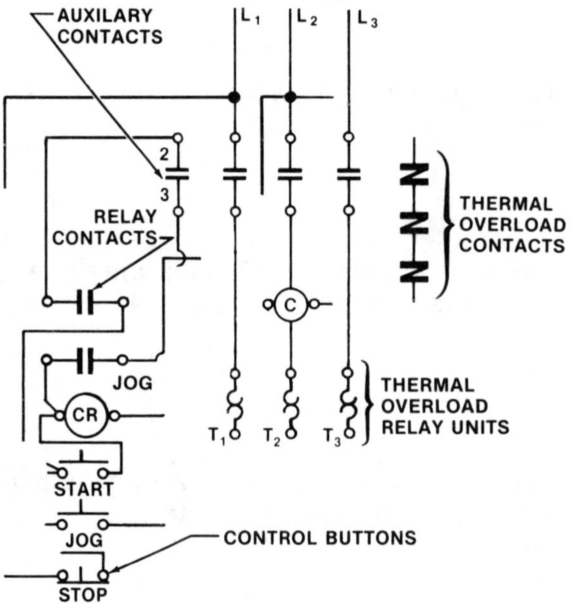

20. Fill in the blanks for the motor system.

A. ____________ size circuit breaker

B. ____________ size time-delay fuse and disconnect

C. ____________ size overload

D. ____________ size THHN copper conductors

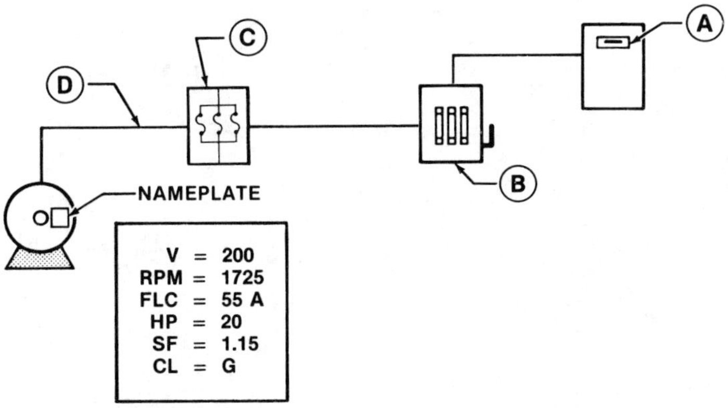

21. Find the values for the motor.

A. ____________ full-load torque

B. ____________ starting torque

C. ____________ locked-rotor current based on code letter F

D. ____________ maximum size time-delay fuses

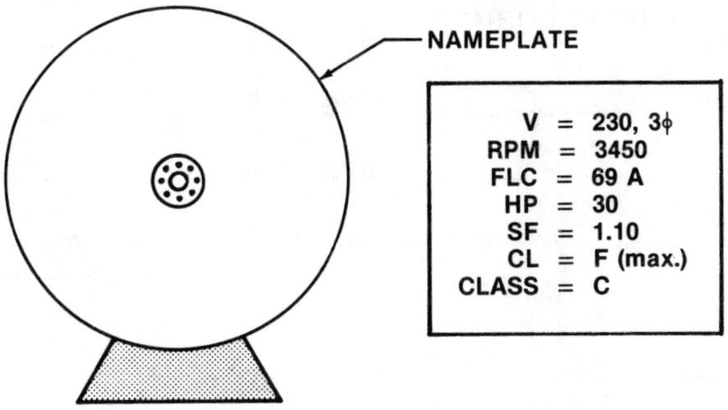

22. Connect the three-phase induction motor for 240-volt delta operation. Match the leads to the connection of the triangle.

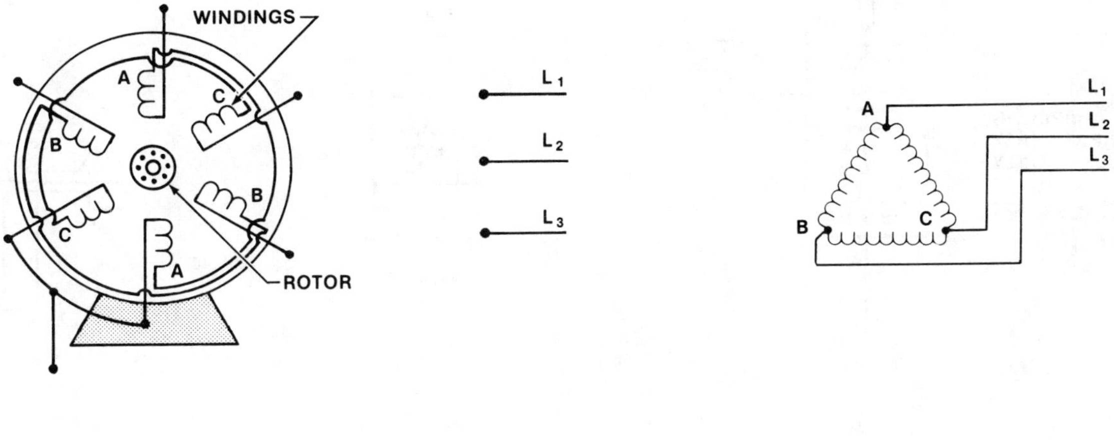

23. Connect the windings of the capacitor motor for two speeds using a switch.

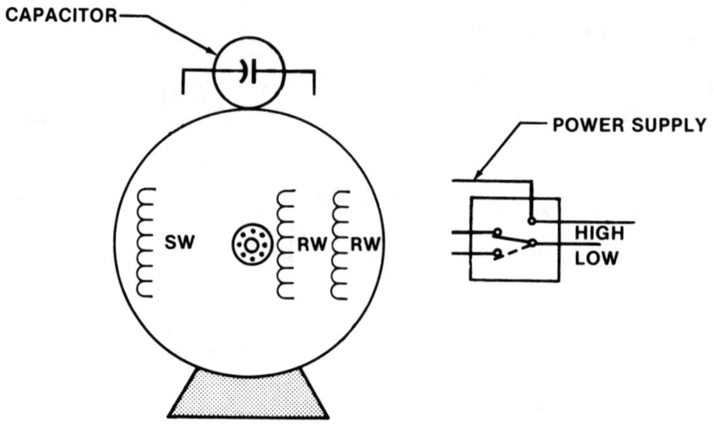

24. What size lighting and power transformers are required to supply a three-phase load of 85 amps and a single-phase load of 62 amps? Connect the secondary windings of the transformers to the panelboard.

A. __________ size kVA

B. __________ size kVA

C. __________ size kVA

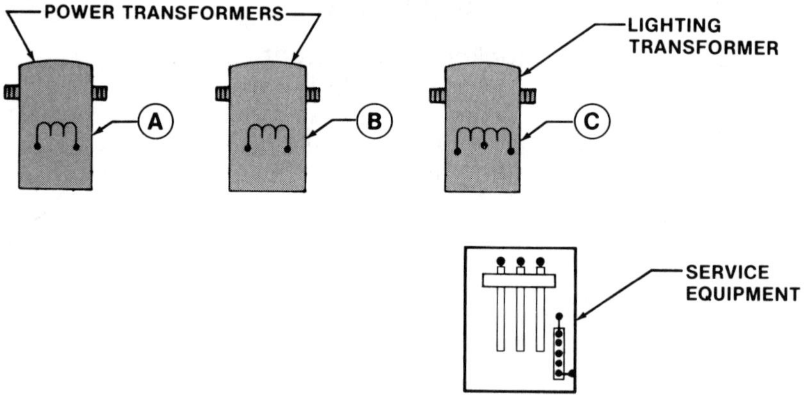

25. What is the minimum size grounded conductor required for the service panelboard where only three-phase and single-phase 480 volts are used? Connect the power supply to the equipment.

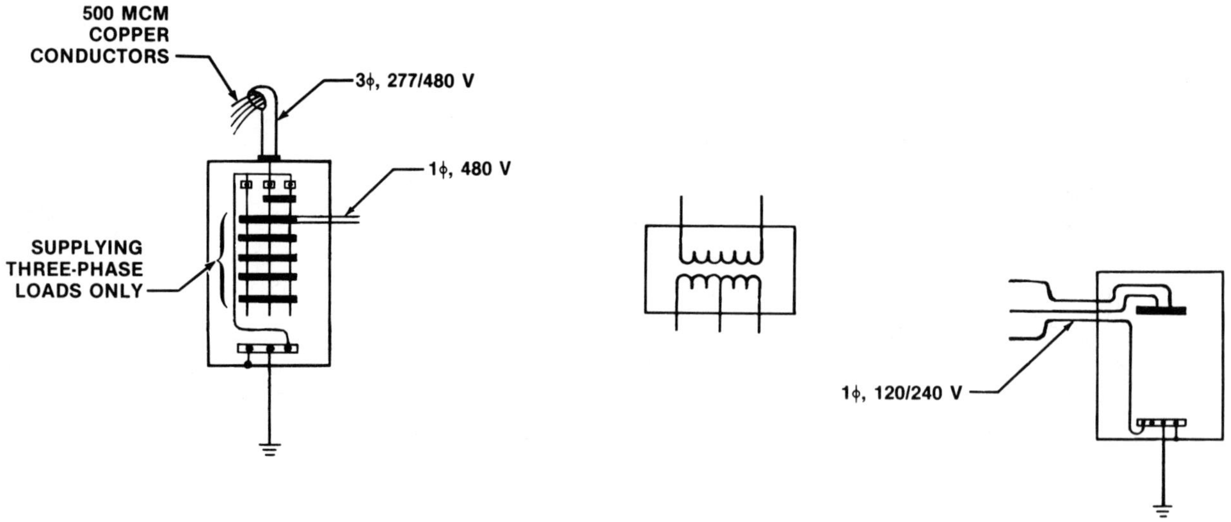

TEST 5—CHAPTER 16

Name ______________________________ **Date** ______________

True-False

T F **1.** Motors must produce enough torque to start and drive a load.

T F **2.** The horsepower of motors is determined by the load they are required to start and drive.

T F **3.** One horsepower equals 550 foot-pounds of energy per second.

T F **4.** Motors at full speed and under full-load conditions will not develop full-load torque.

T F **5.** The polarity of an electromagnet cannot be reversed by changing the flow of current through its coil.

T F **6.** The strength of an electromagnet can be increased by adding turns to its coil.

T F **7.** The magnetic poles of electromagnets reverse their polarity 120 times per second.

Completion

__________ **8.** The starting winding has fewer __________ than the running winding.

__________ **9.** The starting torque of split-phase motors is obtained by adding another high __________ winding.

__________ **10.** Two-phase motors have two windings that are located __________ apart on the stator.

__________ **11.** The starting winding in a split-phase motor has angular phase displacement that rarely exceeds ______° to ______° in time.

____

__________ **12.** The __________ switch opens the circuit to the starting winding at about 80% of the synchronous speed of the motor.

__________ **13.** The __________ protector in split-phase motors protects the motor windings from overheating and burning out.

Multiple Choice

__________ **14.** Transformers rated at 600 volts or less can be protected in the primary side at a maximum of __________% or less.

A. 125
B. 150
C. 175
D. 250

________ 15. Loads tapped from the secondary side of transformers must be balanced as close as possible to prevent ________ any one phase conductor.
   A. underloading
   B. overloading
   C. either A or B
   D. neither A nor B

________ 16. The voltage-to-ground will ________ when the grounded connection is lost at the transformer and service equipment.
   A. fluctuate
   B. remain constant
   C. either A or B
   D. neither A nor B

________ 17. The overcurrent protection device in the primary side must not exceed 167% for transformers rated from 2 amps to less than ________ amps.
   A. 5
   B. 7
   C. 9
   D. 10

________ 18. Transformers rated over 600 volts and having overcurrent protection devices in the primary and secondary sides are selected on the percentage of ________.
   A. power factor
   B. impedance
   C. either A or B
   D. neither A nor B

________ 19. A ________′ tap may be made from the secondary side of transformers in the same manner as feeders.
   A. 10 or 25
   B. 15 or 20
   C. 20 or 35
   D. 30 or 50

Problems

20. Give the size of the conductors for the motor circuit.

A. ________ TW 60°C CU

B. ________ THWN 75°C CU

C. ________ THHN 90°C CU

D. ________ XHHW 90°C ALUM

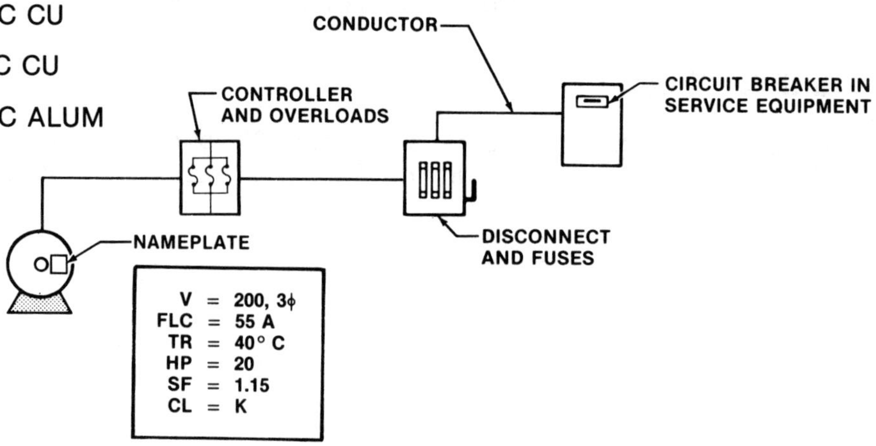

21. Fill in the blanks for the motor circuit.

A. ____________ nontime-delay fuses

B. ____________ time-delay fuses

C. ____________ instantaneous circuit breakers

D. ____________ inverse circuit breaker

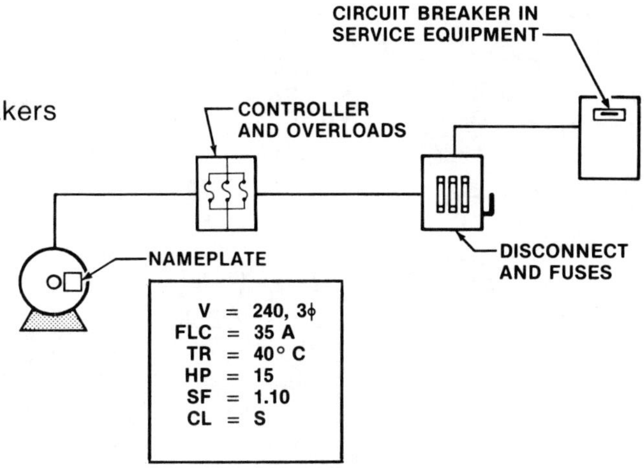

22. Fill in the blanks for the motor circuit.

A. ____________ size minimum overloads

B. ____________ size maximum overloads

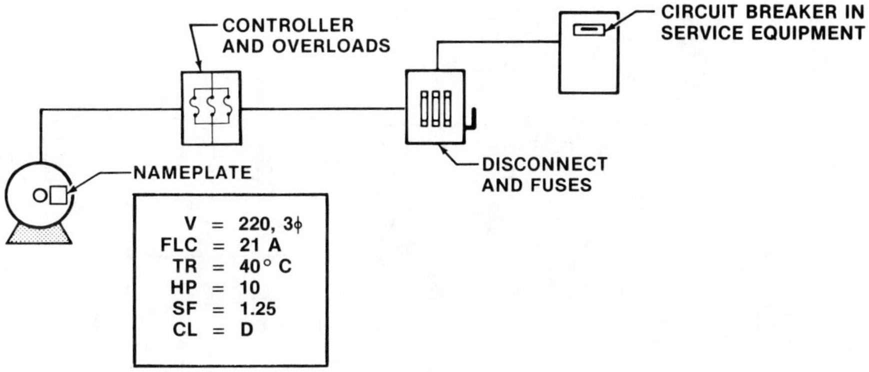

23. Fill in the blanks for the motor circuit.

A. ____________ size nonfusible disconnect

B. ____________ size fusible time-delay fuses

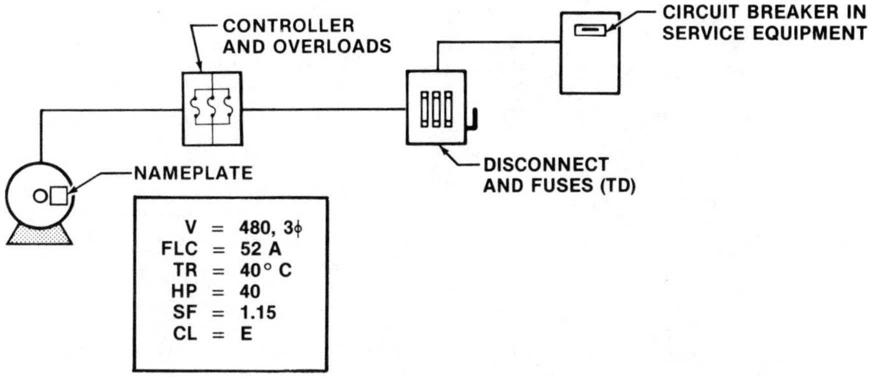

24. Fill in the blanks for the size controller of motor circuit.

A. __________ hp

B. __________ A or greater

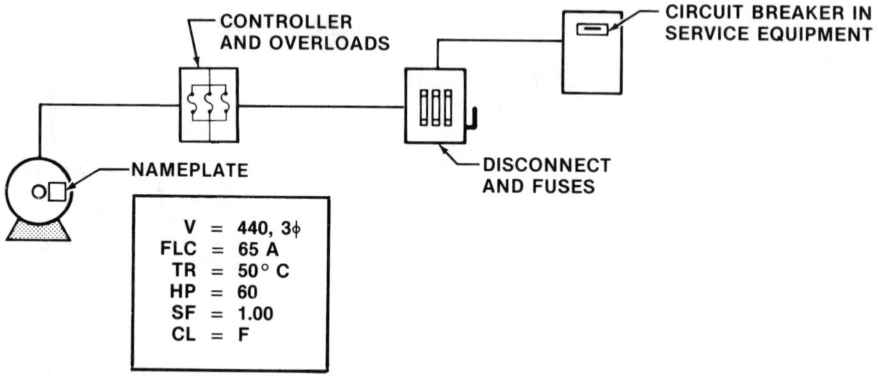

25. Fill in the blanks for the motor circuit.

A. __________ full-load torque

B. __________ starting torque

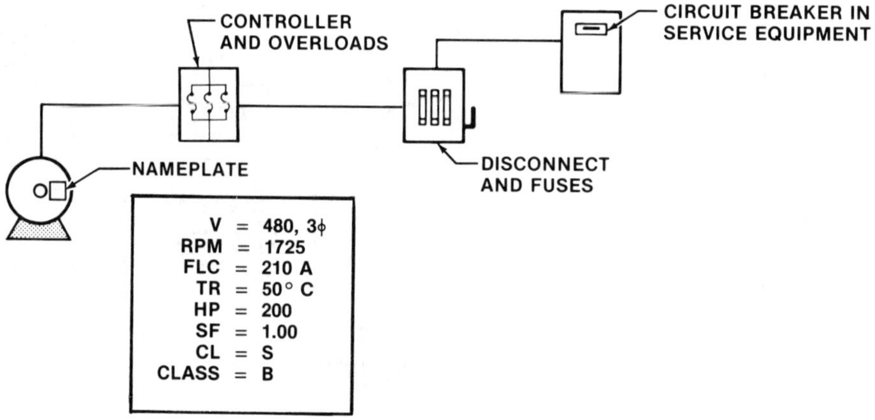

TEST 6—CHAPTER 16

Name ______________________________ **Date** ______________

True-False

T F 1. A piece of soft iron carries less magnetism than air.

T F 2. Each of the three currents in a three-phase system goes through its own cycle independently.

T F 3. In induction motors, the rotor is pushed and pulled through the rotating magnetic field of the stator poles.

T F 4. The class or design of a motor is determined by the way the copper or aluminum bars are embedded in the rotor.

T F 5. Alternating currents change the poles of the stator from north to south and push and pull the rotor through the field.

T F 6. The magnetic field of the rotor is the same as that of the stator field poles.

Completion

__________ 7. If the starting switch fails to disconnect the starting winding from the circuit, the __________ resistance winding will burn out.

__________ 8. Standard direction of split-phase motors can be determined by viewing the motor from the __________ end.

__________ 9. Split-phase motors can be reversed by interchanging the connections of either the __________ winding or running winding.

__________ 10. __________ motors are equipped with a starting capacitor to provide more starting torque.

__________ 11. The line current in split phase motors is __________% greater than that in capacitor-start motors.

__________ 12. Transformers can be __________ to provide multiple speeds for shaded motors.

__________ 13. Two speeds can be obtained by providing __________ windings with one winding having smaller wire with many turns.

Multiple Choice

__________ 14. If a 10′ tap is made from the secondary of a transformer to the lugs of a 225-amp panelboard, the conductors must be __________.

A. 1/0 THWN copper
B. 2/0 THWN copper
C. 3/0 THWN copper
D. 300 kcmil THWN aluminum

______ **15.** If the primary of a transformer is 2400 volts and the ratio is 10 to 1, the secondary voltages will be ______ volts.
A. 120
B. 208
C. 240
D. 277

______ **16.** Where the secondary supply of transformers are not grounded, a(n) ______ conductor is not required to be routed to the service equipment.
A. grounded
B. equipment grounding
C. ungrounded
D. all of the above

______ **17.** Transformers with an ungrounded supply conductor must have ______ grounding conductors routed with all feeder and branch-circuit loads to ground all metal.
A. ungrounded
B. equipment
C. either A or B
D. neither A nor B

______ **18.** The fault current that a 150 kVA, 480-volt, three-phase transformer delivers at at its terminals with 3% impedance is ______ amps.
A. 4960
B. 5016
C. 6016.9
D. 8090.9

______ **19.** The minimum size fuses required for the primary side of a 112 kVA, 480-volt, three-phase transformer is ______ amps.
A. 125
B. 175
C. 200
D. 225

Problems

20. Give the current rating for phases A and B with a blown fuse in phase C.

A. ______ phase A

B. ______ phase B

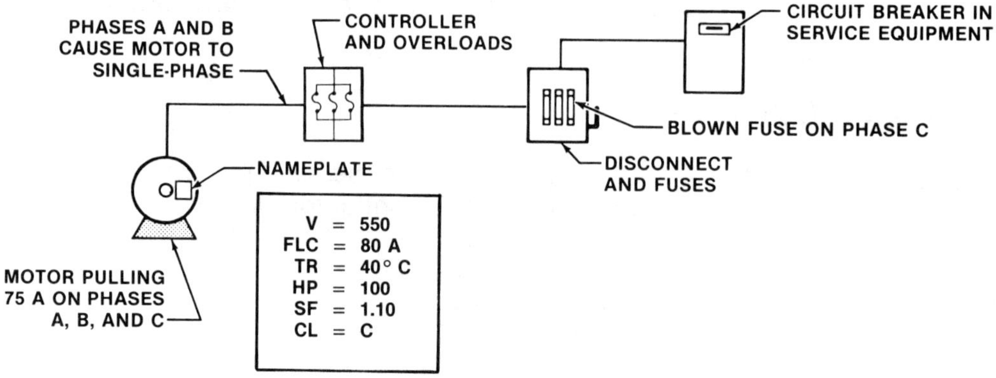

21. Fill in the blanks for reduced starting of the motor.

For resistor and reactor starting
A. reduced starting current = __________ A

For autotransformer starting (65% tap)
B. winding current = __________ A

C. line wire current = __________ A

D. transformation current = __________ A

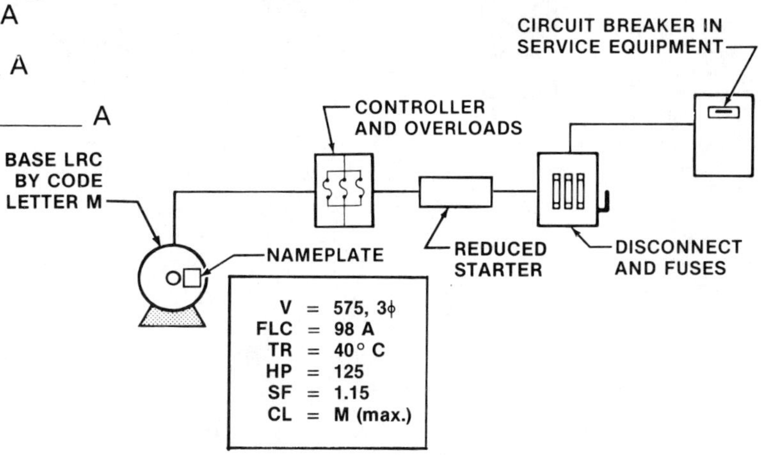

22. Fill in the blanks for the motor circuit using motor windings to reduce the starting current.

A. part-winding starting = __________ A

B. wye-delta starting = __________ A

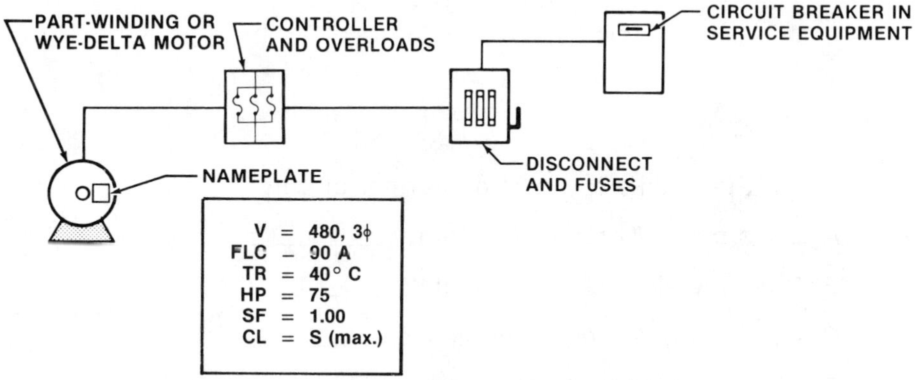

23. Connect the leads of the windings for reverse rotation using the drum switch.

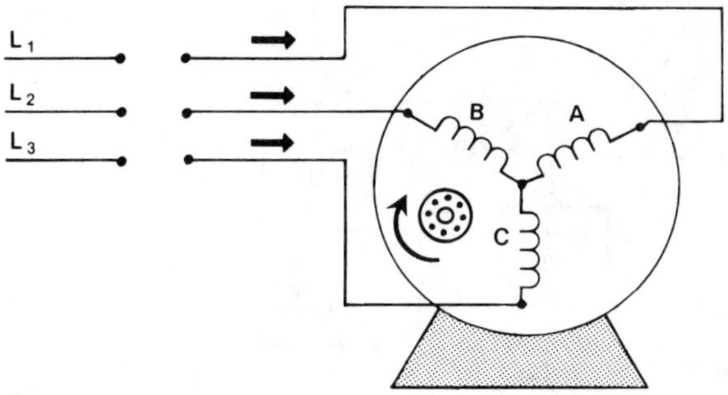

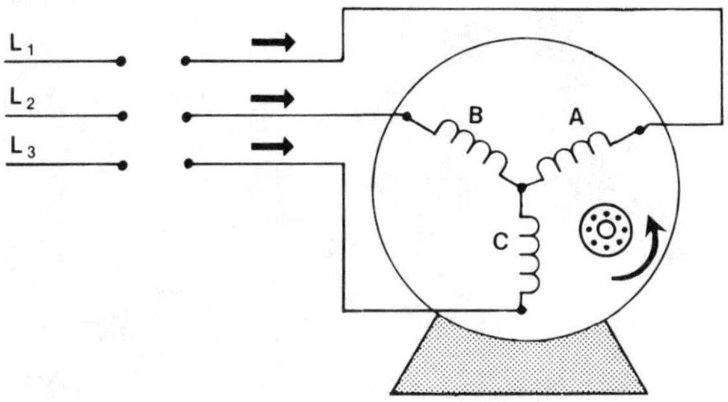

24. Connect the transformer's windings and terminate the leads to the meter and service equipment.

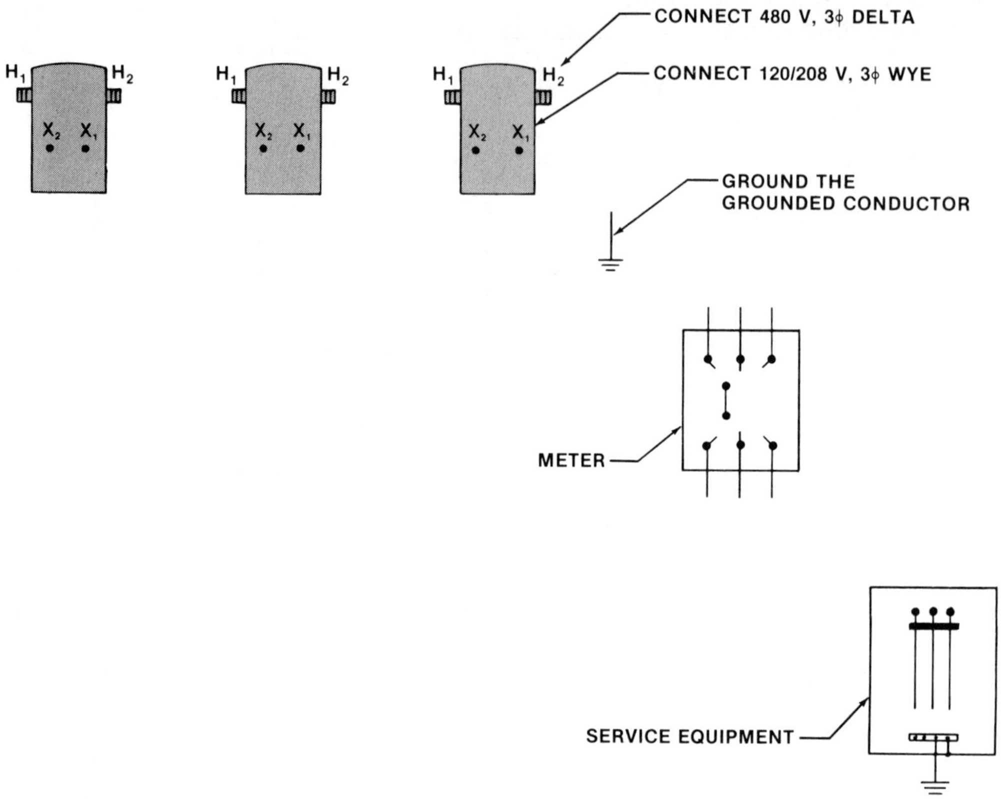

25. Fill in the blanks for the transformer circuit.

A. Does panel #1 require a main? ___________

B. Does panel #2 require a main? ___________

C. Does panel #3 require a main? ___________

D. What size transformer is required? ___________

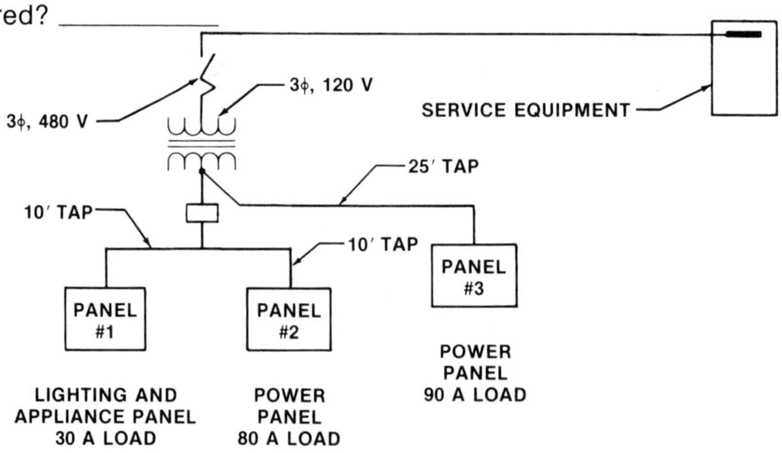

TEST 7—CHAPTER 16

Name ______________________________ **Date** ______________

True-False

T F **1.** The copper bars that are close to the surface in the rotor have high resistance windings.

T F **2.** A rotor turning rapidly through the magnetic field cuts fewer lines of force.

T F **3.** Class D motors do not provide high starting torque to start heavy loads.

T F **4.** Motors are usually designed to operate in a range from 0°C to 4°C.

T F **5.** Motor insulation must withstand very high temperatures without burning up.

T F **6.** There are four basic classes of insulation used for motor windings.

Completion

__________ **7.** Dry-type transformers rated at 600 volts or less and not over 112½ kVA must clear combustible material at ______________″ when located indoors.

__________ **8.** Two additive transformers and one subtractive transformer can be __________ for subtractive operation.

__________ **9.** Where transformers are protected with an automatic sprinkler system, the door of the vault is not required to have a(n) ______________ hour rating.

__________ **10.** Ventilation openings must be located as far away from doors, windows, and ______________ material as possible.

__________ **11.** Autotransformers used in a zigzag configuration for grounding three-phase systems must be rated for ______________ current.

__________ **12.** When the primary full-load current rating of transformers are multiplied by 167%, the next ______________ size fuse must be used where the ampacity does not correspond to a standard size device.

Multiple Choice

__________ **13.** The grounded phase of a corner-grounded delta system must not be __________ at the service equipment disconnect.

A. terminated
B. fused
C. none of the above
D. all of the above

__________ **14.** A ______________-amp circuit breaker is required to protect the secondary side of a 300 kVA, 2400-volt, three-phase transformer with 2% impedance (unsupervised location).

A. 200
B. 250
C. 300
D. 400

__________ **15.** The grounded conductor cannot be ____________ when the load of the service is electrical discharge lighting.
A. bare
B. reduced
C. increased
D. left out

__________ **16.** Transformers connected in parallel must have as near the same percentage of ____________ as possible.
A. impedance
B. voltage
C. both A and B
D. neither A nor B

__________ **17.** Where the 25′ tap rule includes the primary transformer plus the secondary, the size of the primary tap must be ____________ the feeder size.
A. one-half
B. one-third
C. three-fourths
D. none of the above

__________ **18.** Where a 10′ tap is made from the secondary side of a transformer and a disconnect is installed, the feeder circuit must have a(n) ____________ grounding conductor if routed in PVC.
A. equipment
B. grounded
C. none of the above
D. all of the above

__________ **19.** Separately derived systems can be grounded either at the service equipment or at the source of ____________.
A. supply
B. power
C. both A and B
D. neither A nor B

Problems

20. Fill in the blanks for the transformer circuit.

A. __________ size THWN copper primary tap

B. __________ size THWN copper secondary tap

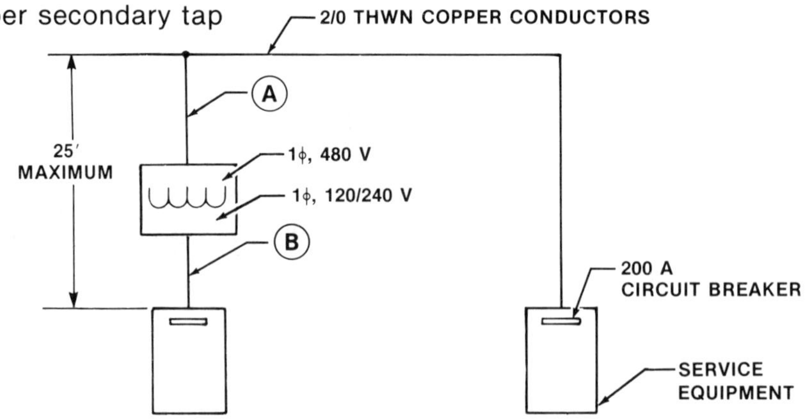

21. Fill in the blanks for the transformer circuit.

A. __________ size main circuit breaker

B. __________ size 10′ tapped conductors

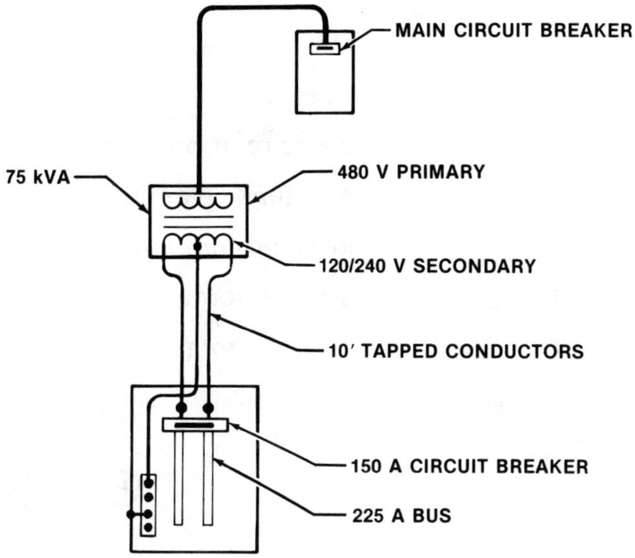

22. Fill in the blanks for the motor system.

A. __________ full-load torque

B. __________ starting torque

C. __________ locked-rotor current

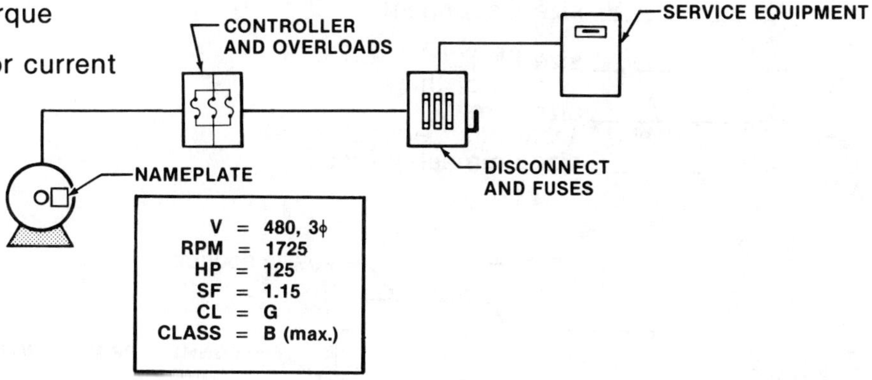

23. Fill in the blanks for the motor circuit.

A. __________ size wire

B. __________ size overload

C. __________ size circuit breaker

D. __________ size disconnect and time-delay fuses

E. __________ size controller

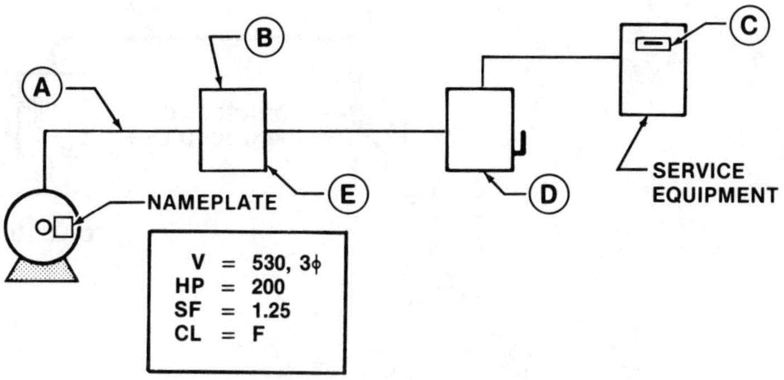

24. Fill in the blanks for the motor circuit.

A. __________ size overloads

B. __________ size wire

C. __________ size circuit breaker

D. __________ size time-delay fuses

E. __________ locked-rotor current

F. __________ starting torque

G. __________ full-load torque

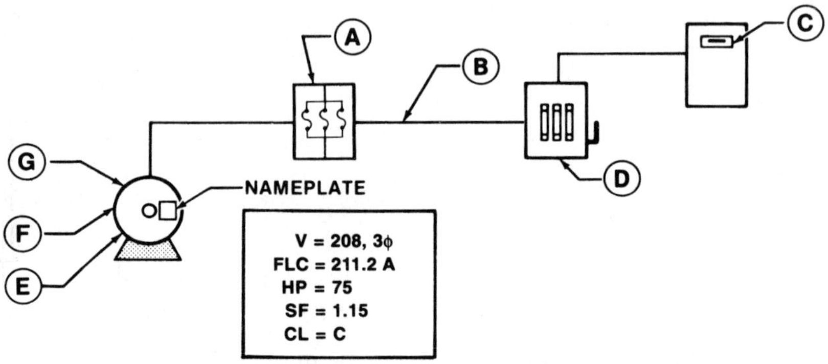

25. Connect the transformer circuit to the disconnect and motor. Fill in the blanks.

A. __________ size disconnect

B. __________ size EMT

C. __________ size wire

D. __________ size time-delay fuses

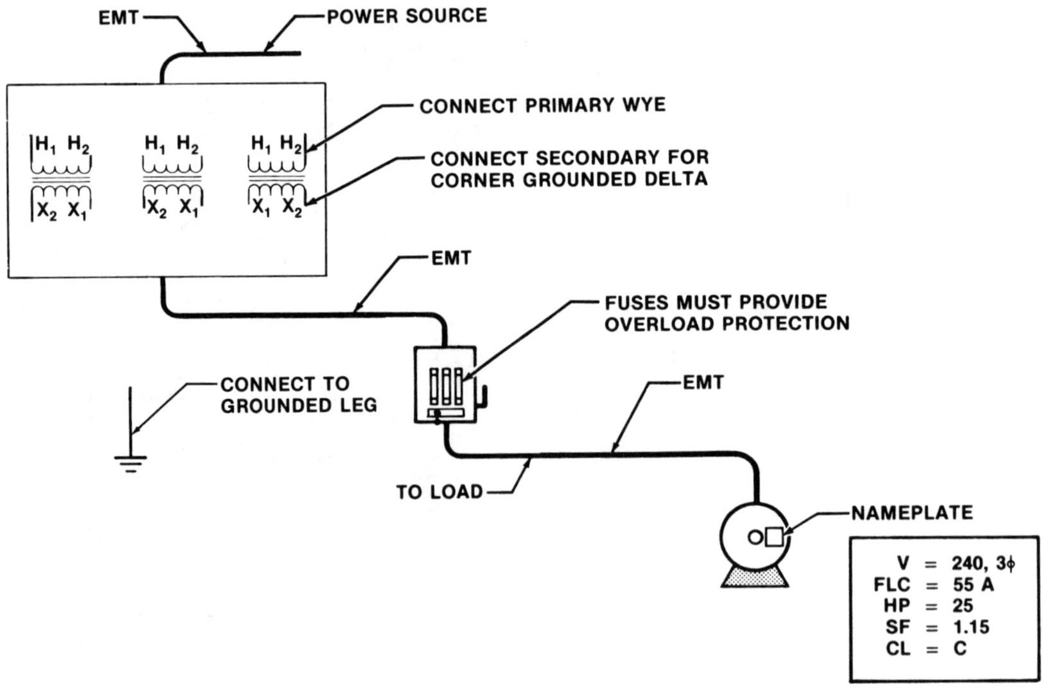

TEST 8—CHAPTER 16

Name Date

True-False

T F 1. The rotor of an induction motor is designed with solid iron cores to limit eddy currents.

T F 2. Conductors moving through a magnetic field in a closed circuit do not measure current flow.

T F 3. Six-pole induction motors provide three different speeds for motor operation.

T F 4. One of the main advantages of connecting a bank of transformers closed delta is that if one winding burns out, the other two can be connected open delta.

T F 5. Transformers located close to the building deliver as much fault current to the service equipment as transformers located far away.

T F 6. Transformers supplying service equipment located in buildings must be grounded at the transformer.

Completion

________ 7. The power and control conductors to an individual motor can be ________ in the same conduit under certain conditions.

________ 8. When motor power conductors and Class I control circuits are pulled through the same conduit and used at continuous duty, they must be ________ according to the number.

________ 9. When capacitors are connected on the load side of magnetic starters, the ________ must be sized for the rating of the improved power factor.

________ 10. The grounded leg of corner-grounded delta systems carries the same amount of ________ as the other phases.

Multiple Choice

________ 11. The size of the grounding conductor connected to the building steel for a transformer that supplies a panelboard with 4/0 THWN copper is #________. copper.

A. 2
B. 3
C. 4
D. 6

________ 12. A bank of wye-to-wye transformers should not be used unless the system is a ________ grounded system.

A. three wire
B. four wire
C. none of the above
D. all of the above

_______ **13.** The current in _______ systems chase each other.
A. delta
B. wye
C. both A and B
D. neither A nor B

_______ **14.** The current in _______ systems are opposite each other.
A. delta
B. wye
C. both A and B
D. neither A nor B

_______ **15.** Open delta-connected transformers should supply buildings where the loads are predominantly _______.
A. single-phase
B. three-phase
C. ungrounded
D. all of the above

_______ **16.** The NEC® requires the grounded conductor between the transformer and the service equipment in the building to be at least _______% of the largest phase conductor.
A. 10
B. 12½
C. 15
D. 20

Problems

17. Connect the leads of the motor for the higher voltage. Fill in the blanks.

A. _______ size wire
B. _______ size wire
C. _______ size EMT
D. _______ size circuit breaker (min.)
E. _______ size full-load torque
F. _______ size EMT
G. _______ size wire
H. _______ size time-delay fuse
I. _______ size circuit breaker
J. _______ size EMT

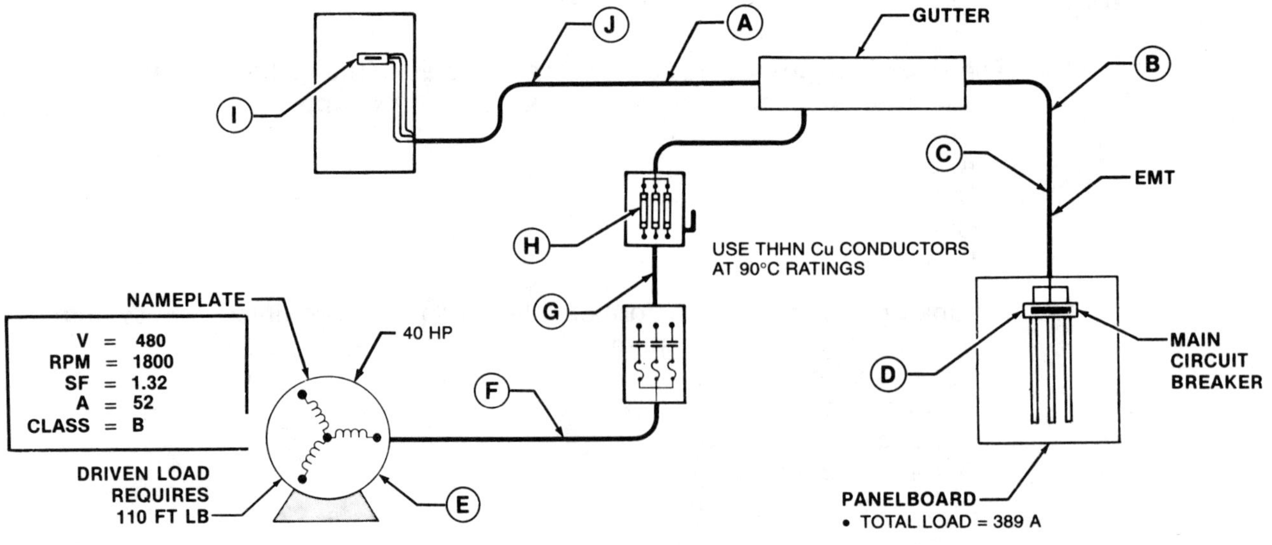

18. Fill in the blanks for the electrical system.

A. __________ size EMT

B. __________ size transformer

C. __________ size circuit breaker

D. __________ size panel

E. __________ size conduit

F. __________ size wire

G. __________ size circuit breaker

H. __________ size wire

I. __________ size EMT

J. __________ size wire

K. __________ size wire based on transformer size

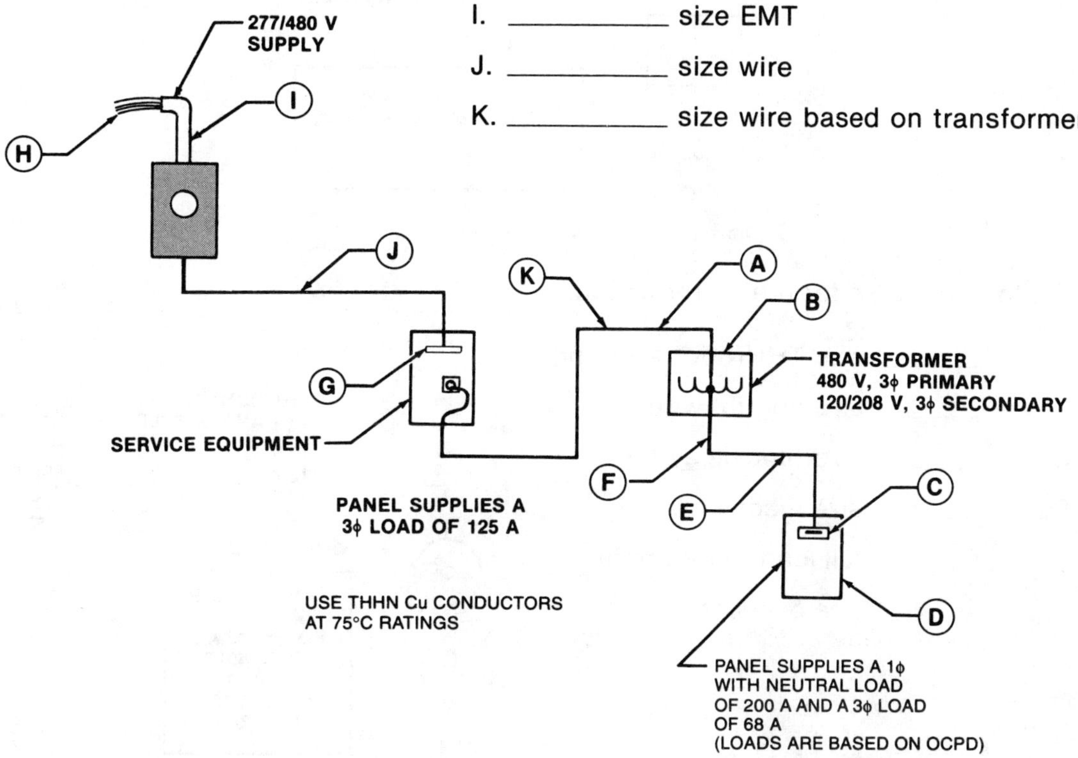

19. What size conductors and overcurrent protection device are required for the primary of the following transformer used at continuous duty?

A. __________ size THWN conductors

B. __________ size primary overcurrent protection device

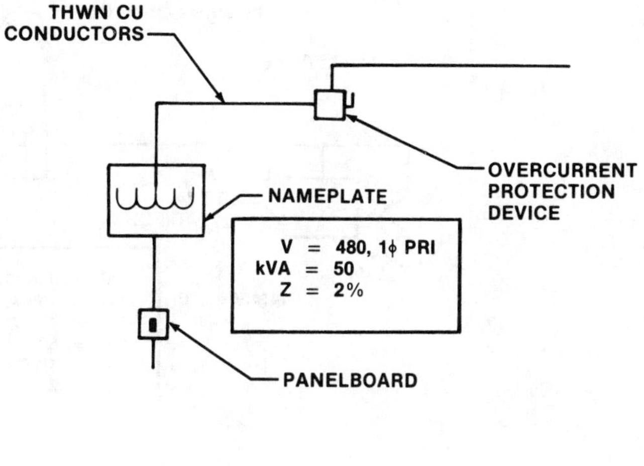

20. Size the following overcurrent protection devices for the transformer circuit.

A. primary protection = __________ A circuit breaker

B. primary and secondary protection = __________ A fuses

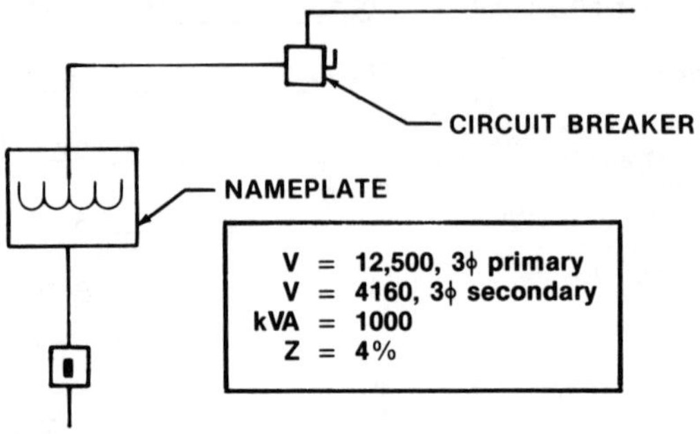

21. Fill in the blanks for the motor circuit.

A. __________ size THHN copper conductors

B. __________ size control wire

C. __________ size time-delay fuses

D. __________ size disconnect

E. __________ full-load torque rating

F. __________ size overloads

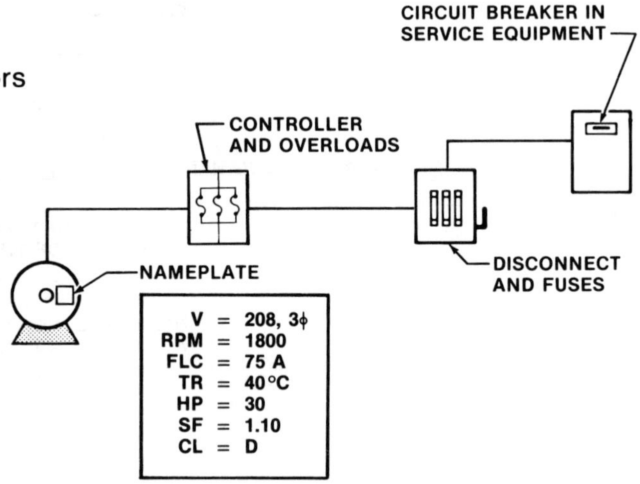

22. Fill in the blanks for the motor circuit.

A. __________ size THHN copper conductors to capacitors (use 90°C ratings)

B. __________ size disconnect for capacitors

C. __________ size overcurrent protection for capacitors

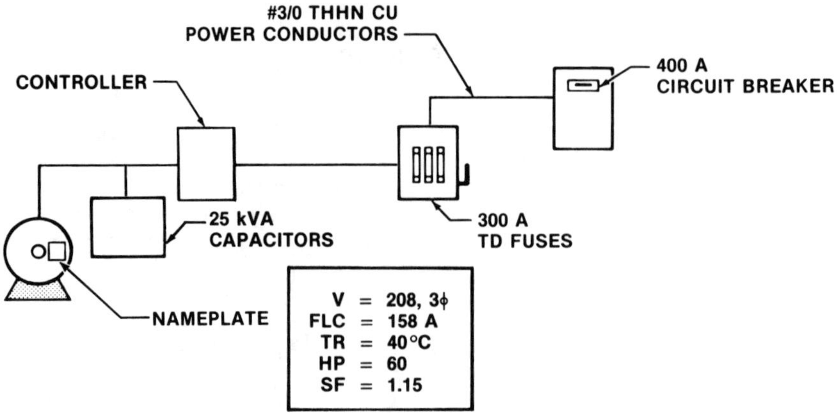

23. Fill in the blanks for the transformer circuit.

Select percentage for overcurrent protection devices.

A. __________ nontime-delay fuses

B. __________ time-delay fuses

C. __________ instantaneous circuit breaker

D. __________ inverse time circuit breaker

Select percentage for disconnect.

E. __________ nonfused

F. __________ fused

Select percentage for overloads.

G. __________ minimum

H. __________ maximum

Select percentage for conductors.

I. __________ single motor

J. __________ two or more motors

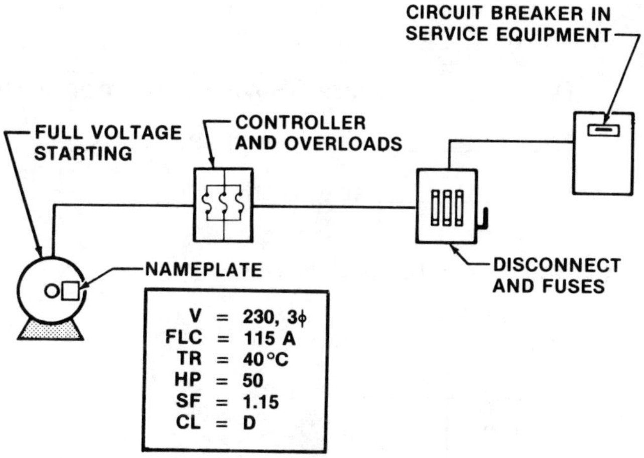

24. Select the percentage of overcurrent protection devices and fill in the blanks for the transformer and motor circuits.

control circuits less than 2 amps
A. __________% of FLC

9 amps or more
B. __________% of FLC

2 amps or more and less than 9 amps
C. __________% of FLC

less than 2 amps
D. __________% of FLC

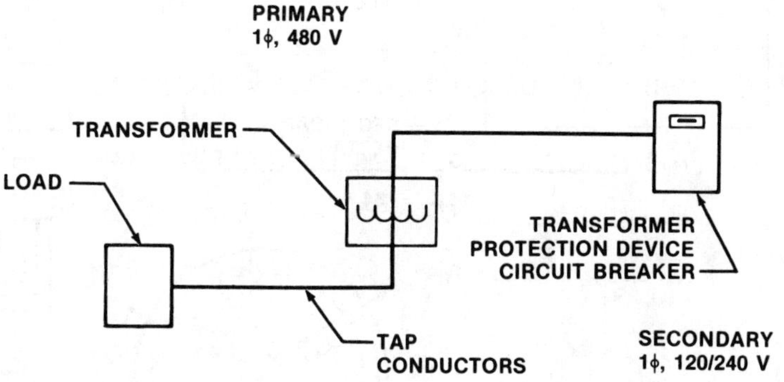

Using the 10′ tap rule, is a main required for the following?

E. __________ lighting and appliance panel

F. __________ power panel

Using the 25′ tap rule, is a main required for the following?

G. __________ lighting and appliance panel

H. __________ power panel

25. Using 15 horsepower motors, fill in the blanks for the transformer and motor circuits. Connect the motors for proper operation.

15 Hp, 3ϕ, 240 V motor

A. ___________ size THWN copper conductors

B. ___________ size time-delay fuses

C. ___________ size overload

15 Hp, 3ϕ, 208 V motor

D. ___________ size THWN copper conductors

E. ___________ size time-delay fuses

F. ___________ size overload

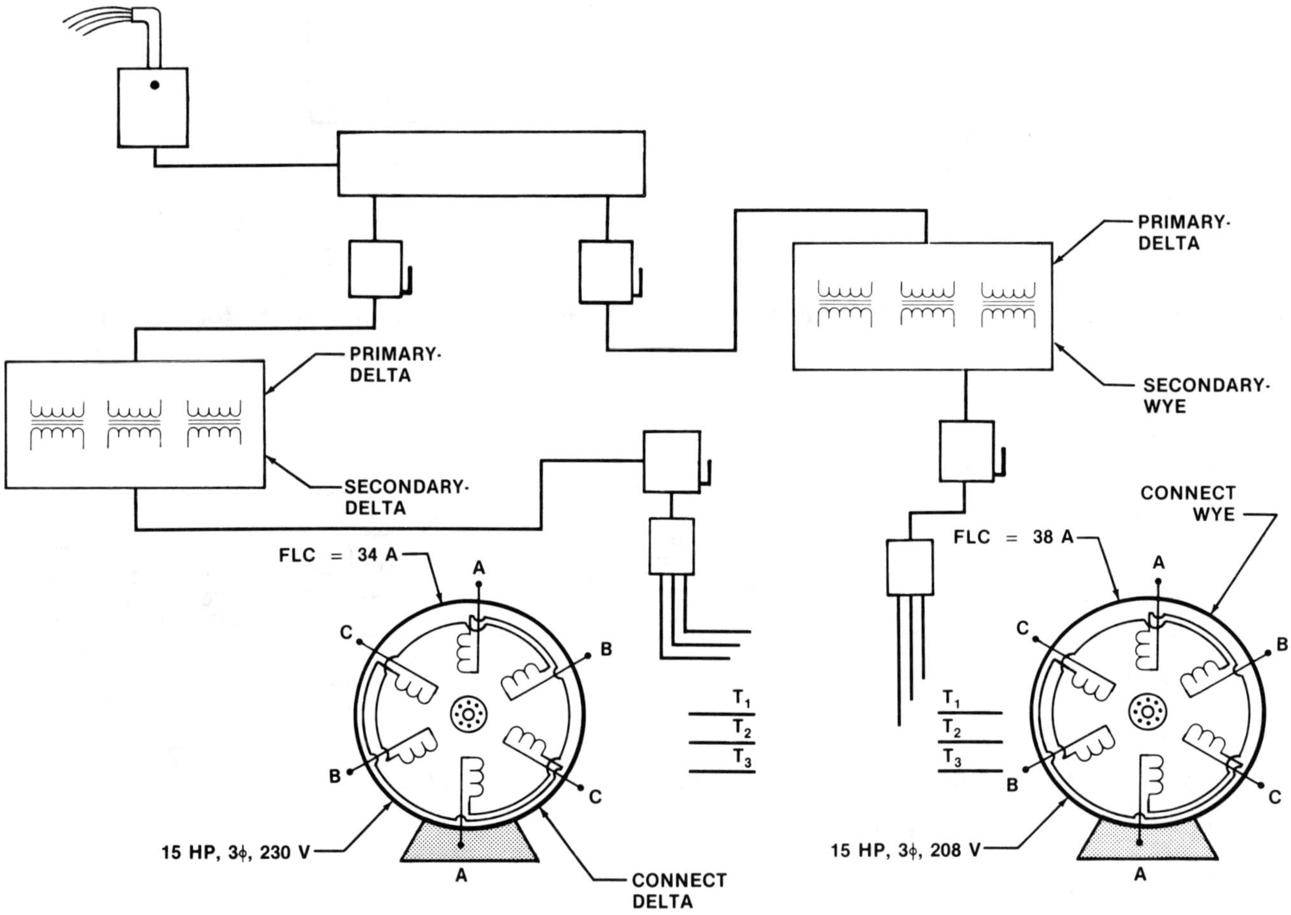

INDEX

A

B

C

D

E

F